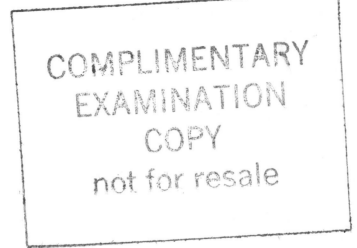

THE SOCIAL
CONDITION
OF HUMANITY

Irving M. Zeitlin

University of Toronto

THE SOCIAL CONDITION OF HUMANITY:

An Introduction to Sociology

SECOND EDITION

New York   Oxford
OXFORD UNIVERSITY PRESS
1984

Library of Congress Cataloging in Publication Data

Zeitlin, Irving M.
  The social condition of humanity.

  Bibliography: p.
  Includes index.
  1. Sociology.   I. Title.
HM51.Z44   1983   301   83-2286
ISBN 0-19-503350-7

Printing (last digit):   9

Printed in the United States of America

For Esther, Ruthie, Michael, Bethie, and Jeremy

Finally, the new chapter deals with the most urgent question of all: The entire human race and, indeed, all of earth's inhabitants today live in the shadow of the danger of a nuclear holocaust. Can this frightful menace be eliminated? More and more people now recognize that there is only one way to answer this question if humanity is to survive.

I.M.Z.

Contents

Contents

THE SOCIAL
CONDITION
OF HUMANITY

1

Sociology: Its Scope and Central Concerns

If one were to ask the majority of sociologists to define the scope and substance of sociology in a few sentences, they most likely would reply in something like the following terms.

Sociology is a social science devoted to the study of human groups of all kinds and all sizes. A group may be defined as two or more interacting individuals. The interaction of individuals gives rise to a variety of social relations and social processes such as cooperation, competition, conflict, and domination. The full range of human behavior and relations is the subject matter of sociology.

We call sociology a *science* because our main intellectual aim is to comprehend and explain the workings of the social world. Like all sciences, sociology pursues truth and knowledge by employing methods of inquiry based on logic and evidence and by subjecting theories and findings to an ongoing critical examination.

However, the statement I have placed in the mouths of my colleagues tends to exaggerate their unity of outlook. The differences among sociologists are often as significant as their common ground. Some sociologists imitate the methods of the natural sciences and see themselves working in close parallel with physicists and chemists; other sociologists believe they share the aims of novelists and poets. Between those extremes there is a wide range of research methods, techniques, ideas, and just plain notions. Nevertheless, the several different sociologies do have something in common that may be called the sociological *perspective* or *approach*.

3

The essentials of the sociological approach are best conveyed by introducing the student to the pioneers of sociological theory and analysis who wrote in the late nineteenth and early twentieth centuries. Their works are now regarded as the classic tradition of sociological thinking because they have stood the test of time. We refer to these writings again and again; we rely on pioneers' theoretical ideas; we employ their concepts; we continue to investigate questions they raised; and, finally, we emulate their intellectual craftsmanship. In a word, the classic tradition provides the theoretical foundations of the sociological perspective.

In this introduction we cannot fully convey the intellectual riches of that tradition, nor can we dwell on the specific works and ideas of the thinkers who contributed to it. Those are among the tasks of this book as a whole. All we can do here is to provide a general idea of the tradition's significance.

Not all sociologists would agree on the thinkers to be included in the classic tradition. Some of my colleagues would have longer lists than others. However, most lists would have to include at least three of the masters who receive detailed consideration in this book: Emile Durkheim, Karl Marx, and Max Weber. Although the names of these men are not likely to mean anything to the beginning student, I mention them here and postpone introducing them until the appropriate contexts in the following chapters.

The founders of the discipline which we today call sociology did not necessarily look upon themselves as professional sociologists. One finds in their writings an exploration of the full scope of the human condition. They freely availed themselves of the intellectual resources of anthropology, psychology, history, economics, and philosophy. Today those fields of study are distinct academic disciplines with jealously guarded borders. But for the classical sociological thinkers, academic boundaries were either nonexistent or quite fluid. They were more interested in addressing significant questions than in remaining faithful to this or that academic area.

It is therefore in the spirit of the classical tradition that sociology is conceived of and introduced in this book. Following in the footsteps of the pioneering masters, we shall try to focus attention on key questions. In the remainder of this introduction, I merely raise some of those questions in the hope that the student will agree on their importance and look forward to addressing them.

QUESTIONS ADDRESSED
SOCIOLOGICALLY IN THIS BOOK

What are the similarities and differences between humans and other living creatures? It is true that human beings share many physiological characteristics with some other members of the animal kingdom. But is it not also true that the human possesses something *more* than they do? What does that "more" consist of?

A related issue is the nature of human nature. Is the human being driven by instincts, by internal, genetically determined biopsychic forces? Are humans innately aggressive, as Konrad Lorenz and other prestigious students of animal behavior have alleged? Or to take an opposing but equally influential view, are humans wholly determined by external, environmental factors? The psychologist B. F. Skinner, for example, has claimed that human beings are passive entities totally shaped by external stimuli. For Skinner and his followers, "autonomy," "will," "mind," "self," and other forms of consciousness are just so many illusions. Finally, there is the question raised anew in recent years by sociobiologists: How do the principles of human social organization compare with those on which animal societies are based?

As one shifts attention to large human organizations and societies, new theoretical problems emerge. How does a society establish and maintain order, that is, ensure a measure of internal peace that will enable its members to attend to their vital affairs? What is the connection between order and justice? That question was skillfully explored by the renowned French sociologist Emile Durkheim, who noted that modern industrial society is largely based on formal contractual relations. Contracts are often made between social unequals, where one party dominates and the other has no choice but to serve or to suffer worse consequences. Durkheim's reply, as we shall see, raises the critical question of whether a social order resting on basic social inequalities can be stable.

However, the social theorist who raised that question most dramatically was Karl Marx, the well-known nineteenth-century writer whose ideas have had so much political influence in the present century. Sociologists have increasingly recognized that Marx made significant contributions to sociological theory and analysis, especially for the study of social classes and social change. Though Marx and Marxism are highly controversial subjects, I shall try to discuss his work objectively, with the aim of highlighting his sociological contributions.

Social stratification is the term sociologists employ to describe the study of inequalities based on wealth, power, prestige, and other social conditions. Key questions are: How do basic social inequalities come about? How are they perpetuated? Are modern societies still structured along social-class lines? Some sociologists favor a social-class type of analysis; others do not. Some sociologists believe that basic social inequalities can be reduced and even eliminated. Others insist on the necessity of social inequality, arguing that a society's social positions must be stratified and that the rights and duties associated with those positions must be unequal.

Similarly, some theorists hold that throughout history, from the dawn of civilization to the present, two classes of people have been evident: One that rules and another that is ruled. That state of affairs has prevailed whether government has been despotic, aristocratic, democratic, or what have you. Moreover, the division of society into rulers and ruled, those theorists maintain, is inevitable and bound to persist so long as there are human societies on earth. We shall subject several such theories to close scrutiny.

One very clear manifestation of social inequality is the status of ethnic minorities. Ethnic minorities the world over have continued to suffer from a variety of disabilities, including prejudice, discrimination, segregation, exploitation, and persecution. The vastness of the subject constrains us to focus attention on minorities in the United States. Our analysis begins with a historical overview of ethnic immigration from the colonial era to the 1930s. A sound generalization may be made concerning all the European minorities that immigrated in the course of those three centuries: They eventually improved their economic circumstances considerably. However, the lot of the groups that arrived since the 1930s has been quite different. The major influx since the 1930s has consisted of Spanish-speaking minorities from Central and Latin America, people of Mexican, Puerto Rican, Cuban, and other Spanish-speaking origins. The Mexican Americans are the most numerous group, and they also constitute the largest ethnic minority in the United States after black Americans. The vast majority of the Spanish-speaking people in the United States find themselves at the bottom of the socio-economic pyramid or in what we call the *underclass*. They thus share economic circumstances similar to those of the majority of black Americans. Finally, American Indians, though native to the continent, are also among the most disadvantaged. It is therefore with good reason that we give several ethnic groups more sustained consideration than others.

6   Social institutions have been central to the concerns of sociology from

its very inception. In the language of sociology, an institution is a firmly established social practice together with its accompanying form of organization. Examples of major institutions are the family and religion. In our discussion of the family, we shall concern ourselves with such questions as these: Is the nuclear family (two parents and at least one offspring) found in all human societies? What are the conditions underlying monogamy and polygamy? What impact has the women's-liberation movement had on the family? How has the liberalization of divorce laws affected the family? What are the effects of experiments with new forms of sexual liaison? Is the family in a state of decline?

In studying the institutions of religion, sociologists concern themselves with the connections between religious phenomena and other facets of social life. What are the social roots of the idea of the divine? Why is a belief in the divine found in all societies? Why is the sacred everywhere sharply distinguished from the profane? Is religion a positive phenomenon in that it meets certain basic social and individual needs? Or is it injurious to humans, blinding them to reality and forestalling a rational understanding of their actual circumstances? Is the influence of religion declining in modern society? To shed light on those and other intriguing questions, we employ the classical writings of William James, Emile Durkheim, Bronislaw Malinowski, Sigmund Freud, Karl Marx, and Max Weber.

Besides institutions, another central concern of sociology is to understand social movements and other forms of collective behavior. Why and how do some small groups become mass movements? Throughout history we see evidence of concerted social action on the part of large numbers of people. Religious, political, and other social movements appear again and again. Often a movement begins with a single leader of striking qualities and a small band of followers. At first one group is scarcely distinguishable from all the other similar ones in the vicinity, but in time it does in fact distinguish itself by its remarkably dynamic qualities. Its utterances and practices begin to appeal to ever-larger numbers of people, and the originally minute circle finds its doctrine spreading far and wide. Somehow it has fostered a movement sweeping across borders and firing the imagination of broad masses of converts. How shall we account for such a remarkable phenomenon? To answer that question, we take a close look at three twentieth-century movements: (1) the Nazi movement in Germany, which led to the Second World War; (2) the Chinese Communist movement, which resulted in Communist rule in China; and (3) the American civil-rights movement, which has profoundly affected the consciousness of all Americans.

Another major area of sociological concern is deviant behavior or the question of why some people deviate from the norms (rules, standards, and laws) of society. Crime is an example of deviant behavior. What makes a person become a criminal? Traditional *nonsociological* replies to that question are that a criminal is an abnormal human being afflicted with some defect that accounts for his deviance. A criminal is an atavism, a kind of throwback to an earlier evolutionary stage who possesses ferocious instincts. That was in fact the doctrine promoted by a well-known nineteenth-century criminologist, Césare Lombroso.

However, Lombroso's younger contemporary, Durkheim, had a quite different view of the matter. In his pioneering sociological analysis, he put forth the proposition that deviance, like conformity, is firmly rooted in social conditions. Building on the work of Durkheim, we address such matters as how an individual becomes a criminal, the nature of lower-class criminal subcultures, the higher incidence of certain types of law violations in the ghetto, and white-collar crime. Finally, the question is raised whether homosexuality may be considered a form of social deviance.

How do societies change? What processes account for the transformation of society and culture? Urbanization, for instance, had a beginning and can be traced to the ancient Near East. On the site of biblical Jericho, archeologists have uncovered in the oldest strata the remains of a community dating back to the eighth millennium B.C. Those strata appear to be evidence of the earliest human settlement on record. Another of the earliest human settlements is Jarmo, located on the foothills running along the Tigris-Euphrates plain between Iraq and Iran. Archeological methods (carbon-14 tests) indicate that farming communities existed in that region by about 6750 B.C. However, neither of those ancient settlements, though large and proto-urban in some respects, were true cities. Rather, they were tangible evidence of the Neolithic Revolution, which prepared the way for genuine urbanization. Beginning with the earliest true cities of Mesopotamia, we follow the urbanization process through the *polis* or Greek city-state, the medieval town, and the modern metropolis. We then examine the trend which has become evident since the 1950s: the movement from city to suburb.

In addition to urbanization, sociologists study the social conditions of economic development, modernization, and other forms of social change. The urgent theoretical and practical questions are these: Why are there rich and poor nations? Why is it that some countries have successfully industrialized themselves while others have remained economically backward? More specifically, why have most societies of the so-called Third

World (Asia, Africa, and Latin America) thus far failed to achieve the kind of modern industrial development which has been characteristic of Western Europe, the United States, and Japan? Is the answer to be sought primarily in climatic, geographic, and biological factors, or is it rather to be sought in the concrete social and *historical* circumstances of the countries concerned?

I have emphasized the word *historical* because I believe this is the place to say a word about the relation of sociology to history. One aim of sociology, conceived of as a science, is to develop sound generalizations about the variety of human groups. In the study of social change, for example, we note that some societies are industrially advanced while others are comparatively backward. That fact prompts the sociologist to search for what it is that the advanced societies have in common which the backward societies lack. If we should discover the presence of one or more factors among the advanced and the absence of those factors among the backward, that would help explain the difference between them. A hypothesis might then be formulated to the effect that all societies possessing conditions x, y, and z will successfully industrialize, whereas all societies lacking those conditions will fail to do so. Eventually, if the hypothesis is adequately tested, it may acquire the status of a sound generalization.

Contemporary sociological theories often turn out to be unsound when confronted with available historical evidence, for many are set forth without due regard for the history of the groups or societies in question. It is not going too far to say that much of contemporary sociology is nonhistorical. Yet it would have been unthinkable for the leading classical theorists to ignore history. As we shall see in due course, though the classical theorists were certainly interested in grasping the general and universal, they concerned themselves with specific periods, particular events, and concrete circumstances. They clearly recognized that an adequate explanation of social facts requires a historical account of how the facts came to be what they are. When the classical theorists generalized, it was always *from* the historical record.

To my mind there can be no doubt that we must follow the classical tradition in that respect, as in many others. Only by doing so can we hope to produce truly sound sociological theories. That is the reason why the student will find more history in this book than in other introductions to sociology. In four chapters in particular (those dealing with ethnic groups, urbanization, social movements, and social change), I have furnished considerable historical background and detail which, I am confident, can only enrich the sociological analysis.

9

2

The Concept of Culture

THE HUMAN SPECIES IN THE ANIMAL KINGDOM

In 1859 Charles Darwin published his theory of biological evolution in an epoch-making book called *On the Origin of Species by Means of Natural Selection*. Darwin's theory effectively challenged the prevailing biblical and Aristotelian views. According to those views, the various species of the animal kingdom were created once and for all and forever remained in their original forms. Darwin, however, presented massive evidence to show that *all* forms of animal life have undergone and continue to undergo evolutionary changes. Evolution is accomplished through a process he termed *natural selection*. That term means that some individual members of a species are better suited than others for survival in a particular natural environment. Those who as a result of their biological equipment are best suited to the existing conditions will survive and reproduce; the others will ultimately perish.

Each individual member of every species inherits certain biological traits from its ancestors. Some of those traits are more advantageous than others for adaptation to a specific environment. A deer, for example, in its natural habitat has several predatory enemies. It depends for survival on the ability to detect those enemies and to flee rapidly. Hence the deer with the keenest sense of smell and/or greatest speed is the one most likely to survive and reproduce.

Darwin did not systematically explore the hereditary process itself. He mistakenly believed that the process was somehow rooted in the

11

blood. It was his unknown contemporary Gregor Mendel who first satisfactorily explained why the offspring in every species tend to resemble their parents in physical characteristics. Mendel thus laid the foundation for the scientific theory of genetics.

According to the theory of genetics, heredity operates through *chromosomes* and *genes*. Chromosomes are complex structures. They are the nuclei of the many cells of which a living organism consists. Chromosomes occur in pairs in all cells except the sex cells, and the number of pairs varies from one species to another. (In human beings there are 23 such pairs.) Every chromosome is made up of genes, which are the actual determinants of an individual's hereditary potential. Genes occur in pairs, and each individual has no fewer than two genes for any given biological characteristic.

As for the sex cells, they differ from all others in that they contain single chromosomes and genes rather than pairs. When the male sex cell (the sperm) unites with the female sex cell (the ovum), the single chromosomes from each cell come together to form the normal paired arrangement. Thus every individual receives half of his* genetic structure from each parent. In human inheritance many thousands of genes are involved, combining in an almost infinite number of ways. Genes, it should be emphasized, only provide an individual's *potential* for the development of physical and other characteristics. The *actual* characteristics depend on the interplay of genetic potential and the environmental circumstances affecting that potential. For instance, an individual may inherit the possibility of becoming big and strong but may actually turn out to be small and weak because he has suffered from malnutrition or from some debilitating disease.

If offspring tend to inherit the traits of their parents, where do new and different characteristics come from? They result from chemical changes occurring in the genes and are called *mutations*. Mutations occur at random, most of them being unfavorable. However, some may prove to be advantageous to an organism in a specific environment. Since such mutations are passed on through reproduction, they may eventuate in a new and different biological form. Thus adaptive mutations which are "naturally selected" may result in evolutionary change.

* Hereafter I employ the pronouns *he, his,* and *him* to refer either to a male or to a female of the human species. Similarly, I am inclined not to refer to an animal as an *it*. So, for example, I later speak of a dog and "his" master, fully intending by that term a dog of either sex. I am aware that the practice of using only masculine pronouns to denote an individual may offend some people. But the alternative practice would result in such recurrent and cumbersome phrases as "his/her" or "his or her" and would only impede the flow of the text.

So we see that the two originally independent scientific theories of Darwin and Mendel are compatible and complementary. Natural selection and genetics are the foundations of modern evolutionary theory.

In Darwin's *On the Origin of Species,* it was only implied that the human being, like all other species, was also subject to the laws of evolution. More than a decade elapsed before Darwin decided to publish his *Descent of Man* (1871). There he presented his evidence demonstrating the evolution of humans from earlier animal forms. Systematically and at length, Darwin reviewed the parallels in structure and development between humans and earlier animals, highlighting similarities in muscles, sense organs, bone structures, reproductive organs, etc. Then he went on to consider in detail a number of arguments that insisted on the uniqueness of the human being, especially his mind. In the end Darwin concluded that when all is said and done, "the difference in mind between man and the higher animals, great as it is, certainly is one of degree and not of kind."[1]

That conclusion has been largely rejected by modern sociologists and anthropologists. These scholars fully accept Darwin's general theory but they maintain that the difference between the human being and even the most intelligent animal is fundamental.

It is clear that since Darwin was intent upon demonstrating the evolutionary emergence of the human species, he was bound to stress the continuities between humans and other members of the animal kingdom. Given the purpose of his inquiry, he was reluctant to acknowledge any difference in kind which might be taken as an "impassable barrier" between animals and humans in the evolutionary process. Throughout Darwin underscored the common and parallel structures and processes. Animals, he argued, share the same emotions as we do. Terror causes the muscles to tremble, the heart to palpitate, the sphincters to relax, and the hair to stand on end. Love, affection, sympathy, jealousy, courage, etc. are experienced by animals no less than by humans. A dog shows great love for his master, to whom he is fiercely loyal. He is jealous of any affection his master might lavish on another creature. The same is true of other higher mammals as different as monkeys and porpoises. Darwin maintained that animals not only love but desire to be loved in return. They enjoy being praised for the skillful accomplishment of some feat, and they also exhibit pride or dignity. On the other hand, they may feel shame upon misbehaving and modesty when begging too often for food. They also exhibit something akin to a sense of humor: A dog instructed to fetch an object thrown by his master may squat beside it instead, waiting for his master to retrieve it. Then when his master comes quite close, he will seize it and run away, evidently enjoying the "practical joke."

In Darwin's view animals also show wonder and curiosity and even some power of what we call reasoning. Dogs and cats appear to pause, deliberate, resolve, and act. Darwin tells us that he saw an elephant in a zoo obtain an object beyond his reach by blowing with his trunk on the ground beyond the object so that the reflected current drove the object closer to him. Darwin suggested that that action is a form of reasoning without language and that animals appear to grasp words and even short sentences. In the end, however, he acknowledged that the human being possesses certain unique qualities.

There can be no doubt that the difference between the mind of the lowest man and that of the highest animal is immense. An anthropomorphous ape, if he could take a dispassionate view of his own case, would admit that though he could form an artful plan to plunder a garden—though he could use stones for fighting or for breaking open nuts, yet that the thought of fashioning a stone into a tool was quite beyond his scope. Still less, as he would admit, could he follow out a train of metaphysical reasoning, or solve a mathematical problem, or reflect on God, or admire a grand natural scene. Some apes, however, would probably declare that they could and did admire the beauty of the coloured skin and fur of their partners in marriage. They would admit, that though they could make other apes understand by cries some of their perceptions and simpler wants, the notion of expressing definite ideas by definite sounds had never crossed their minds. They might insist that they were ready to aid their fellow-apes of the same troop in many ways, to risk their lives for them, and to take charge of their orphans; but they would be forced to acknowledge that disinterested love for all living creatures, the most amiable attribute of man, was quite beyond their comprehension.[2]

For Darwin, uniquely human qualities were made possible by the evolutionary process. Humanity's early progenitors, like all other animals, must have tended to multiply beyond the available means of subsistence. They were exposed to a struggle for existence and therefore subject to the law of natural selection. Individuals with favorable variations survived and reproduced; individuals with less favorable or injurious variations perished. Ultimately that process enabled the human being to become the dominant animal form on earth. Humans owe their dominant position to their intellectual faculties, social organization, and bodily structure. The supreme importance of those factors has been proved by human success in the struggle for life.

14   To understand how humans achieved their position of dominance

and how they surpassed the other primates, we must review the similarities and differences between them. Then we can more intelligently attempt to reconstruct the process by which humans emerged from previous primate forms.

HUMANITY'S CLOSEST ZOOLOGICAL RELATIVES

All primates have *prehensile* hands and feet, that is, they are capable of grasping objects. The grasping is made possible by the thumb or great toe, which can be opposed to the other four digits. Humans lost the prehensility of their feet when they became bipeds, standing erect and supporting their entire weight on their feet. The human being is the only biped on earth. All other primates move about on all fours and stand on their hind legs only momentarily.

Primates can turn their forearms and hands in any direction. Their movable fingers and toes are equipped with flat nails rather than claws. Primates have well-developed clavicles (collarbones) supporting the muscles of the arms, which are capable of versatile and vigorous movements. Because primates use their hands for many of the tasks for which other animals use their teeth, the teeth and jaws are less developed in primates. The sense of vision is considerably developed in higher primates. Most of them have their eyes on the front of the face, so that they can look forward rather than to the sides as do most other mammals. Primates also have stereoscopic vision: The eyes focus together and transmit to the brain a single image with depth perception.

Most primates are *arboreal,* spending much of their time in the trees. They are well-adapted tree dwellers owing to the effective use of their prehensile hands and feet and their excellent vision. There are two suborders of primates. Those furthest removed from man are classified by zoologists in the suborder of *Prosimii.* That suborder includes tree shrews, lemurs, and tarsiers—small tree-dwelling creatures that lack some of the characteristics of the higher primates. The other suborder, *Anthropoidea,* includes monkeys, apes, and man. Monkeys must not be confused with apes. Monkeys have tails, apes do not. Furthermore, there are significant differences between New World and Old World monkeys. While New World monkeys have prehensile tails, Old World monkeys do not, and are therefore more closely related to apes. The fact that there are no ape species native to North or South America suggests that evolution in this hemisphere never advanced beyond the stage of the monkeys.

The anthropoid apes are humanity's closest zoological relatives. That is evident from structural similarities and from common habits and practices. But the anatomical differences are more important in that they give us insight into the process of divergence between humans and apes. The most important differences stem from the fact that the characteristic form of locomotion for apes is *brachiation* (swinging from branch to branch with hands and arms), while humans are the only creatures who move about on two feet. As a result human feet have lost their prehensile character. The great toe is no longer opposable, but rather in line with the other toes, all proportionately smaller than in other primates. The bones at the rear of the foot are enlarged and shifted backward, thus creating a heel. That helps to balance the body and to lift it in walking. Human legs are longer and the arms shorter, in relation to body length, than they are in the apes. That too is a consequence of the greater strength required by human bipedal locomotion unaided by the arms.

The erect human posture is made possible by the short, broad, and powerful pelvis, which sustains the heavy muscles required to support the body in an upright position. The human backbone is shaped like the letter S and is not an almost continuous curve from top to bottom as in the ape. Correspondingly, the *foramen magnum* (the opening in the skull through which the spinal cord enters) is directly below the skull in humans, while in apes it is more at the rear of the skull. Thus the human head is balanced on the top of the spinal column, with comparatively light muscles holding it in position. Human teeth and jaws are much smaller than the ape's. In apes, there is a marked *prognathism,* that is, the jaws protrude beyond the level of the rest of the face. The massive jaws are operated by heavy muscles enclosing nearly the entire skull. And in some larger apes, there is even a crest of bone along the top of the skull providing additional support for the jaw muscles. Finally, the substantial difference in brain size between apes and humans should be noted. In apes it rarely exceeds 500 cc., while in humans it averages 1,350 cc.

The similarity in physical characteristics between humans and apes suggests that they shared a common ancestry. However, the very substantial dissimilarities indicate that humans diverged millions of years ago from the evolutionary path of the great apes—i.e., the chimpanzee, gorilla, gibbon, and orangutan.

PROTO-HOMINOIDS

It is probable that humans and great apes ultimately had common ancestors not shared by any other living species. The common ancestors may be called *proto-hominoids*. It is with those creatures that we must begin our attempt to reconstruct the evolutionary process which led to the emergence of that new and distinctive species called *Homo sapiens*.

The proto-hominoids were predominantly tree dwellers.[3] It was in their arboreal way of life that they developed their exceptionally keen vision and their freely movable arms with manipulative hands. Like most of their descendants, they were hairy and tailless. They lived in bands of ten to thirty members, typically consisting of one or several adult males, their females, and their offspring.

Since they spent most of their time in the trees, they were of course expert climbers, quite comfortable living in their tree nests and sleeping there. Like other animals they climbed up a tree head first; but unlike others, such as rodents, they climbed down stern first. They slept at night and were active during the day. Their main activity, a constant search for food, compelled them to leave the trees regularly. On the ground they had a semi-upright position which enabled them to raise their heads above the tall grass and to look about. Sitting or standing, they could employ their arms and hands in manipulative motions. They walked and ran on all fours, using their forearms as crutches much like present-day apes; and only infrequently did they walk on two feet like humans. Proto-hominoids picked up sticks and stones and might even have reshaped the sticks somewhat with their hands and teeth. They carried such "tools" short distances together with other materials employed in nest building. Although they were mainly vegetarians, supplementing their diet with worms and grubs, they were also meat eaters, scavenging the remains of carnivores whenever possible.

The males had the pendulous penis characteristic of the primates, and copulation was accomplished exclusively with the dorsal approach common to all land mammals. The period of gestation was about 30 weeks. The mammary glands were pectoral, and a nursing female held the infant to her breast. The females had a year-round menstrual pattern rather than an estrous cycle, and mating remained within the band.

The proto-hominoids did not have articulate speech, though they might have had a call system not unlike that of the gibbon. Such a system might have consisted of a few distinct signals appropriate to several important situations, e.g., discovery of food or the detection of   17 danger. Or a signal might simply have communicated an individual's whereabouts.

THE EMERGENCE OF THE HUMAN SPECIES

How did some of the proto-hominoids diverge from the others and thus become, among other things, erect hominid bipeds? The single most important condition that accounts for the beginning of this process is the fact that they were forced to leave the trees and to make their way permanently on the ground. Most likely it was the drastic climatic changes of the Miocene period that brought them to the ground. Continuous tropical forest was transformed into open savannah with scattered groves of trees. Some bands of proto-hominoids remained in the diminishing and widely separated groves, never abandoning their arboreal mode of existence. However, other bands, whose physical structure facilitated survival in the open country, thrived and multiplied while those who could not cope with the new conditions became extinct. Those hominoids who remained in the trees were the ancestors of the great apes of today; those who became fulltime ground-dwelling bipeds increasingly diverged from their apelike ancestors.

As humanity's progenitors became more and more erect, their hands and arms were modified for skillful manipulation, and their legs and feet were adapted to bipedal locomotion. Those changes were accompanied by other profound anatomical alterations. The pelvis broadened, the spine acquired its distinctively human S shape, and the head shifted its position. The free use of arms and hands, partly the cause and partly the effect of the erect human posture, led to still other structural modifications. As the progenitors gradually acquired the habit of employing sticks and stones as tools, they used their jaws and teeth less and less, and these accordingly diminished in size. Since it is the massive jaw muscles that largely account for the peculiar shape of the ape's skull, we can surmise that as the hominid jaws and teeth gradually diminished, the adult skull came more and more to resemble the skull of *Homo sapiens*.

The exodus from the trees and the permanent descent to the ground presented hominids with a host of new problems and challenges. They now had to cope with radically new circumstances and to develop new mental powers. Accordingly, those individuals with larger and heavier brains adapted more easily and survived. The growth of the brain affected the shape of the skull, while the increased weight of the skull influenced the development of the supporting spinal column. Thus the earliest humans diverged from their apelike ancestors. There is a paradox here. For although humans soon became the most powerful beings on earth, they entered the world as the weakest of nature's creatures. The human being comes into the world without great teeth and claws for defense, and his physical strength and speed are by no means the greatest. His biological equipment in and of itself would prove quite ineffectual in the struggle for survival. It is therefore clear that the "secret" of human dominance lies elsewhere.

It is true that all animals struggle and cooperate with nature in order to survive. But with the emergence of the human species, a new mode of struggle and cooperation came into being. Humans were the first creatures that did not have to rely on their bodily equipment alone for survival. Humans were not only occasional tool users as were their hominoid ancestors, but *toolmakers* as well. Tool using and toolmaking were the normal mode of human existence. No other creature had ever invented tools by which to appropriate its means of life. But this is precisely what became characteristic even of the earliest humans: They produced the means by which to appropriate their means of subsistence. They *labored* and *produced*. It was that unique mode of interacting with nature that made humans dominant despite their natural weakness. As Darwin observed over 100 years ago:

An animal possessing great size, strength, and ferocity, and which, like the gorilla, could defend itself from all enemies, would not perhaps have become social: and this would most effectually have checked the acquirement of the higher mental qualities, such as sympathy and the love of his fellows. Hence it might have been an immense advantage to man to have sprung from some comparatively weak creature.

The small strength and speed of man, his want of natural weapons, etc., are more than counter-balanced, firstly, by his intellectual powers, through which he has formed for himself weapons, tools, etc. . . . and, secondly, by his social qualities which lead him to give and receive aid from his fellow man.[4]

The hominids, as weak and gregarious beings, struggled to establish themselves in their new habitat by means of *labor* and *production*. The fundamental role of labor was recognized by Darwin's contemporary, Frederick Engels. "Labor," he wrote, "is the prime basic condition for all human existence, and this to such an extent that, in a sense, we have to say that labor created man himself."[5] Clarifying this idea further, he added:

19

The mastery over nature, which began with the development of the hand, with labor, widened man's horizon at every new advance. He was continually discovering new, hitherto unknown, properties of natural objects. On the other hand, the development of labor necessarily helped to bring the members of society closer together by multiplying cases of mutual support, joint activity, and by making clear the advantage of this joint activity to each individual.[6]

Humans now had the need to say something to one another. Cooperation gave rise to oral communication, which in turn facilitated further cooperation:

First labor, after it and then with it speech—these were the two most essential stimuli under the influence of which the brain of the ape gradually changed into that of man.[7]

Thus the human brain with its unique faculties evolved together with the development of the human hand itself through the medium of labor and production.

Along with the extension of cooperative labor, the development of articulate speech, and the growth of the brain, humans acquired an increasing capacity for abstract thinking and reasoning; and reasoning in turn imparted a new impulse both to the labor process and to speech. Human labors became more and more diversified, and tasks were divided among individuals. And the more the activity of these early humans assumed the character of premeditated, planned action directed toward definite ends known in advance, the further removed they became from their hominoid forebears.

CULTURE: A UNIQUELY HUMAN CONDITION

Thus we see that toolmaking, social organization, and the productive process, mediated by language, are distinctively human. *Culture* is the all-encompassing term which, since the time of E. B. Tylor, has been used to denote the totality of uniquely human creations. Culture refers to "that complex whole which includes knowledge, belief, art, morals, law, custom, and any other capabilities and habits acquired by man as a member of society."[8]

Culture has been called *superorganic*.[9] That term does not mean that culture in any sense replaces nature or renders its laws inoperative. What *superorganic* is supposed to convey is that humans have something more than nature. In the words of Matthew Arnold:

20

Man hath all that nature hath, but more,
And in that more lie all his hopes of good.[10]

Let us then clarify what this "more" consists of.

The development of culture has neither suppressed nor supplanted natural selection. Rather the cultural process interacts with the genetic system to produce both genetic and cultural changes. However, human adaptation and genetic changes now largely occur in connection with the cultural changes human beings produce. Culture affects the adaptive value of genes. Under conditions of culture, natural selection has increasingly placed a premium on genes enhancing cultural competence. Human fitness is no longer a matter of possessing fixed responses to a specific and limited environment. It is instead a matter of developing a *generalized* capacity to adapt to a very wide range of circumstances.

Culture therefore stands in an interactive relation to the biological nature of humans. Culture is rooted in the human organism but contributes at the same time to its development. Culture is transmitted nongenetically, that is, through the process of social interaction and *learning*. And the main form of human learning, in contrast to that of all other animals, is *symbolic communication and conceptualization*. Culture is humanity's social inheritance. Though human nature, which makes culture possible, is based on genetic factors, culture itself is a nongenetic, superorganic process.

It follows that every human being has an immense amount of learning to do before he can become culturally competent. In humans the learning period extends far beyond childhood and in fact continues throughout life. That condition is related to the fact that the period of dependency and immaturity of the human infant is far more prolonged than that of any other mammal. What is the significance of that prolonged dependency?

It is known that the newborn elephant and deer, for example, are able to run with the herd soon after they are born. The infant seal learns to swim by the age of six weeks. Those species have long gestation periods. Zoologists have accounted for this by suggesting that animals giving birth to small litters are less able to protect themselves from predators and therefore must give birth to offspring who are in a fairly mature condition. Such maturation is made possible by a long gestation period.

Apes, like humans, also give birth to immature young and have long gestation periods.[11] For the gorilla it is about 252 days, for the orangutan 273 days, and for the chimpanzee 231 days. Typically, one infant is

21

produced, as in humans. But the young ape's physical development is much more rapid than the human's. The infant ape becomes capable of lifting his head, rolling, crawling, sitting, standing, and walking in only a fraction (one third to two thirds) of the time that it takes the infant human to achieve the same things. Nevertheless, the infant ape is cared for by his mother for several years. It is not unusual for the ape mother to breastfeed her baby for three or more years. Hence human immaturity in infancy may be viewed as an extension of the infant immaturity common to all hominoids.

There is, however, a very important difference between humans and the apes in this respect. Although the gestation period is almost the same for anthropoids and humans, the growth of the fetus is markedly different in the two species. The human rate of growth is greatly accelerated toward the end of the gestation period. That is most evident in the growth of the brain, which has acquired a volume of 375 cc. to 400 cc. by the time of birth. The total body weight of the newborn human averages seven pounds, which contrasts with 4⅓ lbs. for the body weight of the chimpanzee and 200 cc. for his brain. The body weight of the newborn gorilla is 4⅓ lbs., and his brain size is not much greater than that of the chimpanzee. The human being's comparatively large body size and especially the large head size at 266½ days of gestation make it necessary for the child to be born at that time.

To pursue that point, childbirth seems exceptionally painful for the woman as compared with the female of other species, including the female anthropoids. Unusual pain at childbirth is probably a consequence of the major changes in the human pelvis that have accompanied the development of the erect posture. One such change is a significant narrowing of the pelvic outlet. In adapting to that condition, the skull bones of the human infant grow considerably slower than those of the ape infant of the same gestation age. The slower growth is extraordinarily important since it allows for the movement, flexibility, and overlapping of the membranes in the human infant's skull, thus accommodating the various pressures upon them in the process of birth. Given the rapid growth of the brain and skull in the last month of intrauterine development, it appears that the human infant is born when it is because its head size has attained the limit compatible with viable birth. Anthropoid apes have no such problems owing to the ape mother's ample pelvic outlet.

As a result of the peculiar pelvic arrangements of the human being, the human infant, it appears, is born before the gestation period has fully prepared him for life in the world. The rapid developmental rate

22

of the human fetus in the late stages of its uterine growth suggests that it should enjoy a much longer period in the womb. As Ashley Montagu has explained:

> The growth of the brain of the fetus is so rapid during the latter part of gestation, and the mother's pelvis is so narrow, [that] the fetus must be born no later than the time at which its head has reached the maximum size congruent with its passage through the birth canal. . . . The fact appears to be that the human fetus is born considerably before his gestation period has been completed. The rate of growth of the brain is proceeding at such a pace that it cannot continue within the womb, and must continue outside it. In other words, the survival of the fetus and the mother requires the termination of gestation within the womb when the limit of head size compatible with birth has been reached, and long before maturation occurs.
>
> Gestation, then, is not completed by the process of birth, but is only translated from gestation within the womb to gestation outside the womb.[12]

Montagu also calls attention to the fact that by the time a child is three, his brain (960 cc.) is almost as large as it is (1,200 cc.) when the child is fully grown at age twenty. Doubtless, the comparatively large size of the brain at an early age is related to the immense amount of learning which the very young child must do. That learning is accomplished in a social relation with his parents and others who pass on to him the knowledge and experience accumulated over many generations. The child thus becomes heir to a social tradition which includes an enormous array of skills.

The human child, then, is uniquely helpless and vulnerable because he has so much to learn, so much culture to acquire. If the child is to survive and reach maturity, he needs to be in the care of a social group such as the family. In the human species, the family of parents and children is a more durable group than among species whose young mature faster.

The human child learns some things by following the example of the older generation. In animals learning entails imitation. But what the human infant has to learn could never be accomplished by imitation alone, and indeed only a very small fraction of the child's knowledge is acquired in this way. The rest, the greatest part by far, is acquired through *symbols* and *concepts*. All animals learn; but only humans learn by means of a complex system of symbolic communication called *language*. Only humans have developed a symbolic equipment, which is scarcely less essential for their survival than their material equipment.

The human larynx and tongue enable humans to emit a wide array of sounds. Many animals can also make distinct sounds as, for example,   23 the parrot that can be taught to say, "Polly wants a cracker." But the human being not only has a large repertory of articulate sounds; he has the ability to invest those sounds with conventionally understood meanings. Thus he can carry on a conversation with any member of his linguistic community and with members outside his community by means of an interpreter. That is something the parrot can never do. He is limited to imitating and repeating such sounds as "Polly wants a cracker." If upon hearing that sentence, his master should fail to provide the cracker, we would hardly expect the parrot to engage him in conversation to persuade him that he deserves his reward after all.

Herein lies one major difference between animals and humans. Animals possess a small repertory of sounds and gestures by which they communicate danger and the like to one another. They can also learn to imitate specific sounds; and they can be taught to respond in a definite way—as when a dog rolls over or runs to fetch an object as instructed by his human master. It is no accident that it is the human being who trains the dog, lion, or elephant to perform certain feats at his verbal command, and not the other way around. This is the point. Humans have *created* language. They communicate by means of a very large number of words and other symbols that are mutually comprehensible. And words and symbols are learned not by imitation but by a *conceptual grasp of their meaning.* Only the *sounds* of the words can be learned by imitation, not their meanings. With the exception of a number of onomatopoeic words, the human vocabulary consists of words whose vocal sounds bear no intrinsic connection with the objects they denote. Hence words are strictly conventional symbols, and meaning is imparted to them by the members of a society.

Language provides the human being with a method of learning fundamentally different from that of all other animals. The absence of language in all other creatures limits their mode of learning to the here and now. They must learn by concrete example. The human being, in contrast, can learn about a situation before it occurs: We can be verbally instructed to anticipate and beware. By means of language we are also given the opportunity to acquire the socially inherited knowledge and experience of our predecessors—without having to repeat their trials and errors. Language, then, is the vehicle by which the accumulated wisdom and experience of numerous generations is transmitted to each and every one of us.

Language is basically abstract. It consists of *general,* symbolic concepts. Words are *universals of discourse. Chair, table, apple, orange, man,* and *ape* are all words referring to *classes* of objects and not to any

24

real and particular chair, table, etc. When we say *chair* or *table,* what happens in our heads is not the projection of an image of the particular chairs or tables standing in our homes or offices. Instead, we grasp the general meaning of the word, which entails no concrete image whatsoever. Although this makes for occasional ambiguities in communication, it has its advantages, for it leaves room for interpretation and creative thinking.

Animal learning, in contrast, is restricted to the concrete. Confronted with a situation in which a bunch of bananas are out of his reach, a chimpanzee may discover how to use a stick in order to obtain one. He employs a combination of trial and error and nonlinguistic reasoning. Perhaps he imagines the banana and stick in various positions as he strives for it. His "reasoning," evidently, never goes very far beyond the concrete and actual situation. But what is unique to humans is that they do get beyond the concrete.

We do not know, of course, what goes on in the head of an animal. However, the impression is unavoidable that all higher animals, such as apes, dogs, cats, and horses, have vivid dreams. Those of us who have a pet dog have no doubt observed him while asleep, yet moving his limbs, rolling from side to side, barking, yelping, or growling. It is entirely possible that an element of the dog's dream consists of visual images of some sort, though we should not suppose that a dog "pictures" objects —other dogs, cats, humans, etc.—precisely as we do. Human visualization is not only a product of perception but of conception as well. Besides, given the fact that the dog's olfactory organs are so much more developed than the human's, we may assume that the dog has keen scent memory, and that whatever visual images he might have while dreaming are accompanied by scent "images."

In any event, if dogs, apes, and other animals are in fact capable of conjuring up a visual image or mental picture, that is most likely a picture of a *particular* object. For human beings, however, the visualizing of particular objects in the mind is quite difficult. When we try to picture a specific object, we succeed only momentarily, if at all. The object is often blurred, vague, and elusive. We naturally have no trouble in recognizing our relatives, friends, pets, and houses, but thinking about them does not so much involve concrete images as it does concepts. Thinking with concepts and words is so fundamental in human beings that we find it difficult to bring concrete particulars to mind. When we hear the word *banana* it is the general and abstract class of objects that comes to mind. But the chimpanzee's "thinking" lacks generality and abstractness, and he never gets beyond the particular banana

25

with which he is concerned. Recent research suggests that primates may have a rudimentary capacity for symbol use. That capacity, however, is not a *primary* mechanism of adaptation in primates as it is in humans. Hence, the animal-human distinction with respect to symbolizing ability is not contradicted by the existence of rudimentary primate "language."

Culture As Ideology

Human thinking (or reasoning) is therefore unique in that it consists of operating with symbols. In a sense symbols represent *ideas,* not concrete objects. When we say *man, dog, bone,* or *banana,* those words refer to ideal objects, though real and substantial men, dogs, bones, and bananas of course exist. However, humans can also think and talk about ideas that cannot be seen, smelled, or handled, such as justice, freedom, or immortality. In fact, human beings are often more motivated to pursue and attain ideal objects than they are substantial and edible ones. All ideas are social products, and insofar as they have meaning for humans and inspire them to act, they are essential elements of their cultural world. Having spiritual as well as material needs, humanity has created a cultural equipment with which to satisfy those needs.

Humanity's cultural equipment includes first and foremost those ideas which can be translated into tools and other means of acting upon nature. But it also includes what may be called society's ideology: superstitions, magic, religious beliefs, art, morals, law, customs, fashions, and so forth. No human society has ever existed without ideology in that sense. Even the earliest and most rudimentary societies show a profound concern with the ideas of immortality, magic, and the divine, and have corresponding rituals and ceremonies. Since other members of the animal kingdom cannot form abstract ideas which are passed on from generation to generation, ideology is nowhere to be found among them.

It has been said that "man lives not by bread alone." That was true even of earliest "man." Archeologists inform us that flints and other crude stone tools fashioned more than 50,000 years ago have an artistic as well as a utilitarian aspect. The earliest toolmakers wanted a beautiful implement and not merely a useful one. The earliest of human societies expended considerable labor not only on finding their subsistence but on painting their bodies and ornamenting themselves with necklaces, beads, and shells. Undoubtedly, humans are motivated by abstract ideas and concerned with matters extending far beyond the rudimentary urges of hunger and sex.

26   Ideology may or may not serve the biological and economic needs of

most members of a society and may serve some members better than others. Ideology may contribute to the maintenance and smooth working of a society, or it may split and divide it. In a word, ideology, no less than material equipment, constitutes a crucial determinant of the biological and social condition of humanity. It acts upon society's material equipment, technology, and social organization and is in turn affected by them. The tools humans invent, employ, and perfect, together with the extent and quality of their social cooperation, are clearly conditioned by prevailing ideologies; and ideologies are reciprocally affected by those other all-important cultural elements. Thus ideologies may either facilitate the cooperative social process by which humans produce the means of satisfying their needs or they may hinder and disrupt that process.

The Cultural Variety

So far, we have been talking as if there were one universal human culture. Theoretically it is true that the human beings living today are heirs to the traditions, knowledge, and experience of all their forerunners. Furthermore, since all human beings are members of one species, *Homo sapiens,* they share a basic human nature and basic human needs. The common human nature has given rise to common aspirations and concerns. In *all* human societies, most individuals value life over death, health over sickness, material well being over misery, happiness over pain and suffering.

In practice, however, common aspirations express themselves in a bewildering array of cultures, of which the differences are as significant as the similarities. Humanity is divided into a host of societies distinguished from one another linguistically, religiously, ethnically, nationally, and otherwise. Today we live in national and multinational societies with highly significant differences among them; and within each of those societies there exist innumerable local, regional, ethnic, religious, and social-class distinctions. Those distinctions are all *cultural* in the technical sense in which we have been using the term. They are the products of historical development.

Fossils of early humans going back hundreds of thousands of years have been found in places as remote from one another as Europe, Java, China, and South Africa. The earliest humans lived in small, scattered, and mutually isolated groups. They grappled with diverse climatic, geographic, and other environmental conditions and therefore developed different cultural ways of dealing with their specific circumstances. Every

27

group developed its own historically distinct social traditions, comprising everything from toolmaking to the conception of the supernatural. Each band, horde, or clan was the repository of specific historically evolved traditions.

Often, but not always, those culturally distinct groups also differed in the languages or dialects they spoke. That is a fair inference from what we know about the economically primitive peoples of North America and Australia. Anthropologists inform us that in the early twentieth century, the Australian aborigines, for example, numbering some 200,000 in all, spoke no fewer than 500 languages; and among the native peoples of California, some 31 families of languages and at least 135 dialects have been distinguished. In the Near East several divergent linguistic traditions are evident in the earliest written documents: Egyptian, Sumerian, Semitic, Elamite, and others. With the diffusion of writing, ever new tongues appeared, so that today one can hardly enumerate all the languages and dialects of humanity.

Language and culture vary independently of each other, with language changing much more slowly than culture and social organization. The English language spoken today is scarcely to be distinguished from the English spoken just prior to the Industrial Revolution. Yet English culture and society have changed radically since that time. Similarly, the United States has been transformed from an agrarian to an industrial society during the past 200 years. But the American-English language has changed hardly at all during that time.

If culture is said to be a mode of human adaptation to a natural environment, that should not be taken to mean that culture is determined by geographic, climatic, or other physical conditions. Culture refers to the totality of human designs for living that have evolved historically. In the space of a few hundred years, European and North American societies have undergone basic sociocultural changes. But the physical conditions of those continents have remained, for all practical purposes, virtually unchanged in that time. It is clear that culture is a *variable,* while the physical environment is a constant, relatively speaking. Therefore the condition of the physical environment of, say, Europe cannot account for the historical changes on that continent over the past few centuries. The human species is not physiologically adapted to any specific physical environment—culture has given humans a capacity for general adaptation.

The same logic applies to the relation of culture to the human biological makeup. In the early centuries of the first millennium A.D., China was far more advanced in science and technology than Europe.

By the nineteenth century, however, China had become a relatively backward society, dominated by the industrially powerful countries of Europe. In the late twentieth century, Chinese social structure and culture are quite different again. Clearly, those substantial sociocultural changes cannot be explained by reference to genetic, physical, or "racial" factors—all of which have remained relatively constant.

The human natural capacity for culture is the *necessary* condition for the emergence of culture in general. But the alleged biological or physical nature of some ethnic, linguistic, national, or "racial" group can never be made to account for the state of that group's culture. At birth all human beings unafflicted by mental or physical handicaps are capable of learning any language and acquiring any culture. The validity of that assertion may be seen in a mental experiment. It was employed by the noted anthropologist Alfred Kroeber some 60 years ago and remains as persuasive today as it was then. "Let us take a French baby," wrote Kroeber,

born in France of French parents, themselves descended for numerous generations from French-speaking ancestors. Let us, at once after birth, entrust the infant to a mute nurse, with instructions to let no one handle or see her charge, while she travels by the directest route to the interior heart of China. There she delivers the child to a Chinese couple, who legally adopt it, and rear it as their son. Now suppose that three or ten or thirty years passed. Is it needful to discuss what the growing or grown Frenchman will speak? Not a word of French; pure Chinese, without a trace of accent and with Chinese fluency; and nothing else.[13]

Substitute for the French baby and the Chinese parents persons of any other national-cultural background, and the outcome will be equally predictable.

Culture and Human Nature

The human being is of course a part of nature. Yet the fact that the human species has something *more* than nature to work with gives the species a measure of self-determination and autonomy unique in the organic world. Although human nature will receive detailed attention in the next chapter, which discusses socialization, we need to say a word about that subject in the present context.

Perhaps a good way to appreciate human uniqueness is to compare humanity with both the plant and animal world. We have every right to assume that plants have impulses of some sort that guide their vital processes. But to the best of our knowledge, they lack sensation and consciousness. They have no psychic center, as do animals, to which to report their organic states. They are exclusively oriented to the outside and exist in a state that is beyond self-control. Lacking also the communicative ability of animals, plants can neither free themselves from the immediate environment nor, of course, teach or learn. Hence if plants possess a "psychic" function regulating their interaction with both the inorganic environment and the animate world of insects and birds, they nevertheless have no inner life, no nerve center.

In contrast, all members of the animal kingdom, including humans, do in fact possess an inner life: nerve centers and the ability to communicate. All animals have vital feelings, urges, or impulses, which are at work whether the organism is asleep or awake and which account for the attention given to external stimuli. Here we need to underscore a point in opposition to behavioristic theories. Often such theories tend to portray a living organism as a highly passive entity that merely responds to external stimuli. Even human beings are conceived of in those terms. Thus B. F. Skinner in his book *Beyond Freedom and Dignity* denies all autonomy to humans and argues that they are totally conditioned by external forces.[14] That conception is not only false in relation to humans but to animals as well. No animal is so passive that he merely responds to external stimuli. Even the simplest sensations of an animal, and all the more so his behavior, are always a function of *internal,* drive-motivated attention.

In fact, internal drives and impulses appear to be so strong in animals that they have been called *instincts* and treated by psychologists and ethologists as *the* motivating force of animal behavior. The problem of whether or not there is such a thing as instinctual behavior in animals and humans will be addressed systematically in the next chapter. There we shall try to show that the instinctualist theorists have gone to an extreme. For the present, however, it will be helpful to explore the meaning of "instinct" in order to bring out the main differences between animals and humans. –

1. Instincts facilitate the nourishment and reproduction of the organism in question, *or of another organism,* as in the case of the mother providing and caring for the animal infant.

2. Instinctual behavior manifests itself in a definite pattern or rhythm. The pattern, in the main, is not acquired through learning or habit. Yet it may nonetheless be directed to present and even future needs and states (e.g., provision of food for the winter "as if" the organism actually anticipated the future).

3. Instinctual behavior not only serves the needs of the individual members of the species, but ultimately contributes to the survival of the species itself. Instincts provide a given species with a typical way of coping with the recurring situations which are significant for life. Instincts are fixed and "rigid" in the sense that they guide the organism well in the specific environmental circumstances to which it has adapted. But the same rigidity causes the organism to display error, confusion, and disturbance when the particular environmental conditions change.

4. Instinctual behavior is innate and hereditary. Individual species members inherit not only a general capacity, but specific modes of behavior. "Innateness" does not necessarily mean that the instinct is already present at birth; it may be coordinated with definite stages of growth and development. If we therefore speak of a *hunting instinct,* we should bear in mind that what is inherited is not the skill to do so successfully. For such skill presupposes a certain stage of maturity and a considerable amount of learning.

One should not suppose that because animals are subject to strong internal drives, they are lacking in creative behavior. Everything we know about animals persuades us that they are active beings. The spontaneous explorations of their habitats yield both gratification and the learning of useful patterns. Animals learn through self-training and not merely through human conditioning of them. Imitation and associative memory also make cumulative experience possible among animals, although of course this must be distinguished in the sharpest possible terms from the cumulative cultural traditions that are characteristic of humans. In the case of herds, flocks, packs, and other forms of gregarious animal life, the young learn from the mature. But each generation must acquire this learning anew since, in contrast with humans, animals have no way of preserving and transmitting what they have learned to more than one or two generations.

Finally, animals most certainly possess what might be termed *practical intelligence,* a capacity for solving concrete problems not encountered earlier in their life histories and for making choices, whether among material things or among fellow members of the species. Intelligence involves the ability to employ *experience* and to cope with new situations. It should be stressed that an animal's practical intelligence entails a genuine act of discovery; for it is neither the result of instinct nor of habit. Animals at the highest stage of their psychic life can choose and "deliberate" in the nonsymbolic and nonlinguistic sense. And in   31 their emotions they are even closer to humans, showing the capacity for helping others, for sympathy, and the like.

THE UNIQUENESS OF HUMANITY

From a strictly biological point of view, the human being is just another species of the animal kingdom. Because the human being is an organism, some students of human behavior have attributed human motivations and conduct to various instincts not dissimilar to those found in other animals. However, even such basic activities as ingesting food or engaging in sex cannot be explained in humans as the results of biologically innate and fixed patterns. Humans do not necessarily eat when they are physiologically hungry. They eat culturally selected foods, at socially appointed times, in socially approved ways. Nor are human sexual relations dictated by an estrous cycle. Humans choose their mates on the basis of a variety of social and cultural criteria such as religion, ethnicity, nationality, and social class. Although humans everywhere are quite similar physiologically, their patterns of conduct are everywhere different. The immense range of distinct activities which constitutes the histories of biologically similar humans demonstrates that the goals and objects of human conduct are not biologically given. Both the ends and the means of human activity are diverse and changeable.

Thus with humanity a new quality of being emerges. The human being's place in nature is unique in that he is less subject than all other beings to the processes of his internal physiological nature as well as to the processes of external nature. The animal is immersed in his natural environment and comparatively bound by it. But humanity transforms the environment into a *historically changing cultural world*. Human beings never leave the realm of nature; but as a result of their conscious practical activity they live and act in a world of their own making, a world in which they are both authors and actors. That is the sense in which humans have a greater capacity for self-determination than all other beings.

The creation of humanity's material and symbolic world is made possible by the unique human capacity for conscious and self-conscious reflection. Unlike all other creatures, the human being develops a symbolic center to his being, which we may call a *social self*. The self is "spiritual" in that it cannot be located anywhere in the human anatomy. The self is not a tangible entity that can be touched or handled. It is not even observable. But it is no less real on that account. Consciousness of self and others, emerging in the course of social interaction, is the basis of all distinctively human activities. To understand how the self emerges and how an individual acquires culture, we must explore the process of socialization.

3

Socialization:
Becoming Human

As the newborn human infant emerges from its mother's womb, it enters a world which is at once social, cultural, and historical. It is now on the road to becoming a human social being through a process called *socialization*. That term refers to a person's gradual acquisition of his cultural inheritance: The beliefs, practices, morals, laws, customs, and other capabilities which each and every one of us comes to possess as members of a given society. *Enculturation* is sometimes used interchangeably with *socialization*.

But no sooner has one offered a definition of socialization than one realizes that it raises a fundamental question: What is the nature of the being that is undergoing socialization? For it is clear that the really important question where socialization is concerned is this: What elements of human conduct does the child learn and acquire from his society and culture, and what elements does he bring into the world with him as a part of his genetic endowment? That has been a central issue from the time of the earliest reflections on human nature; and the issue continues to divide students of human behavior to this day. Clearly, an adequate understanding of socialization, its depth, scope, and consequences, requires a prior clarification of this all-important question: What is the nature of human nature?

In response to that question two highly influential doctrines have emerged. On the one side are those whom we might call the *instinctual-ists* and on the other are the *behaviorists,* who comprise not only the modern behavioristic psychologists but all other environmental and sociological determinists as well. Let us examine each of these doctrines in turn.

THE INSTINCTUALIST DOCTRINE

The instinctualists, as the term suggests, conceive of the human being as driven by internal, genetically determined biopsychic forces. Whether it is fear, anger, rivalry, the desire for mastery over others, maternal love, sexual desire, aggressiveness, or what have you, each of those conditions is said to be determined by inherited and innate biological processes. There was a time in the history of human psychology when the list of such instincts was seemingly endless. Today psychologists are much less inclined to employ the term *instinct* in an effort to explain patterns of human conduct. Freudian psychology is a special case which we shall consider in a later context.

In recent decades the instinctualist doctrine has enjoyed a marked revival, primarily among ethologists, that is, students of animal behavior. If we examine the writings of Konrad Lorenz, Niko Tinbergen, Desmond Morris, and other distinguished ethologists, we shall find that no matter how their views may vary in detail, they all share a common body of assumptions: That there exist innate determiners of some kind which when affected by specific stimuli call into operation certain neural, glandular, and muscular processes resulting in particular patterns of behavior or psychological states.[1] That conception of "instinct" has invariably been derived from the study of animals, not humans.

In their scientific writings, the outstanding ethologists have been careful to note that though they believe the existence of instincts has been proved in "lower" animals, it has yet to be proved in humans. Moreover, even with regard to animals, important qualifications are introduced. Tinbergen, for instance, has observed that while many species have a parental instinct, others never take care of their offspring and probably lack the corresponding neurophysiological mechanisms. Not only that, a species can lose an instinct, retaining the nervous mechanism but not the motivational one. Tinbergen cautions us to refrain from generalizations until we know more about such matters.

But despite such properly cautious qualifications, the doctrine of instinct *has* been generalized to animals and to humans as well. Such generalization is mainly a result of the popularizations of modern ethological theories; but in some cases it is due to extrapolations from animal 35 to human behavior made by the ethologists themselves. Hence it is very important to note that the revival of "instinct" as an explanatory principle by Lorenz, Tinbergen, and others has met with widespread criticism and opposition.

The theory that there exist "innate" or "inherited" mechanisms which determine definite types of behavior has been rejected by many if not most ethologists and zoologists. What is especially interesting is that the theory has been rejected for animals as well as for humans. For as we shall see, there does not seem to be anything in the organism of any higher animal, whether in the genes or elsewhere, that determines *fixed* action patterns.

Nevertheless, a particular form of the instinctualist doctrine appears to have gained popular acceptance in recent years. The cause of human aggression, we are told, resides in an instinct which we share with other animals and, indeed, which we presumably have inherited from our prehuman ancestors. According to that doctrine, human beings are by nature aggressive and violent creatures; they are killers. In addition, they have an instinct for territorial defense. Thus if men are cruel, commit murder, and fight wars, those actions are all to be traced to the basic human biological makeup.[2]

To come to grips with such claims one must first have an adequate grasp of what is meant by *aggression* or *aggressive behavior*. Take the phenomenon of war, and especially modern warfare. When a young man in the United States is drafted and sent across the ocean to war, is this an expression of the instinct for aggression? Or when a pilot flying 30,000 feet above the ground drops his bombs on the target below, is he necessarily filled with hatred? He knows that his deliberate actions will result in the death and injury of many human beings. But does he perform those actions in response to some biological urge? Reflection on those questions strongly indicates that the attempt to explain why men go to war by reference to some innate urge is quite misleading.

That humans are aggressive by nature is not a new idea. In his book *Leviathan,* the seventeenth-century philosopher Thomas Hobbes argued at length that humans are naturally inclined to be warlike. In his attempt to grasp what humans would be like in a presocialized condition, Hobbes created an imaginary state of nature. In that state, where "men live without a common Power to keep them all in awe, they are in that condition which is called Warre; and such a warre, as is of every man, against every man." But not long afterward the famous philosopher Jean Jacques Rousseau countered this view with the incisive observation that

"man becomes a citizen before he becomes a soldier."[3] What Rousseau meant to convey in this statement may be expressed in modern socio-logical terms: Before men are ever sent off to war they are taught to play a variety of social roles and they are indoctrinated with certain ideologies. War is a socially organized activity; and before a man can engage in warfare, he must be socialized into the role of soldier. He must be trained as a warrior and taught to kill. War is a *social,* not a natural phenomenon.

It is worth noting that this sociological understanding of the matter was not new with Rousseau. Isaiah, the great Hebrew prophet of the eighth century B.C., foreseeing a time when wars would be banished from the human condition, had this to say in his renowned vision:

> And they [nations] shall beat their swords into
>   plowshares,
> And their spears into pruning-hooks;
> Nation shall not lift up sword against nation,
> Neither shall they learn war any more. (Isa. 2:4)

There is a remarkable sociological insight here. It is nations that make war, not individuals. If men make war, it is because they have *learned* to do so. War is learned behavior, and not the result of an innate drive.

The sociological view is not taken seriously by such highly influential ethologists as Konrad Lorenz and his followers. Their view, in contrast, is that *all* animal behavior, including human actions, is instinctual. The animals most closely studied by Lorenz are domesticated ducks and geese. The "triumph ceremony" of the Greylag geese, for example, with its series of movements and variety of noises—neck stretching, head tilting, raucous cackling, etc.—results in specific relationships between them. Those movements and sounds are presumably "programmed" in the genes of the creatures. All animals, according to Lorenz, have such genetic programs, which are designed to serve what he regards as the four great drives: hunger, fear, sex, and aggression. Humans too al-legedly share those drives, the last of which is supposed to account for war and other violent conduct.

But is there such a thing as an instinct to make war? To answer that question we must first of all recognize that war is an organized conflict between members of the *same* species. As such, it is extremely rare in the animal kingdom. War is not the same as competition between two different species, even if it should involve conflict. To the best of our knowledge, the only animals who make war are human beings and per-haps certain species of ants. And no one will seriously argue that hu-mans and ants make war because they share common genetic endowments. More important, if "instincts" are presumed to be useful to a species, then it is crystal clear that war cannot be the result of an instinct. For war is not only useless but unquestionably harmful to the human species. In modern warfare massive populations are starved, injured, oppressed, and killed and their resources destroyed. Thus if someone argues that wars accomplish a kind of "natural selection" among humans, it is easy to see why the argument is fallacious. In a state of nature, natural selection eliminates those members of an animal species that are unable to adapt to the specific environmental circumstances. In war, in contrast, it is not the weak, sick, or otherwise "unfit" who are sent to be maimed and killed, but rather the strong and healthy. It is clear that war, far from being favorable to the human species, is the very opposite. There are no grounds, then, for the belief in a human war instinct.

There is good reason to believe that Lorenz and his followers have even misunderstood the "aggression" of animals. Field studies conducted over the past 20 years on many different types of animals have shed considerable light on fighting between members of a species. The results of such studies have been summarized by the zoologist V. C. Wynne-Edwards, who has concluded that although

the stakes are sometimes life or death, serious fights and bloodshed are uncommon. Convention restricts the contestants very largely to displaying themselves for mutual appraisal or engaging in a harmless trial of strength, and from these actions they predict what the outcome of real combat would be without needing to fight it out. What they do is to threaten or impress one another, at the crudest extreme by exposing or even briefly using their fighting weapons—butting with their horns or baring their teeth. In the most refined examples, the victor overrides the self-confidence of the loser by sheer magnificence and virtuosity.[4]

And, as Adriaan Northland has observed:

The goal of fighting in many species is not so much fighting in itself but rather to establish a social organization which makes fighting superfluous.[5]

In the same vein, J. L. Cloudsley-Thompson noted that

threatening gestures and ceremonial displays frequently replace actual fighting. In this way conflict tends to become ritualized and adapted, so that its function may be achieved without harm to the rivals.[6]

Finally, as Professors Ueli Nagel and Hans Kummer have succinctly stated,

38   Aggression in animals is primarily a way of competition, not of destruction.[7]

What those studies indicate, then, is that the effect of fighting is to reduce and eliminate bloodshed and killing. The displays of fighting and aggression, far from leading to the destruction of other members of the group, lead to mutually accommodating and even cooperative social relations.

In primate societies, for instance, the display of an aggressive posture by some members of the group appears to be designed to forestall injury to others. It also serves to prevent the disruption of the group's vital activities. Combat is for the most part ritualized, with the individuals in question going through the motions of fighting rather than actually doing so. In that way the "aggressive" encounters between species members preserve peace within the group and contribute to the preservation of the species.

Furthermore, the "aggressive" behavior of animals is clearly contingent upon definite circumstances and not rigidly determined by genetic endowment, as the innate aggressionists would have us believe. Change the developmental conditions of a so-called wild, carnivorous beast, and he will behave quite differently from his "normal" patterns. Lions, when raised as domestic pets, grow up incapable of hunting and are anything but aggressive. When they are returned to their usual habitats, they find themselves unable to provide for their own livelihoods. Although the potentiality for hunting is present in the lion, he must undergo a definite training and acquire experience in order for the potentiality to develop. Wolves, like lions, also have to learn how to hunt. If after being raised domestically they are returned to the wild, they will observe caribou with only casual interest; even when they are hungry it never occurs to them that the caribou are potential prey.

A FALSE INTERPRETATION OF DARWIN'S THEORY

Darwin himself never portrayed the animal world as engaged in a ferocious, dog-eat-dog struggle for existence. Nor did he ever suggest that because they have evolved from "lower" animals, human beings have created a world of hostility, conflict, and aggression. Nevertheless, both of those ideas have become widespread. Darwin's theory, as we have seen, is based on the fact that offspring often differ physically from their parents in significant ways. Those changes, which he termed *variations,* are today called *mutations.* Darwin assumed that the new traits best suited to the circumstances in which the species lives will be most likely to appear again in the next generation. In that way the most successful   39 variations will be transmitted from generation to generation until the species gradually evolves, through the process of natural selection, into a somewhat different species. That is also the way humans evolved out of earlier animal forms.

To illustrate the process, let us assume that the existence of a thinly furred species is gravely threatened as the climate grows colder owing to the advent of a glacial period. As a result of a mutation, some members of the species acquire a thicker growth of fur. They are more likely to survive and reproduce than the less hairy animals. Some of their offspring, in turn, would adapt more easily to the increasing cold and would transmit the thick-fur genes to the next generation. Thus gradually, over a period of many generations, an entire species might change from being thinly to thickly furred.

Clearly, that process of natural selection does not entail a fierce struggle for survival among members of the same species. It is a *struggle for existence,* as Darwin termed it, only in the sense that every organism strives to maintain itself. Darwin himself emphasized that he employed this term in a large and metaphorical sense including *"dependence of one being on another,* and including . . . not only the life of the individual but success in leaving progeny."[8] Let us note the underscored phrase. It is evident that for Darwin the "struggle for existence" not only involved no actual war among species members, but even required mutual dependence. That important element of his theory has been almost entirely overlooked.

In *The Descent of Man* Darwin pointed out that in numerous animal species, the struggle between individuals for the means of existence disappears and is replaced by *cooperation.* He also showed that cooperation results in the development of social and other faculties which afford the species the best conditions for survival. In such cases the "fittest" are not the physically strongest, most cunning, or most aggressive, but rather those who learn to help and support each other. "Those communities," wrote Darwin, "which included the greatest number of the most sympathetic members would flourish best, and rear the greatest number of offspring."[9] Yet Darwin never elaborated that important part of his theory, and his followers ignored it entirely.

It remained for Prince Petr Kropotkin, the Russian geographer and humanist, to explore the overlooked element systematically. Having come under the influence of the French philosopher Alfred Espinas and the Russian zoologist Karl Kessler, Kropotkin spent several years in Siberia and Manchuria studying animal life under natural conditions. He wrote a series of articles presenting the results of his studies, and the articles were published as a book in 1902 under the title *Mutual Aid: A Factor of Evolution*. Although Kropotkin's conclusions were largely ignored at the time, they are increasingly recognized today as an extraordinary scientific contribution.

For Kropotkin, mutual aid, mutual support, and mutual defense—in a word, sociability and cooperation—play at least as important a role in evolution as does mutual struggle. Indeed, cooperation may be an even more important factor. That may be readily seen by asking this question: Who are the fittest, those who engage in continual war or those who assist one another? Kropotkin's studies persuaded him "that those animals which acquire habits of mutual aid are undoubtedly the fittest. They have more chances to survive, and they attain in their respective classes, the highest development of intelligence and bodily organization." Furthermore, mutual aid, as a factor of evolution, "most probably has a far greater importance, inasmuch as it favors the development of such habits and characters as ensure the maintenance and further development of the species, together with the greatest amount of welfare and enjoyment of life for the individual, with the least waste of energy."[10] That is Kropotkin's central thesis which he patiently documented in the first two chapters of his remarkable book. (The remainder of the book is devoted to the importance of cooperation in human development.)

Kropotkin also showed that what is often described as interspecies competition is nothing of the kind. If a species succumbs, it is not because it is exterminated or starved out by another species, but simply because it cannot adequately accommodate itself to new conditions. As with Darwin, a "struggle for life" exists here only in the metaphorical sense, and no actual fighting or conflict occurs between the species. Even under conditions of extreme scarcity, Kropotkin saw no evidence to support the theory of interspecies competition resulting in the extermination of one by another. "If," he wrote,

the physical and the biological conditions of a given area, the extension of the area occupied by a given species, and the habits of all the members of the latter remained unchanged—then the sudden appearance of a new variety might mean the starving out and the extermination of all the individuals which were not endowed in a sufficient degree with the new feature by which the new variety is characterized. But such a combination of conditions is precisely what we do not see in nature. Each species is continually tending to enlarge its abode; migration to new abodes is the rule with the slow snail, as with the swift bird; physical changes are continually going on in every given area; and new varieties among animals consist in an immense number of cases—perhaps in the majority—*not* in the growth of new weapons for

41

snatching the food from the mouth of its congeners . . . but in forming new habits, moving to new abodes, and taking to new sorts of food. In all such cases there will be no extermination, even no competition—the new adaptation being *a relief from competition, if it ever existed.*[11]

Hence from all that we have thus far observed about the animal world, it should be evident why one should either reject all "nature-red-in-tooth-and-claw" theories or at least be very skeptical of them. Not only do theories of innate human aggression and competition remain unproven; so do theories of innate animal aggression and competition.

Before we leave the issue of innate aggression, a word should be said about a related thesis. The instinctualists have also claimed that animals and men share a "territorial instinct," that is, "an inherent drive to gain and defend an exclusive territory." "Instinct" is defined as "the genetically determined pattern which informs an animal how to act in a given situation."[12] But the theory of a territorial instinct has not fared very well. Here is a list compiled by the anthropologist Ashley Montagu of only a few of the mammals that exhibit no territorial tendency whatsoever:

The California ground squirrel, adult male long-tailed field mice, she-wolves, the red fox, the Iowan prairie spotted skunk, the northern plains red fox, the zebra, Grant's gazelle, wild dogs, the Bahamian rodent or hutia, cheetahs, mountain goats, deer, wallabies, rhesus monkeys, langur monkeys, baboons, and in the Hominoidea, the superfamily to which man belongs, together with the orang, the chimpanzee, and the Gorilla.[13]

Montagu also cites Professor Francois Bouliere of the University of Paris who, after surveying the evidence, concluded that "territorial behavior is far from being as important in mammals as in birds, and very often it is limited to the temporary defense of the nest or of certain parts of the home range."[14]

As for humans, the studies we have of food-gathering and hunting peoples reveal no evidence of a territorial instinct. Fighting to defend a territory is practically unknown among such groups. A sense of territoriality emerges among human beings under specific socioeconomic conditions: with the cultivation of plants and the domestication of animals. It is only with the invention of agriculture that sedentary populations came into being. Historically that process occurred for the first time about 12,000 years ago. First villages and then towns came to be regarded as one's "home," and often those settlements were attacked by nomadic marauders. When attacked, the settled inhabitants sought to defend the precious worldly goods which they had produced at so high a cost to themselves, just as the invaders sought to acquire the products of the village in question. But it is evident that neither the defenders nor the invaders of the settlement were activated by a "territorial instinct." The causes of the conflict between them are to be found not in their biological makeup but in their respective social organizations. These causes, in short, are social and have nothing to do with any alleged instincts.

However, some readers might still have nagging doubts. They may acknowledge, in light of the evidence, that human beings have no instinct of aggression. But, they may ask, does that mean that human beings have no instincts whatsoever? We can shed light on that important question by exploring the nature of human sexual behavior.

HUMAN SEXUALITY: IS THERE A "SEX INSTINCT"?

It is highly likely that if most people were asked to name the two most basic instincts in humans and, indeed, in all animals, the answer would be: hunger and sex. Undoubtedly every living organism requires food in some sense if its life is to be maintained. We may suppose that when higher animals such as mammals are deprived of food, they eventually experience certain internal states of activity corresponding to shifts in the physiological balance of their systems. Those internal states prompt them to search for food. And if, say, a carnivore in the wild has properly learned to hunt, and the conditions are favorable, he may succeed in satisfying his hunger. Is there any justification for regarding such behavior as instinctive? Most probably not, if by "instinct" we mean a "genetically determined pattern which informs an animal how to act in a given situation."[15] For we have already seen that a considerable amount of learning is required by wild beasts such as lions and wolves before they can become competent hunters. Without such learning the animal in question remains uninformed as to "how to act in a given situation." So while all animals need food and experience a bodily state which we call *hunger,* that still does not mean that hunger is an "instinct."

The same logic applies to sexual activity in animals. In the preceding chapter we noted that the endocrine and generative systems of a mature female mammal go through an estrous cycle which makes copulation possible in definite periods, and in those periods only. That is the rutting season in deer, for example, when the female is "in heat," which in turn leads to sexual excitement in the male deer.

Even in such cases, however, one should not be too quick to con-   43 clude that deer, dogs, cattle, and sheep have a sexual instinct. For although the estrous cycle may be a necessary condition, it is not a sufficient condition for copulation. Once again, a certain amount of learning seems to be required before the animal is informed "how to act in a given situation." Many of us have no doubt witnessed our own pet dogs engage in indefinite, trial-and-error behavior. That strongly indicates that the sexual urge experienced by dogs and other animals is not so rigid a determinant of behavior as to warrant the term *instinct*.

If there is no good reason for regarding hunger and sex as instincts in animals, there is even less reason for regarding them as such in humans. As we have noted, human beings do not eat simply in response to internal sensations corresponding to the physiological shifts in their systems. An immense amount of learning and socialization is presupposed in human hunger. In all cultures humans eat at socially appointed times which do not simply reflect their physiological states but are conditioned by the given cultural patterns. In all human societies eating is governed by more or less definite cultural norms. It is a social and even ceremonial occasion. What, when, and how one eats, types of cuisine, occasions, and implements for eating are all culturally defined.

As regards human sexuality, we may begin by noting that there is no estrous cycle in humans. There are of course certain neurophysiological organs and processes which make it possible for mature humans to engage in sexual activities. But no human being has the ability to perform sexually before he has learned how to do so in some social context. The requisite knowledge may be acquired from one's family, from one's peers, or from books; but knowledge is absolutely essential if the potential capacity for sexual performance is to be transformed into the actual ability for such performance. In every civilized country there are adult men and women who enter upon marriage without the slightest inkling of how to go about sexual intercourse. Such cases are known to almost every marriage counselor, obstetrician, and gynecologist.

If a person has not learned what to do about his or her sexual feelings, there is nothing in the human organism which automatically leads an individual to sexual intercourse. That is not to deny that there exist in every human being certain physiological processes which are the source of sexual or erotic impulses. But to call those impulses or dispositions *instincts* obscures the fact that the psychophysical conduct of the individual is largely influenced by his cultural context. Thus the object of one's sexual interest may depend on such cultural conditions as standards of beauty and ethnic, religious, and social-class background. Indeed, even the decision of whether the object of one's sexual interest

44

will be a member of the opposite sex, the same sex, or of both sexes will largely depend on the mode of socialization which one has undergone.

In that light it is clear that no instinct governs human sexuality. What we find in human beings instead is a wide range of erotic impulses which may be directed toward oneself, toward members of either or both sexes, and even toward animals and objects. The outcome in the case of every human being is contingent upon socialization. And one very important concomitant of socialization is the process of *repression*.

That socialization necessarily entails some measure of repression may be illustrated this way. As soon as the infant enters the world, the fact that it is either male or female assumes extraordinary importance. From that moment on the infant will be carefully and systematically socialized into a specific gender role. In the culture of the United States, for instance, the male infant will typically be dressed in blue and the female in pink. As very young children, the boy will wear trousers and the girl a dress. The boy will be taught to play baseball and football and to "act like a man," while the girl will play with dolls and model herself after her mother. Throughout childhood males are thus made into boys and females into girls. With time they learn to distinguish men's roles from women's, and "manly" characteristics from "womanly" ones. They also learn that men make love to women and women to men, but that making love to a member of one's own sex is unnatural, deviant, perverse, or the like.

As a result of such direct and indirect inculcation, each male and female consciously or unconsciously represses any impulse he or she might have to behave in a manner unbecoming to his or her gender role. If an adolescent boy or girl should experience an erotic feeling or attraction toward a member of the same sex, it is almost certain that the feeling will be quickly repressed. In those terms heterosexuality is not to be regarded as an automatic behavioral outcome of one's male or female physiology.

In sum, whether it is hunger, sex, aggression, or any other aspect of human conduct, the evidence clearly shows that those aspects cannot be accounted for by any innate predeterminants.

THE BEHAVIORISTIC DOCTRINE

If the instinctualists have fostered an erroneous image of the human being, so too have the behaviorists. The founder of the behaviorist doctrine, John Broadus Watson, was opposed from the very beginning to any psychology that speaks of consciousness.[16] Writing in the early decades of the twentieth century, Watson looked upon the concept of consciousness as a mystical survival of medieval times, a superstitious idea about the soul utterly unworthy of scientific consideration. Taking a naive empiricist view of science, according to which something must be directly perceived in order to be studied scientifically, Watson and his followers vehemently rejected the concept of consciousness and all other subjective states. One can no more prove the existence of consciousness, they asserted, than one can souls and ghosts. Consciousness and all other phenomena of the "mind" are "mere assumptions," for they cannot be seen, touched, or exhibited in a test tube. Even if mind exists, it cannot be studied scientifically and might just as well be ignored.

The behaviorists, like the instinctualists and other physiological determinists, have been interested in studying animals. In fact, one of the main elements of the behaviorists' scientific program has been the application of the methods and standpoint of animal psychology to human psychology. Watson and his followers eventually concluded that instincts, native intelligence, and native talents have no real existence at all. What we ordinarily call native "gifts" and abilities are strictly the product of environment and training.

For Watson, abilities, in animals and humans alike, emerged through the process of *conditioning*. Abilities are simply conditioned reflexes or conditioned responses. In that basic tenet of his theory, Watson was profoundly influenced by the Russian physiologist Pavlov, who in his famous experiments with dogs first aroused a salivary response in them by placing food on their tongues. Food was the "adequate stimulus" to the response of salivation. In the experiment a bell was rung every time the food was presented. After several repetitions the food was withheld, yet the bell alone aroused the response. In other words, a stimulus (the sounding of the bell) that originally did not elicit the response eventually came to do so because it was one of the conditions under which the response was made. Although that experiment involved a very simple case of the conditioned reflex, it illustrates the principle of all conditioning, which became the cornerstone of Watson's studies of humans.

Conditioning, Watson argued, is the simplest form of learning, and all other forms of learning are reducible to that elementary process. Conditioning begins in earliest infancy; and out of the few simple responses which the infant has in its repertory, all the complex activities that an adult exhibits are built up. The so-called native gifts and powers are all products of that process. Through conditioning, complex integra-

46

tions of reactions have been knit together and attached to certain stimuli.

In the behaviorist scheme of things, even emotions are learned or conditioned reactions. For Watson, only three emotions—"fear," "rage," and "love"—could be aroused in the infant, prior to learning, by appropriate stimuli. Of course, in the behaviorists' view the terms *fear, rage,* and *love* have nothing to do with the "mental." Emotions are not matters of feeling or affect; they are simply bodily reactions. They differ from other bodily reactions in that they are predominantly visceral, involving glands and involuntary muscles like those of the intestinal walls. Similarly, Watson vehemently opposed the theory that "pleasure" or "satisfaction" tends to build up successful reactions while "annoyance" or "dissatisfaction" eliminates unsuccessful ones. His main objection to the theory stemmed from the fact that "pleasure," "satisfaction," etc. implied the intervention of a mental process. Hence human learning for Watson was a thoroughly material and mechanical affair.

Following the lead of classical physics, behaviorism conceived of the human being as an organic mechanism and sought to analyze all human conduct into a series of specific stimulus-response connections. Everything human was reduced to definite physiological processes, to particular muscular and glandular movements. Anything human which could not be so reduced was unreal, or at least less real. Classical behaviorism thus conceived of the human being as a passive, machinelike entity, wholly determined by the external stimuli of the environment.

B. F. Skinner

In all essential respects the present-day behaviorism of B. F. Skinner and his followers is hardly to be distinguished from the original doctrine of J. B. Watson. "We can follow the path taken by physics and biology," wrote Skinner,

by turning directly to the relation between behavior and the environment and neglecting supposed mediating states of mind. Physics did not advance by looking more closely at the jubilance of a falling body, or biology by looking at the nature of vital spirits, and we do not need to try to discover what personalities, states of mind, feelings, traits of character, plans, purposes, intentions, or the other prerequisites of autonomous man really are in order to get on with a scientific analysis of behavior.[17]

Skinner went on to say that the stimulus-response model never solved the basic problem of behavior and was never very convincing "because something like an inner man had to be invented to convert a stimulus into a response." But upon reflection, it becomes evident that Skinner's   47 new terminology of *operant conditioning, positive reinforcers,* and *negative reinforcers* achieves no greater success in eliminating an "inner man." He wrote, for example, that "negative reinforcers are called aversive in the sense that they are the things organisms 'turn away from.' "[18] Skinner seems to have overlooked the fact that his theory contains the same contradiction that he noted in classical behaviorism. For while he denied that organisms have any autonomy, he nevertheless conceived of them as turning away from negative reinforcers. Clearly, there remains an "inner man" of some kind in this scheme too, for otherwise how would one account for the organism's turning away? Or did Skinner mean that the turning away is like that of a mechanical doll?

Skinner wished to get rid of the "inner man" and all that this implies, an organism with a will and consciousness of its own. Thus he wrote that "man's struggle for freedom is not due to a will to be free, but to certain behavioral processes characteristic of the human organism, the chief effect of which is the avoidance of or escape from so-called 'aversive' features of the environment."[19] But again it is clear that Skinner succeeded no better than his predecessors, for even in his conception, man has sufficient *consciousness* to distinguish between positive and negative features of the environment; sufficient *will* to desire to escape from the aversive features; and sufficient *autonomy* to in fact escape. But Skinner seems not to have noticed his implicit reliance on the traditional concepts of consciousness, will, and autonomy.

For Skinner as for his predecessors, "freedom," "dignity," "consciousness," "autonomy," "will," etc., were pure fictions. All of those terms imply that the human is an active, creative being. But that is precisely what the behaviorists wish to deny. "A person does not act upon the world," wrote Skinner, "the world acts upon him."[20] The human being is a wholly determined and passive object. We lose nothing, Skinner argued, by stripping away the functions previously ascribed to that fictional entity called "autonomous man," and transferring them to the controlling environment.

Thus Skinner followed closely in the footsteps of Watson. The image of the human being held by both theorists was that of an object totally controlled by external conditions. In Skinner's "Utopia" humans will be no less controlled than before; but in this new society the environment will no longer be the unplanned, nonscientific determiner it has been throughout history. Instead it will be an environment governed by what Skinner regarded as advanced, scientific behavioristic principles.

Several questions about both the theoretical and political aspects of Skinner's proposal suggest themselves. If humans are wholly passive and determined creatures, then who will serve as the active conditioners in Skinner's new society? If everyone learns only through conditioning, then who will condition the conditioners? Skinner asked us to remember that the human environment is largely of our own making.[21] But how do passive, wholly determined beings make their own environment? Because such problems, contradictions, and ambiguities are nowhere resolved in Skinner's theory, we may justifiably conclude that the behavioristic doctrine is no less misleading and erroneous than the doctrine of instinctualism. A need therefore exists for a conception of the human being and his socialization which avoids the errors of both of those extreme doctrines. A giant step in the right direction was taken by George Herbert Mead, a contemporary of J. B. Watson.

When Mead began to develop his distinctive understanding of human conduct, there were several influential theoretical currents in psychology. Two of them have already been examined critically: (1) that of the instinctualists and other physiological determinists, and (2) behaviorism. A third, which we have not yet touched upon, may be called *idealism*. The idealists had no hesitation in employing the concepts of "mind" and "consciousness." But from Mead's standpoint their use of those concepts was unsatisfactory. For they treated mind, self, and consciousness as spiritual givens and thus failed to recognize that those qualities are not bestowed at birth but rather emerge under specific social circumstances. Like the behaviorists, Mead asserted that human behavior arises largely as a function of environmental experience. Mead differed from the behaviorists, however, inasmuch as he believed that humans are active determiners of their social environment and not simply the passive recipients of external stimuli. What follows is a summary exposition of Mead's distinctive social psychology.[22]

MIND, SELF, AND SOCIETY

There are aspects of human behavior and interaction that are mediated by little or no thought and which therefore bear a marked resemblance to the interaction of animals. Two boxers or fencers, for instance, respond to each other's movements and gestures reflexively, without deliberation. Such interaction is not unlike that observed in a dog fight. There each dog's gesture, such as the baring of a fang, is a stimulus for the other's response, which in turn becomes a new stimulus. Both the humans and the dogs in those examples are engaged in a *conversation of gestures.* They interact with each other reflexively, and their gestures therefore carry no symbolic meaning.

On the basis of our present knowledge we may presume that animals remain at this level of interaction—a conversation of nonsymbolic gestures—although they may have a very small repertory of symbolic sounds and utterances. In the case of humans, in contrast, nonsymbolic gestures constitute a small proportion of their total interactions. A man can strike another without intending to do so or recoil from something before he knows why. But he can also shake his fist in anger, displaying a deliberately hostile attitude. Now there is an *idea* behind his gesture. Indeed, when a gesture with a specific meaning arouses in other individuals the same idea it arouses in the first, we have a significant symbol. That is what is commonly meant by *language:* communication by means of symbols. Among meaningful gestures the vocal ones are the most important. It is through symbolic communication that human beings call forth commonly understood acts and responses in groups, communities, and societies. It is also by means of symbols that thinking takes place. Thinking is the conversation through meaningful symbols that one carries on with oneself. It is an internal conversation not unlike the external one we carry on with others.

Thinking or "mind" thus clearly presupposes a social process. Minds have no existence before and independently of social process. If one begins with mind, then its origin becomes a mystery. But if one assumes the priority of social process and communication—priority in the sense that they are antecedent to any individual mind or self and in the further sense that neither mind nor self can emerge except in a social context—then the mystery ceases.

In a conversation of *nonsymbolic* gestures, the stimulus and response *differ:* One animal's threat leads to another's flight; an infant's cry leads to the mother's care; one boxer's jab leads to the other's block. However, in *symbolic* communication, the symbol has to mean the same thing to all the individuals concerned. There is no way of getting from the animal condition in which stimulus and response differ to the human condition in which gestures and sounds carry shared if not identical content for all members of the linguistic and cultural community.

It is within the social process of interaction that humans communicate and form their thought objects. That is done with abstract concepts and universals. Animals can only relate to specific objects and gestures (vocal or physical); but humans can relate to concepts and categories. *Dog,* as a general concept, simply has no meaning for dogs; but *dog* and *man* in general are precisely the kind of *universals* that humans communicate with or about. If one individual indicates to himself the same thing he indicates to others, they share a meaning. An individual indicates to the other from his own perspective and to himself from the other's perspective. And since that which is indicated is a shared meaning to all participants, regardless of their perspectives, it must be a universal.

The ability to generalize, to grasp a universal and to communicate it, gives humans their unique ability to think and reflect. By thinking and reflecting humans can explore alternative courses of future action without actually stepping in any of the possible directions. Humans can also convey the character of a future state to their fellows. A dog can pick out a specific scent, but he cannot indicate that scent to another dog. A person can identify a second person to a third; a dog can only follow a scent himself. That distinguishes man from beast as Mead remarked, or the detective from the bloodhound. Animals make their "assessments" and "combinations" in actual trial and error, and by interacting with the specific and concrete in the here and now. Humans can make their combinations symbolically; they can construct and deconstruct by means of reasoning and imagination. Hence there are significant degrees of freedom in human conduct which are not available to animals. Qualitatively greater opportunities for creativity, selectivity, and contingency are unique to the human situation.

Symbolic communication is made possible by the human being's unique capacity for becoming an object to himself—that is, for taking the role of another toward himself. That is the way the "self" originates; by experiencing situations in which one is now subject, now object.

The Self

The self for Mead was a social entity distinct from the physical organism—although the self could not of course emerge except on the basis of the organism. The self emerges in a specific context of social experience and interaction and continues to develop in a social process. In time the self becomes an object to itself and hence the center about which all bodily experiences are organized. I experience this hand, this foot, this back as mine, because I am aware of myself and all that "belongs" to it. The hand, foot, and back belong to the self in a way that the dog can never experience the tail as his own.

The self is both subject and object; it is an object to itself. That decisively differentiates humans from other animals. For the ability to become an object to onself means that one can achieve *self*-consciousness, not just consciousness. And self-consciousness means that one can adopt an objective attitude toward oneself and one's situation. The human capacity for rational action rests squarely on the unique human ability to look upon oneself *objectively*.

In fact, in the development of the social self it is the objective side that emerges first. The self is an object before it is a subject because the first experiences of oneself are from the standpoint of others. An individual's first experience of self is indirect and objective because he first becomes an object to himself by taking the attitudes toward himself of the *significant others* about him—mother, father, siblings, etc.[23] If we may borrow a metaphor from Charles Horton Cooley, whose work we shall discuss in the next chapter, we may speak here of a "looking-glass self." Just as we can only gain a sense of our physical appearance through its reflection in a mirror, so we can only gain a sense of our social self as it is reflected by others in symbolic communication. The very young child begins to relate to himself as others do by means of symbols: He addresses himself, responds to himself, talks and replies to himself, so that a subjective side emerges and he becomes both subject and object.

The self is therefore a social entity which, once formed, acquires a certain autonomy—it creatively and continually organizes its social experiences. We can conceive of an adult hermit who has only himself as a social companion and talks to himself as he would with others. But what is absolutely inconceivable is a self arising outside the context of social experience.

Conversation with oneself is an essential element of communication with others. When one speaks to another, one also speaks to oneself. One affects oneself while one affects the other, so that in the very process of speaking, one checks, controls, and guides one's speech by assessing its effect on others through the effect on oneself. One also takes account of the specific other to whom one is relating and speaks and acts accordingly. Thus one becomes many social selves, so to speak, exhibiting one self here and another there. The self, therefore, far from being a static entity, is a dynamic element of the interaction process.

It is symbolic communication or language that makes the emergence and development of the self possible. In the absence of language and other ways of conveying meaning, neither self nor mind is conceivable. Language, as we have emphasized, consists of symbols, and all symbols are *universals of discourse*. Only after Helen Keller, who was deaf, mute, and blind, grasped the symbol *water* did she begin to acquire language. In the absence of that breakthrough she never would have developed either a mind or a self. She would have remained "subhuman."

The "I" and the "Me"

We now arrive at a somewhat more complex part of Mead's theory. Perhaps the best way to begin is to say that the "I" is a processual side of oneself which one can never grasp except after the fact. The "I" is the actual process of overt action, and the "me" is the reflective process. By taking the attitudes of others, one introduces the "me" to which one responds as an "I." One can never catch oneself as an "I," for one cannot literally observe or be aware of oneself at the same time as one thinks, speaks, or acts. In Mead's words, the "I"

appears only in memory and by that time it has become a "me." The "I" of this moment is present in the "me" of the next moment. The "I" in memory is there as the spokesman of the self of a second, or minute, or day ago. . . . If you ask, then, where directly in your own experience the "I" comes in, the answer is that it comes in as a historical figure.[24]

The "me" is the conventional side of the self: The organized attitudes of others that one assumes toward oneself. The "I" and the "me" interact in an ongoing process.

Still another characteristic of the "I" is its relative uncertainty or unpredictability. What an individual does as an "I" is not precisely known in advance due to the unique qualities of each new situation. The "I" is never entirely calculable because it is characterized by spontaneity, novelty, initiative, and freedom.

The "Biologic I"

Earlier we saw that the instinctualists conceived of human conduct as determined by innate physiological forces; and we saw why that conception is erroneous. But does the absence of instincts in humans mean that their organic characteristics need not be taken into account in an adequate social psychology? It is with reference to such questions that Mead's "biologic I" becomes an extraordinarily important concept.

The "I" is the manifestation of human impulses and needs. It is embedded in man's biologic nature. The "biologic I" exists in a definite state of tension with the social "me." In that respect Mead's conception is not remote from that of Sigmund Freud. A major function of the "me" is that of Freud's censor. "The situation in which one can let himself   53 go," wrote Mead, "in which the very structure of the 'me' opens the door for the 'I,' is favorable to self-expression. [When the 'me'] opens the door to impulsive expression, one gets a peculiar satisfaction . . . the source of which is the value that attaches to the expression of the 'I' in the social process."[25]

The "me" represents the internalized values of the group; the "I," in contrast, represents the biologically rooted tendency to be spontaneous and to resist social relationships and conventions which are experienced as stifling and oppressive. Thus Mead conceived of the human individual as an active being who affects his environment. He rejected the behavioristic doctrine in which man is a passive object—all "me" and no "I." The term *impulse* enabled Mead to avoid both the behaviorist and instinctualist extremes.

The biologic component of the self is spontaneous, impulsive, and *unconscious*. Thus an important question arises: If the "biologic I," or energizing principle, is unconscious, how is rational, intelligent action possible? What good is spontaneity and initiative if one becomes aware of one's acts only after they are accomplished facts? Mead's resolution of the paradox rests on the nature of the self as a *processual unity*, which enables humans to act and reflect on their acts, modifying them accordingly.

To take Mead's own homely example, it is strictly impulsive to tug harder and harder at a drawer that refuses to open. Such impulsive behavior may be observed, say, in a very young inexperienced child or in an excited impatient adult. The process of reflection enters when one begins to think intelligently and analytically about the drawer: It is a wooden thing; it may be swollen here and there; it has contents that may account for its resistance, etc. By acting *and* reflecting, we finally succeed in opening it.

When we act under impulse the drawer is strictly something to be tugged at, and once the knobs come off, all we can say, in exasperation, is "What have I done?" With reflection, in contrast, the "drawer has ceased for the time being to be a mere something to be pulled."[26] Reflection and analysis now guide action, so we can say, "What am I doing? What needs to be done?" Here action is united with reflection.

But reflection never takes us totally out of the field of impulses, for we continue to use our hands, feeling for resistance and trying to overcome it. Reflection is the unique human capacity enabling us to analyze, recombine, and guide our impulses in the face of obstacles and other troubling factors of our world. Reflection is the process in which human actions are no longer direct, unconscious responses to stimuli, whether

54

internal or external, but rather responses to one's object self. The guidance which the impulsive "I" receives from the reflective self gives the otherwise blind "I" the possibility of conscious and intelligent action.

We see, therefore, that Mead has provided the basis of a social psychology that enables us to avoid the errors of the idealists, behaviorists, and instinctualists. Like the idealists he recognized that *mind, consciousness,* and *self* are unique human qualities and that no understanding of the human situation is possible without such concepts. But Mead parted company with the idealists when they merely posit *mind* and *self* as if they are somehow given, metaphysical entities. Mead demonstrated that those qualities emerge in a specific sociosymbolic process.

As for the behaviorists, Mead acknowledged the influence of the environment. But he rejected their conception of environmental conditioning as an invincible force totally molding the human being as if he were clay. Mead also disposed of the behavioristic caricature of man as a passive, mindless, wholly determined object.

Finally, there is the instinctualist doctrine. Mead forcefully repudiated the theory that there exist in humans innate determinants of specific action patterns. But his rejection of that doctrine did not lead him to ignore the roots of the social self in the human organism.

Mead never lets us forget that socialization entails a definite tension between the "biologic I" and the conventional self. However, some present-day symbolic interactionists have failed to follow Mead in that important respect. One might even say that the dominant schools of sociology as a whole have subscribed to what has been called an "oversocialized conception of man."[27] Those schools look upon the human individual as if he were all "me" and no "I." Socialization is regarded as a process by which an individual simply internalizes and absorbs whatever society decides to implant in him. There is no human nature from that point of view that is not wholly a product of society. The result is that the image of humans encountered in much of contemporary sociology is that of disembodied, lifeless "role players." Mead's conception of the "biologic I" helps us to counteract the one-sided oversocialized view. And Sigmund Freud takes us an additional step in the right direction.

FREUD'S THEORY OF REPRESSION

Anyone who has ever read widely in Freud's writings will have noted that the concept of "instinct" figures prominently in his theory. Freud's   55 native tongue was German, and the word *instinct* is one of several possible translations of the German word *Triebe,* which might also be translated as *urge* or *drive.*

So the first step toward an understanding of Freud's theory is to clarify his use of that term. Freud's conception of instinct has nothing in common with that of the instinctualists, whose doctrine we criticized earlier in this chapter. For the instinctualists an instinct is an innate, rigid determinant of a specific type or pattern of behavior. For Freud an instinctual stimulus also arose from within the organism. But that is where the resemblance between the two conceptions ends.

Freud distinguished internal from external stimuli. When a strong light falls upon the eye, it is *not* experienced as instinctual. A dryness of the mucous membrane of the throat, on the other hand, or an irritation of the membranes of the stomach, *is* experienced as instinctual. Furthermore, a stimulus of any kind is usually thought of as something operating with a single impact. If the source of stimulation is painful or unpleasant, the organism seeks to escape from it by motor flight. An instinct, in contrast, never operates with a momentary impact but rather with a *constant* one. And since its source is internal, the organism cannot escape it by flight.

"A better term for an instinctual stimulus," wrote Freud, "is a 'need'. What does away with a need is 'satisfaction'. This can only be attained by an appropriate ('adequate') alteration of the internal source of stimulation."[28] An instinctual need therefore makes certain demands of the nervous system, causing it to initiate activities directed toward changing the external world so as to afford satisfaction to the organism.

Hence for Freud, *instinct* was a synonym for *need*. Rooted in the nature of the human organism, it expresses itself mentally. "Instinct" thus appears as a "concept on the frontier between the mental and the somatic, as the psychical representative of the stimuli originating from within the organism and reaching the mind."[29] The aim of an instinctual need is always satisfaction, but the objects with which a need may be satisfied vary. The object may be external to the body; but that is not necessary since the object might equally well be a part of the individual's own body. The object of a need may change any number of times and may satisfy several needs simultaneously.

Thus what clearly emerges from Freud's delineation of the concept of "instinct" is that it refers to a general somatic-mental need that is anything but a rigid drive determining specific outcomes. In Freud's words, the instincts undergo "vicissitudes." The needs are persistent and enduring though the forms and objects of their gratification are variable.

As is well known, Freud regarded the "sexual instincts" (*libido*) in humans as primal and fundamental. But the libido is not a single, rigid force. Human sexual "instincts" are such that they may undergo significant transformations. Love is a manifestation of the fundamental erotic need. When one loves and is loved in return, that brings great pleasure and satisfaction. However, the process of learning to love may become arrested at the narcissistic stage. Erotic and other feelings of love may be repressed with all sorts of unfavorable consequences for the organism; and frustrated unrequited love may turn into hate. Finally, one's vital erotic energies can be channeled into creative nonsexual endeavors which the organism finds gratifying and pleasurable.

Therefore for Freud, the term *instinct* referred to a basic biopsychic process: A continually flowing source of impulses, excitations, and endogenous stimuli experienced by every human individual. But those excitations affect our conduct in a general and indirect way only. For the instinct itself "can never become an object of consciousness—only the idea that represents the instinct can. Even in the unconscious, moreover, an instinct cannot be represented otherwise than by an idea. . . . When we nevertheless speak of an unconscious instinctual impulse or of a repressed instinctual impulse . . . we can only mean an instinctual impulse the ideational representative of which is unconscious." With that background in mind, we can go on to consider the theory of *repression,* which Freud himself regarded as "the cornerstone on which the whole structure of psycho-analysis rests."[30]

Id, Ego, and Superego

The essence of repression, Freud wrote, "lies simply in turning something away, and keeping it at a distance, from the conscious."[31] To understand how and why that occurs we need to review some of the other basic Freudian concepts.

The oldest and most fundamental "layer" of the mental structure is the *id,* which is the ideational representative of the primary instincts in the domain of the unconscious. Its sole activity is to strive for the satisfaction of the organism's needs, in accordance with the pleasure principle. The newborn infant is presumably all id. With socialization, a part of the id gradually develops into the *ego,* which is the "mediator" between the id and the outside world. The ego represents the perceptual-conscious process of observing and testing reality and striving to alter the social environment in its own interest. The ego thus has the task of both guiding the id and representing the world to it. For "the id, blindly striving to gratify its instincts in complete disregard of the superior strength of outside forces, could not otherwise escape annihilation."[32] Hence the chief role of the ego is to coordinate, alter, organize, and control the instinctual impulses of the id, so as to reduce conflicts with reality. The role of the ego would include any and all of the following: repressing impulses that are incompatible with reality; reconciling impulses with reality by changing their object; delaying the gratification of impulses; changing their mode of gratification; and more. By those means the *pleasure principle,* which holds undisputed sway over the processes of the id, is "dethroned," and in its place rules the *reality principle.*

In the course of socialization and development, another mental process emerges: the *superego.* Its main source to begin with is the parental influence, which remains its core. Subsequently, however, it incorporates the values of the larger society until it becomes the powerful internal representative of established morality. The introjected social values and prescriptions become the ego's censor or conscience. Now the individual begins to experience guilt; that is, the need for punishment for transgressing or even wishing to transgress certain sociomoral restrictions. In an automatic process the superego comes to exercise a severe control over the oedipal and other wish fantasies, ensuring that they never leave the unconscious. As a result not only the wish fantasies but also the impulses represented by them are repressed.

Taken in its most general terms, then, Freud's theory provides essential insights into the process of socialization. At the very center of this theory one finds an unavoidable tension between the "instinctual" needs of the human being and the requirements of the civilizing process.[33] The development of the ego and the superego necessarily entails a subordination of the individual's organic needs (the pleasure principle) to the demands of the socializing agencies (the reality principle). The result is a repression of organic needs with all that it implies—deep unhappiness, mental disorders, and the like. A tragic and permanent contradiction prevails in which humans have paid and will continue to pay an incalculable biopsychic price for the advance of civilization. That is the conclusion Freud drew from his psychoanalytical studies in his late writings, notably in *Civilization and Its Discontents.*

A powerful reaction to Freud's apparent pessimism soon emerged from the "left." Ultimately it led to the basic revisions of Freud's theory associated with such neo-Freudians as Erich Fromm, Karen Horney, and Harry Stack Sullivan. The gist of their critique may be summed up this way: The ego faces not some abstract, static civilization but a historically changing social world. By obscuring that fact Freud's reality principle becomes conservative. It tends to generalize to all social reality the repressive characteristics of specific sociohistorical forms.

To be sure, the neo-Freudian critique has some merit. But as Herbert Marcuse has observed, that critique does not diminish "the truth in Freud's generalization, namely, that a repressive organization of the instincts underlies *all* historical forms of the reality principle in civilization."[34] It is here that the neo-Freudian revisionists have made their fundamental error: They have *flattened out* man's organic needs. In minimizing the extent and depth of the tension between the human organism's needs and any society, the revisionists proclaim an easy but false solution.

In his effort to draw out quite different implications from Freud's theory, Marcuse introduced two terms:

1. *Surplus-repression.* This refers to the "restrictions necessitated by social domination. This is distinguished from (basic) *repression:* the 'modification' of the instincts necessary for the perpetuation of the human race in civilization."

2. *Performance principle.* That is "the prevailing historical form of the *reality principle.*"[35]

Marcuse's point here is that there is an undeniable truth in Freud's view that human beings, in their historical struggle with scarcity (*Lebensnot*), have had constraint, renunciation, and delay forced upon them. To satisfy their needs, humans have had to engage in labor and to endure painful social arrangements. For the duration of work, which occupies so great a portion of the mature individual's existence, "pleasure is 'suspended' and pain prevails. And since the basic instincts strive for the prevalence of pleasure and the absence of pain, the pleasure principle is incompatible with reality, and the instincts have to undergo a repressive regimentation."[36]

However, Marcuse continued, basic repression does not by itself account for the deprivation and pain humanity has historically suffered. The historically known civilizations never distributed scarce resources in accordance with individual human needs. On the contrary, scarce resources have been controlled by the privileged elements of society and distributed in accordance with their interests. Therefore it is not the

59

reality principle, pure and simple, but rather the performance principle, to which much of the human individual's suffering must be attributed. Added to the necessary repression required by any form of human organization, individuals have endured the controls and restrictions of specific institutions of *domination*. Such institutions have not been necessary, strictly speaking, and therefore have resulted in *surplus-repression*. In other words, the pleasure principle has been hedged in and constrained not simply because humanity's struggle with nature has required it, but also, and mainly, because the privileged interests of those who dominate society have demanded it.

Marcuse's reinterpretation is extraordinarily important. It preserves Freud's insights while extending them: The socializing or civilizing process inevitably exacts from the individual a heavy biopsychic price in the form of deprivation and suffering. But he suffers more than is necessary owing to specific forms of social organization that are unnecessarily repressive.

Therefore whether we call human needs "instinctual" with Freud or "impulsive" with Mead, we must recognize that the socializing process always entails some degree of repression. Freud in particular has alerted us to the fact that the human being has vital organic needs striving for satisfaction. And Marcuse has reminded us that in all societies, human beings suffer more than they must.

It is evident from the foregoing discussion of culture and socialization that mind, self, and other distinctive human qualities presuppose some form of *social organization*. In the following chapter we begin our exploration of that subject with small groups and other microcosmic situations.

4

Social Organization I:
Primary Groups
and Other Microcosms

In its most rudimentary sense, the term *social organization* refers to a comparatively stable pattern of interaction between two or more human beings. The smallest unit of sociological concern consists of at least two persons and their reciprocal activities. Even imaginative literature is unable to conceive of a lone individual who retains his human qualities while living in total isolation from other human beings. Robinson Crusoe had his man Friday.

The tendency to associate with one's fellows and to form groups seems to be common to all higher animals. The processes of nutrition and reproduction, for example, are unavoidably social, since the organism's needs require social relations for their satisfaction. The experience and activity of an individual organism are always embedded in a larger social whole. There is no living organism of any kind that can exist in total isolation from all other living organisms. All living beings find themselves in a network of social interrelationships upon which their continued existence depends.

In humans the processes of nutrition and reproduction are of course no less important than they are in other organisms. Nutritive and sexual-reproductive needs account for the formation of the family, the basic and universal unit of social organization in which essential needs are met. However, the family never exists in complete isolation from other social forms. Typically, it finds itself in some larger social context, such as a clan, tribe, or residential community. Every human being belongs to some organized community or other; and his social relations with the other members of the community contribute to the formation of his personality and character.

That the tendency to associate is characteristic of animals and not only of humans is an important fact. It demonstrates that a measure of mutual aid and cooperation is essential for the life of many species. But that important fact has given rise to a mistaken notion.

Some entomologists, for example, have maintained that the structure of organization found among certain insects, notably bees and ants, parallels human social organization. Tempting as such analogies may be, we have to reject them as misleading anthropomorphisms. The principle of organization among insects is altogether different from the principles of human organization. Insect societies are based on the *physiological* specialization and biofunctional differences of their members. Thus what we find among both bees and ants is that the entire reproductive process is carried on for the whole community by a single queen bee or queen ant; and while the queen has evolved enormous reproductive organs, those organs have degenerated in the other members of the community. We find insects capturing other minute forms and keeping them as we do cows, for their exudations. We find warrior castes that carry out raids and bear off slaves, making use of their labor. But in all cases the basis of insect organization is the physiological differentiation of its members.

In human society such differentiation is *not* the principle of organization. There are, to be sure, basic sexual differences among humans as well as marked physiological differences between adults and children. But apart from those differences, there are no physiological distinctions among adult humans that would account for the structure of a human community. Most normal adult humans of either sex share an identical physiological structure. The differences that exist in size, strength, weight, and intelligence never become the sole or even the main grounds for the division of labor in any human society. That means that one must look elsewhere for the principles of human organization.

There is no evidence among insects that the experience of individuals is somehow accumulated and transmitted by means of communication from one generation to another. Among ants, for instance, there is no accruing of experience; and since the principle of their organization is physiological, the same repeatable cycle of activities may be witnessed generation after generation. In human society, in contrast, the differences between the butcher, the baker, and the candlestick maker are not physiologically determined; nor are any of the other occupations and professions in the human division of labor. Women with slight builds may be found performing physically arduous tasks, while men with the stature of football players may be found engaging in mental labor. The human division of labor consists of biologically similar individuals, participating in a historically evolved, sociocultural organization.

If, therefore, one had to sum up the essential difference between animal and human organization and, hence, the uniqueness of the latter, one could do so in two words: *language* and *labor*. As we have seen in chapter 1, the evolutionary transition to the human form entailed adaptation to new life circumstances on the ground. In that transition, *speech* and the evolution of the *hand* went along together in the development of the human being and his social organization. The hand is a unique limb in the animal world. It is not only extraordinarily versatile as compared with the limbs of all other creatures; it mediates man's interaction with nature and with other human beings. The hand is essential for the making of the human world.

Humans therefore determine their environment even while it determines them. Such mutual determination is also true of animals. An individual organism determines its own environment by its sensitivity. The only environment to which an organism can relate is the one revealed by his sensitivity. Every organism, according to its nature, *selects* the environmental circumstances upon which it acts. But humanity has gained a qualitatively greater control over its material surroundings by virtue of what we have called "culture." Thus human sociocultural organization must be fundamentally distinguished from the organizations of all other animals, however intelligent they may be.[1]

PRIMARY GROUPS

The nature of the human individual is formed and developed in his social world. But in the first instance, it is not so much in the world at large that he is formed as in his small social world, which may be called a *primary group*. A group is called "primary" because its role is fundamental in shaping an individual's personality, character, and ideals. Typically, a primary group is characterized by face-to-face interaction and cooperation. It is a situation in which everyone knows everyone else intimately and in which strong mutual sympathy and identification emerge, so much so that the members of such a group think of themselves as "we." A few examples of a primary situation are the family, children's play groups, and small residential communities. Such groups are met with throughout history and in all cultures of the world. Pri-   63 mary groups therefore have a universal quality and may be regarded as the nursery of human nature.

It should not be supposed, however, that a primary group is perfectly harmonious and homogeneous. Differences, self-assertion, and even rivalry are often present, but they tend to be socialized by sympathy and disciplined by the sense of "we-ness." It is in such small, intimate situations that the individual self emerges. Let us, then, look more carefully at this process.

There is something especially intriguing about the question of how the self-idea first arises. In speaking to a very young child, it is possible to point to him while addressing him as "you" and also while addressing him by his name, say, Johnny. In contrast, words like "I," "me," "my," and "mine" change their referents when employed by different persons. For example, when you and I and a third person say "I," wishing to indicate the referent of the word by means of a gesture, each of us must point to himself. Hence the idea of "I," "me," and "mine" can never be conveyed to a very young child by pointing to him. It is therefore remarkable that a child two years of age or even younger should already be able to grasp the precise meaning of "I," "me," "my," and "mine," even though those words, when uttered by others, never refer to the same person as when uttered by the child. That demonstrates that the child's eventual learning of the first-person pronouns is neither a matter of mere imitation nor of conditioning, for these processes would yield no real communication—only a mechanical repetition of words and phrases. So let us attempt to lay bare the process by which the first-person pronouns are grasped.

Anyone observing a newborn human infant will witness an active, spontaneous being. Doubtless the infant experiences a variety of energizing impulses. From the earliest days of its life, he spends his waking hours restlessly turning this way and that, exploring his surroundings with his eyes, attending to objects and sounds, and responding with a cry or a smile to the words and facial expressions of others. Though the infant has yet to achieve self-awareness, he nevertheless exhibits all the signs of *self-feeling* or *self-experience*. With time, definite acts of will become evident. The baby now wants things and cries and strives for them. Self-feeling expresses itself in appropriative activities such as grasping, tugging, and screaming for something. When the child witnesses similar actions in parents and siblings, he recalls and reexperiences the feeling.

In the course of his interaction with others in the primary group, the
child learns to name his self-experience. He does so by witnessing in

others appropriative and other self-indicating experiences that are accompanied by exclamations of "my," "mine," "I want it," "give it to me," etc. Such exclamations clearly indicate a frequent and vivid *experience* with which the child is already familiar and which he learns to attribute to others. Creative reflection now enables him to make the cognitive leap and to begin the proper use of those words himself.

Imitation and conditioning could never produce that result; it is possible only through social experience and reflection. What the first-person pronouns refer to is *not* primarily the child's body or sensations, but to the social acts accompanying them. The child has discovered that his feelings are experienced by others and that there are visible and audible signs of those feelings. Having had the feeling himself, he now imputes it to others and recognizes that it can be expressed and communicated in words.

Thus the *self-idea* emerges out of the *self-experience* with the learning of the personal pronouns. The self-idea always remains rooted in somatic feelings and impulses; but it becomes social as the child identifies the experiences of others with his own and learns to express them symbolically. The child is now able to put himself in the place of others. He has accomplished what is essential to human communication, for he has learned to use symbols that indicate the same meaning to himself that they indicate to others.

The social self thus emerges in relation to others in typical primary-group situations. The self-other process of symbolic interaction was likened to a "looking-glass self" by the outstanding social psychologist Charles Horton Cooley.[2] But he himself recognized that a looking glass is too mechanical a metaphor to convey the full complexity of the process. For Cooley, the self-idea entailed three chief elements: imagining how we appear to the other person; imagining his judgment of that appearance; and, as a result, experiencing some sort of self-conception and feeling—pride, embarrassment, unease, shame, etc. As the child learns to reflect himself in others, it becomes evident that he cares more for the opinion of certain persons. Typically he cares much for the opinion his family members hold of him. And among the new persons he meets, he shows intense interest in some and indifference and even repugnance to others.

It is now the child's self-conception and not merely his body that brings him great joy or grief, depending upon the treatment which the emerging social self receives. He will cry and be "hurt" when he senses the slightest personal affront, and he will smile and laugh in response to attention and encouragement. He now becomes something of a per-

65

former or social actor, constantly striving for the center of the stage. When the audience shows appreciation, he repeats the performance and invents new acts; but when the audience displays disapproval or indifference, he pouts or runs away weeping. He develops a repertory of performances, trying them out on both family members and outsiders. Soon he learns that with some of his performances he gains considerable influence and even power over others. He tries to get his way by employing dramatic devices. He produces a make-believe cry, for example, in the hope that his mother or father will yield. He experiments with such devices to see what effect they are producing. Creative imagination in taking the role of the other has combined with self-feeling to create a social self which has become a principal object of his concerns and actions.

If at first a child obviously and naively does things for effect, he gradually advances to more sophisticated performances. He learns to hide the fact that he is anxious to elicit visible and audible signs of audience approval. As he matures, the entire process of symbolic interaction becomes more and more complex, subtle, and invisible. The fact that terms like *social actor, role, performance,* and *audience* are commonly employed for an understanding of that process suggests that social interaction may be viewed as drama.

The Dramaturgical Perspective

Erving Goffman, one of the best-known modern social psychologists, has elaborated the theories of Cooley, Mead, and others in a series of ingenious studies. In his first book, entitled *The Presentation of Self in Everyday Life,* he provides the rudiments of his perspective: Social interaction may be viewed as a theatrical performance.

In that book Goffman is concerned with documenting what he calls "expressions given off," that is, expressions of "the more theatrical and contextual kind, the non-verbal, presumably unintentional kind, whether this communication be purposely engineered or not."[3] We gather meanings and obtain impressions from the acts of others, often relying more on their actions than on their words. Facial expressions, gestures, and the rapidity and quality of actions may convey true feelings more accurately than verbal behavior. As we speak, we also communicate in nonverbal ways, and we attend to the latter as a check on the sincerity of the former. We use one stream of communication (the nonverbal) to test the second (the verbal).

As one becomes aware that nonverbal behavior is less controllable than verbal, it becomes possible to employ that knowledge to manage and manipulate the impressions one makes. Since others are also sophisticated in such matters, they attempt to detect some nuance that the individual has not successfully managed to control. Thus social interaction becomes a kind of information game in which each individual tries to manage his own impressions while seeking to penetrate those of others in an effort to grasp their true feelings and intentions.

In everyday life such contests proceed sub rosa. Each participating actor suppresses his true feelings and communicates a view of the situation which he senses the others will find acceptable. Striving to avoid embarrassment by protecting the images they project, actors employ strategy and tactics. In his performance an actor may deceive himself as regards his true motives, or he may cynically deceive others. Those are the extremes within which performance motives range. And, as one might expect, the dramaturgical perspective implies that the actions or performances take place on a "stage," replete with the appropriate scenery and props—facial expressions, clothing, posture, gestures, and the like.

Each actor tends to guide his conduct by what he regards as the official values of the circles he moves in. Putting his best foot forward, he presents an idealized version of himself and underplays those aspects of self that appear incompatible with that version. Every actor thus presents what he would like others to regard as the "essential" self; but at the same time he must contend with the fact that different groups and situations demand that he show different selves. In order to maintain his fostered impression, the actor strives to segregate his audiences so as to enable him to play the right part for the right audience. Social interaction is thus analyzed by Goffman in theatrical and artistic metaphors to convey the fact that performances are delicate and fragile and can be shattered by the slightest misstep.

When several actors cooperate to stage a routine, they constitute a *performance team*. Whether husband and wife, an executive and his secretary, fellow workers, doctor and nurse, and so on, teammates depend on each other to present and maintain a given definition of the situation. Performance teams do not necessarily coincide with the authority structure of a group or organization. Some performances require cooperation within a status or rank, as when parents resolve never to side with their children against each other, or when officers never disagree in the presence of their men. However, other performances require cooperation across statuses, as when an officer aligns himself with his men to put on a successful show for his superior officers. In all per-   67 formances a single member of the team can give the show away or disrupt it by inappropriate conduct. For Goffman, we are all members of such teams, conspirators collaborating to conceal certain facts from our audiences.

To conceal successfully, some of our performances require *regions.* The *front region* is the setting in which the team members actually present their play to the intended audience. A *back region,* in contrast, is set off by barriers or otherwise hidden from the audience's perception. As an illustration, Goffman cites Simone de Beauvoir's description of women's activities when the male audience is absent. Backstage, the woman "is getting her costume together, preparing her makeup, laying out her tactics; she is lingering in dressing-gown and slippers in the wings before making her entrance on the stage."[4]

But "backstage" can refer to any of those regions of social establishments that are generally off limits to the audience. Performers wear their masks in the front region and remove them backstage. Therefore, an interesting time to observe impression management is when the performers leave the back region and come upon the stage or when they return "therefrom, for at these moments one can detect a wonderful putting on and taking off of character."[5]

The separation of front and back regions may be found throughout society. In the home, in public and private establishments, and elsewhere there exist regions where appearances are meticulously maintained and other regions in which the front is dropped and mask and costume are removed. If a performance team implies cooperation, it also implies collusion and conspiracy. Every team has its "dark," "strategic," or "inside" secrets. Dark secrets are concealed because if they were revealed, they would blatantly contradict the image the team is striving to maintain. Strategic secrets are concealed because they entail the plans for future actions a team has vis à vis its opposition. Finally, there are inside secrets, which an individual shares simply by being a member of a group. But secrets are never perfectly kept owing to the existence of what Goffman calls *discrepant roles.*

One such role is the "informer." Posing as a member of the team, he gains access to the back region and gives away the information thus gained to the audience. On the other hand, there is the "shill," who pretends he is a member of the audience but actually is an ally of the performers. The "shill," "claque," and "shillaber" are associated with nonrespectable business performances, but those concepts can apply equally to everyday social encounters.

Despite all the precautions taken, performances are frequently thrown off key by unintended gestures, unanticipated intrusions, and faux pas. Even a deliberate "scene" may occur, resulting in a complete disruption of actor-audience rapport. In everyday life no less than in theater, both the performers and the audience have an interest in minimizing minor and major disruptions. Both therefore develop techniques to reduce the likelihood of their occurrence. To save their show, the performers employ "defensive practices": (1) dramaturgical loyalty (the obligation to safeguard performance secrets); (2) dramaturgical discipline (the obligation to learn one's part well and thus to avoid unintended gestures); and (3) dramaturgical circumspection (the need for prudence and forethought in planning how best to present the performance).

On the other side are the measures the audience employs to help the performers save their show. The audience is generally tactful and discreet and does not go backstage uninvited. It also tries to ignore performances which, though occurring in its presence, are clearly not intended for its eyes or ears. Tact is the key word here. The audience tactfully cooperates with the performers in preventing conduct that might create a "scene." Anything short of a scene the audience pretends not to notice.

A similar type of analysis is developed by Goffman in two other well-known monographs, *Stigma* and *Strategic Interaction*. Goffman sees three types of stigmata: (1) physical deformities; (2) shortcomings of character, especially those inferred from one's record as mental patient, prisoner, drug addict, alcoholic, homosexual, etc.; and (3) those associated with "race," nationality, and religion.[6] Whatever the type of stigma an individual carries, it is something he and others know about and take into account in their social interactions.

The stigmatized individual is ill at ease with "normals," just as they are with him. The normals, however, are frequently in a position to control and reduce the opportunities of the stigmatized. The normals are often normals *against* him; and since the stigmatized individual knows that very well, his main task becomes gaining acceptance or favor among them. Sometimes acceptance is gained by removing his bodily deformity or ridding himself of his characterological defect. However, that may result not in imparting normal status, but rather in transforming him "from someone with a particular blemish into someone with a record of having corrected a particular blemish."[7]

What interests Goffman primarily are the face-to-face encounters of normals and stigmatized, the moment when they find themselves in the same situation and must therefore confront the stigma and take it into consideration. Clearly such encounters may require delicate, calculating impression management on both sides. The stigmatized person is anxious in the face of the too sympathetic or unsympathetic concern of the normal; but the latter is also anxious, looking upon the stigmatized as either too aggressive or too humble. The stigmatized person has the perennial problem of managing the impressions he makes as well as the inevitable tension emerging in encounters with normals. That is not all. Frequently an individual's stigma is not conspicuous or evident, so he has the problem of managing information—should he disclose his shortcoming or not? Sometimes if the "defect" is not readily apparent, the stigmatized individual will try to "pass."

The passer's psychic state continues to be one of anxiety. He lives in fear of being discovered. He has found acceptance among normals who remain prejudiced against the category of persons to which, unbeknownst to them, he belongs. He feels himself neither fish nor fowl, for he has abandoned his own kind without gaining full and true acceptance by the other side. More than the person who never attempts to pass, and certainly more than the normal, the passing but always discreditable person must continually employ strategies to present himself in a manner that will minimize risk and reduce anxiety.

However, Goffman in the end rejects the view that there are two mutually exclusive categories—of stigmatized on the one side and normal on the other. We are all carriers of stigmata, and we are all engaged in passing, just as we are all normal. "The most fortunate of normals," he writes, "is likely to have his half-hidden failing, and for every little failing there is a social occasion when it will loom large, creating a shameful gap between virtual and actual social identity. Therefore, the occasionally precarious and the constantly precarious form a single continuum, their situation in life analyzable by the same framework."[8]

In a later book called *Strategic Interaction,* the focus of interest remains face-to-face interaction that derives particularly "from paralinguistic cries such as intonation, facial gestures, and the like—cries that have an expressive, not semantic character."[9] People "inhibit and fabricate" their expressions. They are therefore engaged in "expression games."

Thus Goffman adds to his dramaturgical vocabulary a vocabulary of games. There is the "unwitting move," an act unoriented to an observer's assessment; a "naive move," a subject's act that an observer takes as it appears; and a "control or covering move," a subject's deliberate act designed to produce expressions that might improve his situation. In such forms of impression management, the subject "tends to make use of the observer's use of his behavior before the observer has a chance to do so."[10]

Observer and subject thus attempt to gain information from the expressive behavior of the other. Each participates in a "contest of assessment," controlling his own expressions while assessing the other's. The contest involves "moves," "covering moves," "uncovering moves," and "counter-uncovering moves."

There can be no doubt that Goffman's perspective provides many valuable insights into some of our daily encounters. Yet a few critical remarks are in order. We find in *Strategic Interaction,* for example, that he draws most of his illustrations not from the face-to-face encounters of everyday life, but rather from the world of espionage, spy games, and cops and robbers. The impression one garners from his total corpus of writings is that the condition he calls "degeneration of expression"— taking nothing at face value, constantly suspecting the motives of others, seeing traps everywhere—is one that he regards as typical, not exceptional.

The inevitable effect of the game metaphors and the illustrations drawn from double agentry is to portray the world of everyday life as if it were devoid of truly innocent expression. Innocence, after all, is precisely the expression that "a guilty expert gamesman would give." Goffman himself reveals some ambivalence toward his metaphors but resolves it rather easily. It is true, he writes, that in a number of important respects "agents are unlike ordinary mortals." But the analogy is nonetheless warranted, he believes, because "getting oneself through an international incident involves contingencies and capacities that have a bearing on the games that go on in local neighborhoods."[11] Goffman thus portrays all of us as agents who expose and discredit others, all the while fearing that we ourselves shall soon be discovered and exposed. With that portrayal, he attributes to the whole of everyday social intercourse a fundamental cynicism. The result is a caricature and not a true and balanced portrait of the presentation of self in everyday life.

The Quantitative Perspective: The Significance of Numbers for Social Interaction

The view that the numerical size of a group is highly significant in determining its forms of interaction was first systematically stated by the renowned sociologist Georg Simmel.[12] His best-known analysis of the relationship between numbers and forms is found in his discussion of *dyads* (social interaction of two) and *triads* (social interaction of three).

The dyad, as the smallest unit of social interaction, has certain characteristics which distinguish it from all larger units, for each of the two   71 members interacts with the other in an immediate and direct fashion. The dyad is the only unit in which members feel no social structure looming over them. Both members know that the very existence of the social unit rests directly on each of them and that the withdrawal of either would destroy the whole.

The direct personal interdependence characteristic of the dyad constitutes the basis of intimacy. We say, "Two is company; three is a crowd," and we sense that the admission of a third person would undermine the intimate relationship one has with the other. The larger the group, the less intimate it becomes. That is true not only because additional individuals can now interpose themselves between the members of a dyad, but also because the addition tends to create objective structures rising above the members. That is a point to which we shall presently return.

The absence in the dyad of any impersonal, objective structures means that the delegation of responsibilities is direct and personal. There is no structure behind which either member can hide; nor can either achieve anonymity or facelessness, which is so easily done in large groups and organizations. In the dyad the dependence of the whole on the individual is perfectly clear, and coresponsibility for all collective actions is unavoidable. It is true that one can attempt to "pass the buck" to the other, a phenomenon frequently found among business partners. But a partner can resist and reject such efforts more immediately and decisively than can a member of a larger organization.

The coming together of two persons creates new social qualities that are absent in either individual taken alone. That new qualities emerge from the unity of elements seems to be a law of nature as well as of society. For example, the elements sodium and chloride are each deadly poisons when taken alone; yet when they are brought together they lose their poisonous qualities and are transformed into salt, a food. Sodium chloride cannot be reduced to its component elements without destroying the thing, or quality, we call salt. In those terms we may justifiably speak of salt and other compounds as wholes which are greater than their parts. The idea being described here is called in philosophy the *theory of emergence* and teaches that phenomena must be grasped at their emergent levels. The opposite view is called *reductionist:* To understand something, we must analyze or *reduce* it to its elements. But we have seen that in reducing salt to its elements, we inevitably lose the quality of salt. Hence the analysis of something into its elements may be useful and enlightening for certain purposes. But the separate elements themselves can never tell us anything about the reality that will emerge once the elements are united.

The theory of emergence applies with at least equal force to social processes as well. There are many respects in which the dyad is greater than its parts. For instance, the cooperation of two persons can accomplish considerably more than can two individuals working alone.

Adding a Third Element

The expansion of the dyad illustrates the theory of emergence even better. Simply by adding one single element to the dyad, one changes the social reality in several fundamental ways. The quantitative expansion of the group gives rise to new qualities. New possibilities of both a positive and negative kind emerge for the first time in a triad, owing to the opportunity for *indirect* relationships.

Conflicts between two parties which they themselves cannot resolve may now be mediated by the third party and/or absorbed in the triadic whole. But the indirect relationship is not an unmixed blessing. For as Simmel remarks: "No matter how close a triad may be, there is always the occasion on which two of the three members regard the third as an intruder."[13] The sharing of a mood and experience with one intimate other is always disturbed by a spectator. And even the most harmonious triad is liable to become three groups of two persons each, thus destroying the purely dyadic relationship of any two of the individuals.

With the triad, Simmel notes, three basic types of social formation become possible. The first he calls *mediation*. The relationship of a married couple may be shaky and precarious, but with the birth of a child it is remarkably strengthened. The child as a third element closes the circle by binding the parents to each other, either directly by increasing the couple's mutual love or by creating indirect bonds between them. The unity of the couple is now mediated by the child's presence, and their mutual affection and sympathy simply would not exist without the child as a point of mediation. That is why an unhappily married couple may deliberately refrain from having a child. The spouses intuitively understand that a child would bind them closer to each other than they care to be. Mediation often results in the greater affection and passion of the parties concerned. But there is another variety of mediation in which passion is reduced.

The nonpartisan mediator enables two conflicting parties to reach an accord by breaking the fatal spiral in which the vehemence and passion 73 of one provoke even greater passion in the other until the entire relationship breaks down. The mediator fulfills his role by presenting each party with the claims and arguments of the other. Since the mediator is disinterested, he can adopt an objective attitude toward the conflict. The result is a reduction of the highly emotional and subjective character of the dispute and a cooler examination of the issues. Each party now assumes a more objective attitude than would have been possible without a mediator. The logic of mediation would apply not only to two individuals but to the representatives of two organizations as well—such as labor and management.

The second basic social formation resulting from the addition of the third element Simmel calls *Tertius Gaudens*—i.e., someone who draws advantage from the conflict of the other two. The *Tertius* may make certain gains simply because the other two conflicting parties (or groups) have roughly equal power and hold each other in check. Since they are absorbed with each other, they can do nothing to prevent *Tertius* from deriving certain benefits from their stalemate.

But *Tertius* may gain advantage in another way. One of the two conflicting parties may bestow benefits upon *Tertius* not for his sake but to hurt his adversary. Simmel cites the example of "English laws for the protection of labor [which] originally derived, in part at least, from the mere rancor of the Tories against the liberal manufacturers."[14] Two parties can be hostile to begin with and thus compete for the support of a third, as in the above example. Or their hostility can result from competition, as in the case of two suitors pursuing the same woman. In those cases *Tertius*'s advantage derives from the fact that he is equally independent of the two conflicting parties. But the advantage vanishes the moment the two parties become a unit again.

The third and final basic form that can arise with the addition of a third element Simmel calls *Divide et Impera,* or divide and conquer. That is a variation on the theme of *Tertius Gaudens*. The third element can draw advantage from the already existing tensions between two parties, but it can also intensify those tensions for additional advantage. The classic example would be British rule in India. The British colonial power played off Hindu against Moslem and exploited to the fullest other religious and ethnic differences and antagonisms. The superior power, intent upon dividing and conquering, can do so in several ways. It can turn to its account already existing disunity; it can create and foster disunity where only differences exist; and it can prevent the unification of elements "which do not yet positively strive after unification but *might* do so."[15] Thus we find that all truly autonomous associations that might unite to oppose the central regime are prohibited in totalitarian systems.

The triad, then, is a totally different structure from the dyad. The possibilities emerging in a three-element system are simply not available in a dyad. Yet the quantitative expansion of the triad to four or more members does not necessarily give rise to additional qualitative changes. There is no basic difference in principle between dividing and controlling two potential or actual opponents and dividing and controlling three or more. Even a large mass of individuals can be controlled by a small minority, provided that the mass is sufficiently atomized and the minority sufficiently organized.

The dyad and the triad thus illustrate how Simmel sought to demonstrate the significance of numbers in social interaction. Underlying the great diversity of concrete social phenomena, a number of basic social patterns may be discerned. Most strikingly in the transition from the dyad to the triad, those patterns are made possible by the number of interacting elements.

Social Interaction as a System of Exchange

In recent years several contemporary sociologists have attempted to apply economic concepts of exchange to social interaction in everyday life. One of the best-known representatives of that theoretical approach is George C. Homans.

Homans's most detailed exposition of his theory is found in *Social Behavior: Its Elementary Forms*.[16] He begins by noting that the ideas he wishes to present and test scientifically are already embodied in proverbs and maxims: "Every man has his price. You scratch my back and I'll scratch yours. Do as you would be done by. You can't eat your cake and have it too. No cross, no crown. Fair exchange is no robbery. . . . And so forth." Such are the truths that men apply in everyday life; but presumably they grasp neither their interconnections nor the theory they imply. Homans therefore wishes to state those ideas in a series of propositions and to test their scientific validity. Such validated propositions, he believes, will enable him to explain the elementary forms of everyday behavior.

To explain behavior Homans relies on two bodies of theory: behavioral psychology and rudimentary economics. They may be merged into one, since both theories regard "human behavior as a function of its payoff: in amount and in kind it depends on the amount and kind of reward and punishment it fetches."[17] Both theories conceive the exchange of human activities in terms of reward and cost, and both seek to explain those activities in terms of what certain actions cost as compared with what they gain for the actor. From that standpoint social interaction may be viewed as an exchange of "goods" and "services" in which each actor strives to reduce costs and maximize profits.

Homans draws his favorite illustration of exchange from Peter Blau's *The Dynamics of Bureaucracy.* The interaction of two office clerks is taken as fairly typical of face-to-face, small-group behavior. "Person" is less skilled at the required paperwork tasks than "Other." Office rules stipulate that if a clerk needs help with his work, he should seek it from his office supervisor. But if Person were to follow the rules, he might come to be regarded as incompetent, thereby hurting his chances for promotion and increments and even jeopardizing his job. Therefore, despite the rules, Person requests help not from his supervisor but from Other who, being more skilled, does his work more quickly and has time to spare. "Other gives Person help and in return Person gives Other thanks and expressions of approval. The two men have exchanged help and approval."[18]

To the dyad Person-Other, Homans adds a third party, whom he calls "Third Man." Like Person, Third Man is deficient in skill and in need of Other's help. Person and Third Man are willing to give Other approval for his help; but now Other must divide his limited time between them. Person and Third Man will now receive less assistance than did Person when he was alone with Other. Other's help has become more scarce and valuable so that Person and Third Man are now prepared to "pay" more approval per unit of Other's help. Thus "Other's bargaining position has improved over what it was when he had only Person to deal with even though no conscious bargaining need take place."[19]

For Homans, the concepts of cost, profit, and reward are therefore essential for an understanding of behavior as exchange. Cost refers to "value forgone" and not only to the pain incurred while engaging in a specific activity. An activity incurs cost insofar as it causes one to forgo a rewarding and alternative activity. Thus when Other helps Person, he inevitably incurs a cost. If he refuses to help and continues instead to do his own work, he forgoes Person's approval; conversely, if he extends assistance, "he forgoes the value of doing his own work."[20] Similarly, if Person does his own work, he loses the value of help; and if he solicits help, he pays in loss of self-respect, since in asking for help he acknowledges his inferiority to Other. Hence the formula governing such exchanges is: Profit = Reward − Cost. "We define psychic profit," writes Homans, "as reward less cost and we argue that *no exchange continues unless both parties are making a profit"* (italics added).[21] With that background in mind we can have a look at a few of Homans's applications of his exchange theory.

Exchange and Power

When two or more individuals enter into an exchange relationship, and one of them is able to change the behavior of others in his own favor, it may be said that "one man is more powerful than another."[22] Homans assumes that Person and Other originally gained equally from their exchange of approval and advice, and that now the value of Other's advice has gone up. He has been taking too much time from his own work. He is now able to get more approval per unit of advice and has therefore become more powerful. At work here is "the principle of least interest." That is, the person who is least interested in continuing a relationship is the most powerful. He can dictate the conditions of the association.

Other gains power over Person because Other possesses expert knowledge of office procedures which Person lacks, and because other sets a lower value on approval than Person sets on advice. Other's power, Homans suggests, results from a *change* in situation. Earlier Other accepted approval as equivalent pay for advice. Now he assigns less value to Person's approval. The change results from the entry of Third Man. Like Person, Third Man is relatively unskilled; he therefore values advice more than the approval he is prepared to give in return. Other now advises two individuals, and his costs rise accordingly because he must now take more time from his own work. And though he is now getting more approval, the value of further approval has declined for him "through satisfaction." On the other hand, because two persons are now seeking advice, neither is receiving as much advice as one did earlier. They are liable to become relatively deprived of advice, and thus the value of advice grows for each of them. "Both effects," writes Homans,

> tend to make exchanges in the immediate future less rewarding to Other than to either Person or Third Man, and it is this that gives Other power over both of them. Power, then, depends on the ability to provide rewards that are valuable because they are scarce. In the office, many men can provide approval, but we have assumed that only Other can give good advice.[23]

That is how Homans explains the emergence of power; it results from a *change* in conditions, specifically the entry of Third Man.

The question that suggests itself is whether Other has power over   77

Person even before the entry of Third Man. Homans seems to imply that power emerges in the Person-Other relationship only with the entry of Third Man; and that otherwise equality of power prevails between Person and Other. For he argues that in repeated exchanges between Person and Other, "provided that there is no other change in the conditions, power differences tend to disappear, and neither party unilaterally will change his behavior any further. Person, for instance, will not make his approval any more fulsome, since it would cost him too much in the way of confessing his inferiority."[24]

Homans repeats that argument in his revised edition of *Social Behavior: Its Elementary Forms*. He again maintains that in a two-person relationship of the Person-Other kind, differences in power tend to disappear or equalize. He then goes on to say that his point is often misunderstood, and that he does not intend it to mean "that Other will lose whatever advantages his superior power may have given him in the past. It implies, on the contrary, that the changes that have already occurred will be maintained. But neither party will be able to get any *more* out of the other."[25]

The new formulation remains ambiguous. If the power of the two parties tends to equalize, how can Other simultaneously retain the advantages of superior power? And if he retains his superior power, then why does Homans say that the power differences between them tend to disappear? Evidently, what Homans really means, judging from the above clarification in the revised edition, is that existing power differences in a two-party relationship tend to become *stabilized*. But if that is in fact his meaning, it is not at all clear why he insists on the point. For it is apparent that if Other has more expert knowledge than Person, and if Person must hide his relative lack of competence from the supervisor, then a fundamental power imbalance is inherent in the relationship. So long as Person cannot (1) do without advice, (2) coerce Other into giving him advice, and (3) find an alternative source for advice, it would seem that there is no reason why Other could not, if he wanted to, increase his power over Person. If approval is no longer adequate payment, Other could demand something else. In any event, there are no grounds, under the circumstances, for insisting that Other will not be able "to get any *more*" out of Person.

Exchange and Justice

To test the validity of his theory of justice, Homans draws upon William F. Whyte's *Street Corner Society*, a study of the now famous Norton

Street gang. Several nights each week, we are told in that study, the gang went bowling. One member named Alec was by reason of his general behavior held in low esteem by the gang. But he was a rather good bowler who in individual matches beat other gang members, including Long John, a friend of Doc, the leader. However, when Doc arranged a tournament among all the members, Alec bowled poorly. Why? Because, Homans argues, the gang heckled him and thus undermined his confidence. In so doing "they maintained a sense of justice." Since Alec contributed little to the group and even violated group norms, the members got even by keeping his bowling score down. And Homans adds: "They kept the value of what he got 'in line' with what he gave."[26]

A few critical comments are in order so as to highlight the shortcomings of the behavioristic method as applied to human conduct. We might begin by asking why the gang's conduct should be interpreted as maintaining a "sense of justice" when it might just as easily be viewed as an attempt to enforce conformity. From Homans's strictly behavioristic standpoint, he has no real means of distinguishing between the two interpretations. Whether a group is seeking to enforce conformity or to administer justice is at least in part a question of what *meaning* the members ascribe to their actions. However, in Homans's behavioristic method, there is no provision for getting at the meaning of acts or the motives of actors.

Let us suppose that Homans would reply that the group was in fact enforcing conformity, but that they regarded such enforcement as just. How can he know this? Perhaps they enforced conformity even while believing that their actions were unjust. All that Homans has to work with are two pieces of data: That Alec conformed poorly, and that he also bowled poorly under certain circumstances. How does Homans know what meaning to attribute to those data? On what grounds does he attribute one meaning rather than another?

But let us for the moment accept Homans's supposition that the group members were seeking justice, and that they viewed their punishment of Alec as commensurate with his transgressions. Does it follow that the outside observer or social scientist must accept the group's judgment uncritically? When Homans says that "they kept the value of what he [Alec] got 'in line' with what he gave," how does he know that the two values are "in line"? Would any other reaction of the gang, either more or less severe, also have been in line? That Homans would reply in the affirmative to the last question seems to follow from his theory that persons remain in relationships only so long as they profit from them. Hence so long as Alec remains in the group, *anything* it does to punish him for lack of conformity would, according to that logic, result in an equivalence between what he gave and what he got. So long as Alec has not terminated his relationship with the group in an effort to find a more profitable alternative, the group, presumably, can do him no injustice.

Or would Homans argue that a reaction of a certain severity would have been "out of line," and therefore unjust? If so, we would once again have to ask what criterion he has for determining whether the gang's reaction is in or out of line.

"Justice" in the Bank Wiring Room

Homans's rule of justice is that "a man's rewards in exchange with others should be proportional to his investment."[27] To test that proposition he utilizes another well-known body of research, carried out by Elton Mayo and his colleagues in the Hawthorne plant of the Western Electric Company. The scene is the bank wiring observation room, in which there are two main jobs: wiring and soldering. The wiremen received more pay than the soldermen because they were thought to have more skill and because they also had greater seniority. Among the workers two cliques emerged, the first holding itself superior to the second in accordance with the higher skill, pay, and seniority of its members.

To the first, or connector, clique but in a subordinate position as befitted his inferior job-status, belonged Steinhardt, the solderman for the first three connector wiremen. To the second, or selector clique in the same sort of position belonged Cermak, who had replaced another man as solderman for the three selector wiremen. . . .
All the men ate lunch in the room, sending out one of their number to pick up their food and drink from the plant restaurant. The question was who should be the "lunch boy"—note the word "boy." It was not a very valuable service; anyone could have performed it for himself, and it was menial: the lunch boy was a servant for the others. At the time the men were first assigned to the room Steinhardt had reluctantly agreed to do the job. But as soon as Cermak came in as a solderman for the selector wiremen and was accepted as a member of their clique, Steinhardt was relieved and Cermak took over as lunch boy.[28]

Homans then asks: "Was it appropriate that he should do so?" And he replies with an unqualified affirmative. Why was it appropriate? Because Cermak "was a solderman, and soldermen held the lowest job status in the room; of the soldermen he was the least senior and the last to come into the group. Accordingly the group assigned him the least rewarding activity at its disposal: his menial job was in line with the other features of his status."[29] Cermak's rewards were presumably proportional to his "investments," and hence the treatment accorded him was fair.

Now as earlier, the most evident problem with that argument is that the situation Homans describes might easily be labeled unjust. The members of the group delegated a menial and servile task to the man who was already low man on the totem pole. Why is it just and appropriate that one who already receives least should continue to receive least? To that question Homans might respond: It is just because he has invested less than everyone else.

Homans's analysis of those situations is problematic because it tends to justify and dignify anything and everything the group does. It attributes justice to relationships that might just as accurately be comprehended in terms of exploitation, coercion, and domination.

Why exploitation? Let us remember that we are talking about some reciprocal exchange of services. Cermak provided a definite service for the others; he brought them their lunches. Homans tries to depreciate its value by saying that "it was not a very valuable service; anyone could have performed it for himself." But that is the same as saying: The service the slave performs for the master is not valuable because the master could perform it for himself. It is undeniable that Cermak performed a definite service for the group. What did he get in return? Very little and perhaps nothing—though Homans would insist that Cermak received approval for fulfilling his menial, servile function; or, negatively, that he was not punished for refusing to serve, either by rebuke or ostracism; or, finally, that he fulfilled his task unwillingly and experienced anger, which was his reward. Yet if we compare in time and effort the daily service Cermak rendered the group with what they gave him in return, it becomes evident that Cermak received, both qualitatively and quantitatively, less from the group than he gave, and was to that extent exploited.

Cermak was also coerced and dominated by the group. He could have refused to comply, but that alternative, he probably felt, would be more costly. In any event, group pressure won out. The group effectively coerced him into doing its bidding and successfully kept him in his place. But the group's success hardly warrants our saying that this is the way things ought to be, that this is just a state of affairs.

Homans writes that "we must remember that we are talking about justice as seen by the members of the particular group and not about our own sense of justice, which is of course Olympian."[30] Once again we must ask: How does Homans know that the group members re-

81

garded their actions toward Cermak as just? Homans has described only their overt behavior and knows nothing about how they assessed their acts. From the available data there is no way of knowing what meaning they ascribed to their pressure on Cermak—whether they viewed it as just, whether they questioned its fairness, or, finally, whether they persisted in their behavior even while recognizing its injustice.

The behavioristic method favored by Homans does not enable us to answer such questions. Yet without knowing something about the motives of the actors concerned and the meaning of their actions, there are no real grounds for asserting that the clique gave Cermak his just deserts.

THE PRIMARY GROUP AS A COMMUNITY

Sociologists have attempted to grasp the nature of "community" by contrasting it with "society." For instance, C. H. Cooley, as we have seen, spoke of *primary groups* which he implicitly contrasted with *secondary groups*. That is only one of numerous dichotomies of a similar kind, such as Ferdinand Tonnies's *Gemeinschaft* and *Gesellschaft;* Henry Maine's *Status* and *Contract* societies; Herbert Spencer's *Militant* and *Industrial* societies; and Emile Durkheim's *Mechanical* and *Organic* solidarity. In the balance of this chapter, we shall say a few words about the "community" side of the dichotomy, leaving for the next chapter a consideration of that macrocosm we call *society*.

The term *community* refers to a situation in which all members of the group are in principle capable of knowing one another personally, and where everyone interacts with all others in face-to-face relationships. Those characteristics imply that a community is a comparatively small demographic unit. It is an informal and unspecialized social organization which therefore stands in contrast to such formal and specialized organizations as a professional association or a private or governmental bureaucracy.

Historically the most common form of community throughout the world has been the small village or rural neighborhood. Every such community is characterized by a certain "consciousness of kind," expressing itself in a common language or dialect and common cultural traditions. Socially and linguistically the members of a community may be so close and distinctive as to have little or nothing to do with the adjacent or surrounding communities. The term *parochial* literally refers to a situation of this kind. It is an attitude that restricts community membership to those within the borders of a parish and excludes as aliens all those outside those borders. That attitude of "we" and "they," of insiders and outsiders, appears to be characteristic of communities in many different cultures. In Italy, for example, the term *campanilismo* denotes the very same thing as the English word *parochialism*: Everyone beyond the range of one's own church bell is a stranger and outsider.

Throughout history communities in the above sense have been a major form of human social organization. Had there not been other tendencies at work, parochialism would have precluded any wider social integration. But there were countervailing forces. With time, economic and other relations between communities tended to break down parochial barriers and to result in cultural osmosis and assimilation. Small communities became the components of larger social entities, so that eventually provinces and regions came to develop a "consciousness of kind."

It is only recently that local communities, provinces, and regions have been brought together in larger *national* societies. The nation-state is a modern phenomenon, as is nationalism—in the sense of a larger consciousness of kind that is supposed to supersede cultural and linguistic particularisms. Furthermore, the large national frameworks did not develop organically of their own accord. They were imposed from above by force. French nationalism developed in connection with the revolutionary and Napoleonic wars; and in reaction, nationalist movements emerged in the rest of Europe. Germany was unified by force in the last decades of the nineteenth century, as were Italy, Japan, and the United States. It is sometimes forgotten that the United States was also unified by force, and that national unity was achieved at the cost of a devastating civil war.

Although history texts speak of the unification of, say, Italy and Germany in the 1870s, that is a misleading statement culturally and linguistically. To this day there is no common spoken language in those countries, but rather several dialects, some of them mutually unintelligible to their speakers. Culturally, there are also important local, provincial, and regional differences. Such cultural and linguistic variations are not unique to Italy and Germany but are found in every nation-state of Europe. Indeed, the term *nation-state* is often inappropriate since the states in question are actually multinational. Little Switzerland and giant Russia are cases in point. National and multinational state organizations are to be subsumed under *Gesellschaft,* not *Gemeinschaft.*

Gemeinschaft, or "community," is therefore a matter of kinship, neighborhood, or friendship relations and the corresponding states of   83 mind. Its main historical forms have been family, clan, and rural village life, regulated by folkways and customs. *Gesellschaft,* or "society," on the other hand, is associated with modern urban life, with national organizations and institutions and with a cosmopolitan as opposed to a parochial consciousness.

Gesellschaft, historically speaking, comes into being with the growth of trade and commerce and the increasing expansion of money, contracts, and business. *Gesellschaft* received its greatest impetus from the emergence of industrial capitalism in the eighteenth century and from its continuing development since that time. Finally, *Gesellschaft* is associated with the development of modern science and instrumental rationality. Looked at historically, therefore, *Gemeinschaft* tends toward *Gesellschaft,* toward large and complex social macrocosms.

5

Social Organization II:
Whole Societies
and Other Macrocosms

THE PROBLEM OF ORDER

How does a society establish and maintain order? How does it ensure
a minimal measure of internal peace that will enable its members to
make a living and attend to their other vital affairs? By what means
does a society seek to establish social stability and cohesiveness? It is
such questions that constitute the "problem of order." That problem
has been a central concern of social and political theory from the time
of philosophers' earliest reflections on society and government. Plato's
Republic is an inquiry into the possibility of creating a higher and more
just order, and Aristotle's *Politics* is a comparative analysis of types of
social and political structure: Which is the best type, and how is it to
be established?

For Artistotle, social order presupposed a measure of equality. Inequality, he believed, "is generally at the bottom of internal warfare in
states, for it is in their striving for what is fair and equal that men become divided." The best society is based not on the extremes of rich
and poor but on a large class of the moderately well-to-do. In possessions, "to own a middling amount is best of all." A good society should
"consist as far as possible of those who are like and equal, a condition
found chiefly among the middle section."[1] When most men possess
moderate property, they have a stake in the existing order.

When, in contrast, an entire society consists of some who possess a
great deal and others who possess nothing, we can expect one of two   85

results, or an alternation between them. Either order will be continually menaced by social tensions, conflicts, and revolution; or order will be forcefully imposed by a tyrant from above. The poor, having nothing to lose, may easily become insurrectionist; while the rich, having so much to lose, will do their utmost to prevent overt opposition and uprisings. Hence moderation in property, as in all things, was for Aristotle the firmest foundation of social peace.

After Aristotle the problem of order was often addressed anew. One of the most influential theories was put forth by Thomas Hobbes. In the tumultuous days of the English civil war, Hobbes produced his *Leviathan* (1651), a classic study of power as a precondition of social peace. Because of Hobbes's overriding concern with peace and his systematic analysis of power, his views have come to occupy a central place in discussions of order.

Hobbes defined power as a man's "present means to obtain some future apparent Good."[2] For Hobbes there existed in all humans a *natural* and restless desire for power that ceased only in death. Indeed, an individual pursues more and more power, although not because he hopes for the greater delight that increments of power will bring him nor because he cannot rest content with moderate power. Rather, the reason is that he cannot secure the present power he has to live well without acquiring more.

Hobbes made another assumption about human nature: That the natural condition of humanity is one of *equality*. In the presocial state of nature which Hobbes postulated, humans are equal in the faculties of both body and mind. True, one individual may be physically stronger or mentally quicker than another, "yet when all is reckoned together, the difference between man, and man, is not so considerable, as that one man can thereupon claim to himself any benefit, to which another may not pretend, as well as he. For as to the strength of body, the weakest has strength enough to kill the strongest, either by secret machination, or by confederacy with others, that are in the same danger with himselfe." And in faculties of mind Hobbes saw even greater equality among humans. "For Prudence, is but Experience; which equall time, equally bestowes on all men, in those things they equally apply themselves unto. That which may perhaps make such equality incredible, is but a vain conceit of one's owne wisdome, which almost all men think they have in a greater degree, than the Vulgar."[3]

The condition of fundamental equality among humans gives rise to equality of hope in the attainment of their ends. Therefore, "if any two men desire the same thing, which neverthelesse they cannot both enjoy, they become enemies; and in the way to their End, . . . endeavour to destroy or subdue one another." The result is that humans quarrel and fight for gain, for safety, and for reputation. Hence the *natural* state is one in which humans are engaged in *war*—a war of everyone against everyone. War, for Hobbes, referred not only to the act of fighting itself, but also to the will to contend. The consequences of such war are that humans live without culture and without society; and worst of all, they live in continual fear and danger of violent death; and the life of the individual is "solitary, poore, nasty, brutish, and short."[4]

Hobbes, then, viewed the state of nature as one in which force and fraud prevail. In that state, there is no right or wrong, no just or unjust, for those are social, not natural, qualities, which humans acquire only in society. And the single, most important condition that makes society possible is a "common Power to feare." Wherever and whenever no such common power exists, humans revert to a state of nature and war. The fear of falling back into that state and the will to survive evoke in humans a modicum of reason which leads them to the formation of a *social contract.* Under its terms, individuals agree to give up their natural liberty and to subordinate themselves to a sovereign authority who in turn guarantees them security and protection from force and fraud. It is only the common terror of the sovereign that holds the war of all against all in check. The contract with the sovereign is guaranteed by the sword, for contracts "without the Sword, are but Words, and of no strength to secure a man at all." In thus conferring all their strength on one sovereign, individuals form society, or what Hobbes called a *Commonwealth.* It is "that great Leviathan, or rather (to speake more reverently) . . . that *Mortall God,* to which we owe under the *Immortall God,* our peace and defence."[5]

Thus Hobbes saw human individuals as warlike in nature; he believed it is only in society that their fighting and disposition to fight are restrained. War is natural, and peace is social. We have already seen in chapter 3 how Rousseau replied to Hobbes: "Man becomes a citizen before he becomes a soldier." The use of force and violence is learned behavior, learned in the course of socialization in the context of social organization. We have also seen that the various theories of innate aggression in humans are untenable. We shall not repeat those arguments here but rather limit our attention to the one element of Hobbes's theory that pertains to the problem of order.

Contract theory in Hobbes, as in other theorists whom we shall shortly consider, is sociologically erroneous. All contract theories postulate human societies as arising out of individuals, not individuals out of society. Those theories assume that first it is individuals with rational selves that are in existence, and that it is those individuals who get together to form society. In that view societies are like business organizations formed by a group of investors, who choose their officers and constitute themselves as a society. But we have seen in the preceding chapters that the individual is a strictly social being, and that his existence is unthinkable before and outside of society.

Nevertheless, for certain heuristic purposes the notion of social contract may be very useful; and there can be no doubt that social and political theorists have employed the notion with illuminating results. If, for example, we restrict our attention to Hobbes's explanation of order, we can see an element of truth in his theory.

If we reflect on society, our own and the many others we know about from history and in the world today, we can readily see that the *fear of power* is one important component of any social order. As we shall see, however, it is only one such component and not the whole story by any means. There are some societies in which fear of the rulers looms so large that one would have to say that such fear is the key element. Individuals obey because they are intimidated and terrorized by the state. At the other extreme, there are societies in which fear of the sovereign, his agents, and the police plays a very small role in eliciting the obedience of citizens. In that type of society, individuals obey for quite different reasons. Thus whatever other criticisms we may have of Hobbes, we would have to acknowledge that he rightly uncovered one element of social order. To grasp the other elements, we must consider another contract theory, as well as major criticisms of it.

UTILITARIAN THEORIES

Contract theories of a utilitarian sort attempt to account for order quite differently. If Hobbes rested his entire argument on fear of a central authority, the utilitarians dispensed with a central authority altogether. We shall discuss the utilitarian view as if it were a pure type, ignoring the individual differences and emphases among the thinkers in question. John Locke, the English classical economists (e.g., Adam Smith and David Ricardo), Jeremy Bentham, and Herbert Spencer are among the best-known representatives of that point of view.

For the utilitarians, social order and harmony resulted from the division of labor. Order is an *automatic* consequence of an economic system in which every individual pursues his own interests. That theory expressed itself in the doctrine of *laissez faire*—the belief that the economy works best when left alone. No central regulatory agency is required for its smooth operation. Indeed, the intrusion of such an agency can only disrupt what is essentially a self-regulating system. All economic affairs take place through the medium of free exchange, and if each individual dedicates himself to the pursuit of his own interests, that will lead to the "greatest good of the greatest number."

It is doubtful whether the market system ever worked in the automatic, self-regulating way in which the classical economists conceived of it. It is even more doubtful that it worked for the general good. As early as the second decade of the nineteenth century, the economist Sismondi in his *Nouveaux Principes d'Economie Politique* (1819) demonstrated that the poor suffer most from economic crises and that the utilitarians were therefore simply wrong. Those economic matters aside, the important question for our purposes is whether free exchange and other contractual relationships must lead to social order.

The most telling critique of utilitarian theory was delivered by the noted French sociologist Emile Durkheim (1858–1917). Durkheim observed that if we look carefully into the so-called total harmony of interests of the utilitarians, we can see that it conceals a latent conflict. For it is clear that if an individual's own interests are the sole regulator of his conduct, there is nothing to prevent everyone's relentless pursuit of self-interest from degenerating into a Hobbesian war of all against all. "There is nothing less constant than interest," wrote Durkheim. "Today it unites me to you; tomorrow it will make me your enemy."[6] It is true, argued Durkheim, that interests can bind people together; but they can do so only temporarily and partially. For when the mutual interest in exchange of two or more parties ceases, they will either turn away from each other or even against each other. It follows that *contractual interests* alone no more than *fear of the Leviathan* alone can account for social order. Each of those conditions may constitute one element of order, but neither condition by itself can serve to ensure order. In developing his critique, Durkheim provided the third element which any adequate theory of order must include.

What the utilitarians have overlooked, Durkheim correctly noted, is that every contract contains *noncontractual* or *sociomoral* elements that exercise some regulative control over the parties concerned. In fact, a contract has no validity if it fails to fulfill the conditions required by law. The contracting parties acquire obligations of a moral or legal kind that are not specified in the terms of the contract. Thus contract   89 law, rooted in custom, tradition, and precedent, provides the socio-moral context of every exchange relationship. We may exchange and cooperate because it is in our mutual interest, but the relationship we thereby form is necessarily hedged in by duties and obligations *not* of our own making. "The agreement of parties," said Durkheim, "cannot render a clause just which by itself is unjust, and there are rules of justice whose violation social justice prevents, even if it has been consented to by the interested parties."[7] Thus Durkheim logically demonstrated that contractual relationships alone cannot form the basis of a stable social order.

FROM MECHANICAL TO ORGANIC SOLIDARITY

The shortcomings of utilitarianism led Durkheim to conclude that the increasingly complex modern industrial division of labor must be viewed in a new light. The "economic services that it can render are picayune compared to the moral effect that it produces, and its true function is to create in two or more persons a feeling of solidarity."[8] Durkheim presented his theory by exploring the transition from the preindustrial to the industrial social structure. His main concern was the implications of that transition for order, or what he called *solidarity*. In Durkheim's view the preindustrial society was solidary because it was socially homogeneous: Everyone was socially similar. However, with the advancing division of labor, the original solidarity was undermined. Does that mean that solidarity is now forever impossible? No, for according to Durkheim, a new and *higher* type of solidarity is being generated by the growing complexity of the division of labor. That is the thesis Durkheim developed, beginning with a consideration of what he termed *mechanical solidarity, or solidarity through likeness.*

The solidarity of Durkheim's hypothetical preindustrial society is based upon common collective sentiments, a *conscience collective,* which in French carries the connotation of both a common consciousness and a common conscience. The collective sentiments are strongly engraved on all individual consciences. The best empirical indication of this is the social reaction to crime. In fact, an act was "criminal" for Durkheim precisely because it was carried out in opposition to the collective sentiments. One must not say, he wrote, "that an action shocks the common conscience because it is criminal, but rather that it is criminal because it shocks the common conscience."[9] Any act which violates the common conscience threatens the solidarity—the very existence—of the community. An offense left unpunished weakens to that same degree the community's social unity. Punishment therefore serves the important function of restoring and reconstituting social unity. In that type of society, punishment is expiatory and retaliatory. It is a passionate reaction by the community against those who dared violate its basic rules. Restitution is not enough; the social body must make a passionate reaction, for it is thanks to the intensity of the reaction that the group revitalizes itself.

If repressive and expiatory law is characteristic of mechanical solidarity, it is *restitutive* law which is typical of organic solidarity. Here the aim is not punishment pure and simple, but restoring damaged interests. Since industrial society is a complex of many and diverse groups and interests, law acts through specialized organs. But it is *society* that empowers those organs and acts through them. Even contractual relations that are ostensibly private and individual become binding precisely because society authorizes and sanctions the obligations contracted for. What is presupposed in every contract is that society stands ready to intervene in order to ensure respect for the agreements that have been made.

Durkheim now turned to the positive social consequences of the industrial division of labor: The increasing specialization of tasks, he argued, leads to a growing reciprocity and interdependence. A higher, organic solidarity can be achieved on the basis of the modern division of labor, which "more and more fills the role that was formerly filled by the Common Conscience. It is the principal bond of social aggregates of higher types."[10] Yet Durkheim felt a certain uneasiness with his thesis since it was evident that the modern industrial division of labor was not in fact engendering the solidarity he predicted. How did he deal with that fact?

If the anticipated solidarity was failing to appear, that was a consequence of the *abnormal* or pathological forms which the division of labor assumed. "Though *normally,*" wrote Durkheim, "the division of labor produces solidarity, it sometimes happens that it has different, and even contrary results. Now, it is important to find out what makes it deviate from its natural course, for if we do not prove that these cases are exceptional, the division of labor might be accused of logically implying them."[11] Thus for Durkheim it was not the normal but the *pathological* forms of the division of labor that have forestalled the emergence of the higher, organic solidarity.

The first of the pathological forms he called the *anomic* division of labor. That word comes from the Greek *anomia,* referring to a state of society in which normative standards of conduct are weak or absent. Durkheim employed the term to convey his view that what was lacking or poorly developed in modern industrial society was a moral-legal code appropriate to the new conditions. Such a code was essential in order to mediate among the many and diverse interest groups in society and thus to regulate and moderate the socioeconomic system.

But Durkheim realized that rules and regulations cannot be the whole solution and that often the rules themselves serve to perpetuate certain social ills. That is particularly evident in class conflicts. Socioeconomic classes, an integral aspect of the industrial division of labor, are a source of dissension and conflict. The "lower classes not being . . . satisfied with the role which has devolved upon them from custom or law aspire to functions which are closed to them and seek to dispossess those who are exercising these functions. Thus civil conflicts arise which are due to the manner in which labor is distributed."[12] In this way Durkheim introduced a second major pathological form—*the forced division of labor*.

Order and Justice

Under the heading of the "Forced Division of Labor," Durkheim examined the relationship between order and justice. The higher organic solidarity requires new rules, but if those rules are not inherently just, the solidarity will never materialize. "If one class of society is obliged in order to live, to take any price for its services, while another can abstain from such action thanks to resources at its disposal . . . , the second has an unjust advantage over the first at law. In other words, there cannot be rich and poor at birth without there being unjust contracts."[13] Durkheim therefore saw the task of modern society as a "work of justice."

That issue was further pursued in another book of his, entitled *Professional Ethics and Civic Morals*. A contract is a juridical-moral bond between two subjects that specifies their mutual rights and obligations. Generally, said Durkheim, " a right exists on both sides." He was quick to add, however, that "these mutual rights are not inevitable. The slave is bound in law to his master and yet has no right over him."[14] Thus Durkheim made the important observation that some contracts are made between social unequals, where one dominates and the other serves and where the latter has no other choice than to serve or to suffer even worse consequences. Can such a contract, though sanctioned

92

by "a moral authority that stands higher," be just? To that question Durkheim replied with a resounding No.

Tracing the development of contracts as a social institution, Durkheim showed that a bona fide contract cannot be "one of good faith except on condition of its being one by mutual consent."[15] But consent truly binds only those whose consent has been *freely* given.

Anything that lessens the liberty of a contracting party, lessens the binding force of the contract. This rule should not be confused with the one that requires that the contract be made with deliberate intent. For I may very well have had the will to contract as I have done, and yet have contracted only under coercion. In this case I will the obligations I subscribe to, but I will them by reason of pressure being put upon me. The consent in such instances is said to be invalidated and thus the contract is null and void.[16]

A contract cannot be viewed as just simply because an individual has consented to submit to its conditions. The crucial criterion is how much freedom one has to avoid entering into certain types of contractual agreements. Whether a contract is binding or not depends not merely on one's subjective will, but on the *objective* conditions under which it is made. Durkheim's penetrating analysis is worth quoting at length: If

contracts imposed by constraint, direct or indirect, are not binding, this does not arise from the state of the will when it gave consent. It arises from the consequences that an obligation thus formed inevitably brings upon the contracting party. It may be, in fact, that he took the step that has bound him only under external pressure, that his consent has been extracted from him. If this is so, it means that the consent was against his own interests. . . . The use of coercion could have had no other aim or consequence but that of forcing him to yield up something which he did not wish to, to do something he did not wish to do, or indeed of forcing him to the one action or the other on conditions he did not will. Penalty and distress have thus been undeservedly laid on him.[17]

Such a contract, Durkheim observed, is increasingly regarded as invalid inasmuch as it causes an individual to suffer an unjustified injury. Such a contract is unjust. A contract is moral and just only if it is not a "means of exploiting one of the contracting parties."[18]

What did Durkheim mean by an exploitative or unjust contract? Every object of exchange, he argued, has a "determined value which we might call its social value. It represents the quantity of useful labor which it contains. . . . Although this magnitude cannot be mathematically calculated, it is nonetheless real." According to that principle an unjust exchange is one in which "the price of the object bears no re-   93 lation to the trouble it cost and the service it renders." A contract is uncoerced and just "only if the services exchanged have an equivalent social value," and in order "that this equivalence be the rule for contracts, it is necessary that the contracting parties be placed in conditions externally equal." Public morality must therefore condemn "every kind of leonine contract wherein one of the parties is exploited by the other because he is too weak to receive the just reward for his services. Public conscience demands, in an ever more pressing manner, an exact reciprocity in the services exchanged."[19] That led Durkheim to the conclusion that the institution of private inheritance of wealth and property is a "supreme obstacle" to just relations in society.

Now inheritance as an institution results in men being born either rich or poor; that is to say, there are two main classes in society, linked by all sorts of intermediate classes: the one which in order to live has to make its services acceptable to the other at whatever the cost; the other class which can do without these services, because it can call on certain resources. . . . Therefore, as long as such sharp class differences exist in society, fairly effective palliatives may lessen the injustice of contracts; but in principle, the system operates in conditions which do not allow of justice.[20]

That is the way Durkheim arrived at the conclusion that a true organic solidarity requires the abolition of fundamental inequalities. A higher, consensual order must be based on social justice.

The problem of order is also a major concern of a highly influential theoretical current in contemporary sociology called *functionalism*.

FUNCTIONALISM

In their treatment of the problem of order, functionalists have followed neither Hobbes nor the utilitarians nor even Durkheim. Instead, they have attempted to ground social order in moral consensus alone.

From the traditional functionalist standpoint, societies and other large organizations may be studied as if they are living organisms. Just as all the organs of a living being contribute to its survival, so presumably do all the parts of a society contribute to its maintenance. A society, like an organism, has certain needs, and its institutions and other structures serve to fulfill those needs. Naturally, because sociological functionalists deal with social rather than biological systems, they have had to modify the biological vocabulary somewhat in order to adapt it to their specific purposes. Thus sociological functionalists employ such terms as *struc-*

94

ture, function, functional prerequisites, system, system needs, integration, and *adaptation.*

Structure refers to any comparatively stable social pattern of conduct. Most often that term refers to a major institutional pattern such as that associated with marriage or religion. But the term may also refer to a simple pattern of face-to-face interaction in everyday life.

Function refers to the consequences of a particular structure. Those may be consequences for another structure or for a system as a whole. For a long time functions were regarded as uniformly positive. Then, in response to criticisms, a distinguished sociologist named Robert K. Merton introduced several theoretical revisions, one of which allowed for negative consequences. Those he called *dysfunctions.* We will look at his revisions momentarily.

System refers to any relatively stable pattern of interaction between two or more persons or structures. A social system can be as small and simple as a dyad or as large and complex as a whole society.

System needs, following the organismic analogy, refer to needs which must be fulfilled if the system is to survive and maintain itself. One of the basic needs is called *integration,* a concept that emerged out of a consideration of the "problem of order." Every social system, the functionalists argue, must achieve a minimal degree of integration. Another basic need is called *adaptation.* A social system must not only integrate itself internally, but must also adapt to the external environment. Finally, functionalists speak of *functional prerequisites,* i.e., consequences of structures that fulfill vital system needs.

Those are the key concepts of sociological functionalism. Underlying the functionalist approach are two fundamental assumptions: (1) All social structures, or at least the major ones, contribute to the integration and adaptation of the system in which they operate; and (2) the existence and persistence of any given structure is to be explained by means of its consequences, which are both necessary and beneficial to the society in question. Generally, functionalists have tried to show that a given structure fulfills some vital system need and that that fulfillment explains the existence of the structure—presumably because other structures act back upon the first one and contribute to its maintenance. Thus functionalists have endeavored to produce a scientific sociological theory of societies and social systems. It needs to be stressed that functionalists do not use their concepts as mere metaphors. They rather intend them to be employed as precise scientific terms. And it is on those scientific grounds that numerous critics have met them, delivering some telling criticisms.

One such criticism is that the functionalist terms are not very precise at all, but rather vague and ambiguous. Take the concept *society*. When we use that word in everyday discourse, as, for example, in the phrases "American society," "French society," etc., we readily understand the metaphor. We may disagree with one another's characterizations of those societies, but we readily grasp the meaning of the phrase. We do not pretend that the word *society* is a scientific term, and we can therefore tolerate some ambiguity in it.

It is quite different with the functionalists. They intend their terms to have definite empirical referents. One leading functionalist has defined a society as "a system of action in operation that (a) involves a plurality of interacting individuals . . . who are recruited at least in part by the sexual reproduction of members of the plurality involved, (b) is . . . *self-sufficient* for the action of this plurality, and (c) is capable of existing longer than the life span of an individual" (italics added).[21]

"Self-sufficiency" is the key criterion which that author employs to distinguish a society from any other social system. Evidently he wishes to define a social entity which is larger than the family, local community, city, province, or region—the entity we call the nation-state. But in response to his definition the question has been asked: Is there such a thing as a self-sufficient society? Is there any nation-state or multinational state so self-sufficient that it can dispense with economic, political, cultural, and other relations with other states? If the answer to that question is No, we are compelled to conclude that the functionalist term *society* has no precise empirical referent.

Equally vulnerable is the other master concept, *social system,* which has been defined as "any system of social action involving a plurality of interacting individuals" or "any patterned collection of elements." As critics have observed, that definition is so vague and general as to be meaningless. For it does not enable us to distinguish among the various kinds of social groups we all know about—families, friendship groups, residential communities, formal and complex organizations, etc.—all of which may be regarded, according to the definition, as "social systems."

The fact that concepts as fundamental as "society" and "social system" are scientifically inadequate suggests that all the associated concepts together with the accompanying assumptions are also problematic. If one has no clear and precise idea of what "society" or "social system" refers to, how can one speak of "system needs," which purportedly are fulfilled by various "structures"?

Another criticism is that the functionalists have failed to verify their assertions and instead have ensnared themselves in tautological or circular reasoning. Functionalists have claimed, for instance, that a society cannot prevail unless it avoids all four of the following conditions:

1. the biological extinction or dispersion of its members;
2. the apathy of its members;
3. a war of all against all;
4. the absorption of the society into another society.[22]

Is there any proof, even in the most modest sense of that term, that those are *the four* conditions that terminate a society? The proponents of that theory have provided no proof. They have rather had recourse to a tautology: The very persistence of a society demonstrates that the four conditions are inoperative.

Nor is the notion of the "functional prerequisites of a society" any more satisfactory, either from a logical or an empirical standpoint. The list of such prerequisites includes:

1. provision for adequate relationships to the environment and for sexual recruitment;
2. role differentiation and role assignment;
3. communication;
4. shared cognitive orientations;
5. a shared set of articulated goals;
6. the normative regulation of means;
7. the regulation of affective expression;
8. socialization;
9. the effective control of behavior.[23]

Since the authors again offer no proof, those presumed prerequisites, like the four conditions discussed earlier, resolve themselves into tautologies: The authors posit certain conditions for a society's survival and then point to its survival as proof that the functional prerequisites have been fulfilled. In a similar fashion, functionalists have maintained that structures persist only if they are adaptive, i.e., fulfill system needs. The persistence of a structure is then used to "prove" its adaptive character. One remains neatly enclosed in a circle.

The trouble with functionalist explanations, even the most sophisticated, is that the key concepts are defined so loosely that any and all social patterns may be regarded as adaptive or maladaptive, as the case may be. But what we demand from a scientific explanation is that it define clearly and precisely what is meant by "needs," "adaptive," "integrative," "function," etc. Yet functionalists have defined "system needs" so broadly that, as F. M. Cancian has remarked,

almost any social process can be seen as contributing to at least one of them, and, being a list, there are few constraints on adding more needs. But a limited set of prerequisites is essential to functional explanation; without it, an imaginative investigator can "explain" any social pattern merely by describing some consequence or effect of a pattern and arguing that the consequence is a functional prerequisite.[24]

As we observed earlier, the noted sociologist R. K. Merton has responded to criticisms of functionalism by revising its major postulates. His revisions may be concisely summarized as follows: A society's unity ought not be posited in advance but ought rather to be an empirical question. One must specify the unit for which a given item is alleged to be functional; and one must beware of assuming that all structures fulfill vital functions. Functional analysis must allow for diverse consequences, including dysfunctions as well as functions. Finally, one must not assume that one and only one structure can fulfill a given function. Rather, social scientists should allow for the possibility "that alternative social structures (and cultural forms) have served . . . the functions necessary for the persistence of groups. Proceeding further, we must set forth a major theorem of functional analysis; *just as the same item may have multiple functions, so may the same function be diversely fulfilled by alternative items.*"[25]

Let us examine what is being advocated in that italicized passage. If we accept the assumption that several patterns, x plus others, can fulfill the same function—say, "controlling disruptive behavior"—that is scientifically meaningful only if one can distinguish such patterns from nonequivalent ones. Moreover, that must be done in a manner that will enable other investigators to test the proposition by, for example, finding a society that controls disruptive behavior without x or any of its presumed equivalents. If one does not adhere to those simple rules, discovering "equivalents" will always be fairly easy.

In his "paradigm for functional analysis," Merton writes: "*Functions* are those observed consequences which make for the adaptation or adjustment of a given system; and *dysfunctions,* those observed consequences which lessen the adaptation or adjustment of the system."[26] At the same time he acknowledges that concepts such as "prerequisites," "system needs," "adaptation," and the like are "the cloudiest and empirically most debatable." With the aim of suggesting how to overcome the traditional functionalist ambiguities, he further writes:

We have observed the difficulties entailed in *confining* analysis to functions fulfilled for the "society," since items may be functional for some individuals and subgroups and dysfunctional for others. It is necessary, therefore, to consider a *range* of units for which the item has designated consequences: individuals in diverse statuses, subgroups, the larger social system and cultural systems.[27]

Merton appears to be recommending here that the social scientist should concern himself with the question of whether certain social patterns and institutions operate for or against the interests of specific individuals and groups. If that is in fact his aim, there would seem to be better and easier ways to accomplish the task. Durkheim's assessment of contractual relationships is a case in point. As we have seen, Durkheim put forth rather objective criteria for determining whether a specific agreement may be injurious to one or the other of the contracting parties. Under certain circumstances of inequality, he argued, the social institution of contracts may constrain an individual to act against his own best interests and cause him to be exploited. Thus Durkheim effectively and unambiguously examined the impact of social institutions and structures (e.g., social classes) on individuals and groups. We have also seen that in his analysis of mechanical solidarity, the retaliatory punishment for a crime is interpreted as serving to enhance the solidarity of a group. In a word, though Durkheim most certainly considered the consequences of structures for individuals and groups, he did so without encumbering himself with concepts such as "system needs" and the like. It would seem, therefore, that the functionalist approach is beset with basic problems and ambiguities.

A quite different approach to the study of societies and other large social organizations is derived from the writings of Karl Marx (1818–1883). In what follows we shall concentrate on those ideas of his which appear to be valuable for social science.

THE MARXIAN VIEW OF SOCIAL ORGANIZATION
INTRODUCTION: PHILOSOPHICAL FOUNDATIONS

Marx described his method of studying society as *materialist*. The use of that term has contributed to a widespread misunderstanding of his intention. We must therefore clarify what he meant by "material" before we can gain an adequate appreciation of his theory and method.

The main reason why Marx sought a "materialist basis" for his method can be traced to his dissatisfaction with the state of intellectual life in Germany, the country of his birth. In Marx's youth German social thought was dominated by Hegelian and other forms of philosophical idealism in which social and historical events were reflected in highly abstract, metaphysical terms. Real human activities and struggles were scarcely perceptible behind the Hegelian categories. Intellectual discourse was carried on as if humanity and the world existed only in the realm of pure thought. Thus we find Hegel transforming the Idea, or Reason, into an independent, historical subject. Reason becomes the creative and guiding principle of the real world, while the real world is merely the external form of the Idea.

As a young man Marx was so taken by Hegel's ideas that for a time he joined a group of disciples called the Young Hegelians. But he soon came under other intellectual influences which led to his dissatisfaction with the Old and Young Hegelians alike. Their major defect, Marx believed, was that all of them had failed to inquire into the connection between German philosophy and German reality. More generally, they had failed to inquire into the relationship of their ideas to "their own material surroundings."[28] What Marx here meant by "material" can for the moment be conveyed by using several words in tandem: *social, economic, political, historical,* and *cultural.*

Yet Marx's dissatisfaction with the Hegelian tradition did not lead him to reject it in toto. He discerned a very important "rational kernel" in Hegel's mystical shell. For Hegel (and the idealists generally) had brought out the *active* side of the human individual; Hegel had given the *self* a creative and determining role in his philosophy—though only in a form abstracted from real human history and social activity. Furthermore, Hegel had not only recognized the existing state of things but their *negation* as well, their inevitable transformation and breaking up. For Hegel every social form was in fluid movement. History was a process in which social forms come into being, develop, and then give way to new social forms. It is that rational form of Hegel's dialectical philosophy that Marx wished to salvage and to apply in his own social theory. Thus Marx selected certain elements of Hegel's view which he deemed intellectually valuable and rejected others.

That is the same selective attitude that Marx adopted toward the materialists of his time. He vehemently rejected their mechanistic and reductionist doctrine according to which human mental activity is nothing more than matter in motion. Chemical bodily processes, that doctrine taught, were sufficient to explain human ideas and emotions. Materialists of that kind contended that "Ideas stand in the same relation to the brain as bile does to the liver or urine to the kidneys."[29]

The *locus classicus* of Marx's critique of mechanistic materialism is his famous "Theses on Feuerbach." There he forcefully rejected any doctrine which ignores the active, creative, and determining side of the human individual and which overlooks the dialectical or interactive relationship of the individual to his circumstances. In his first thesis Marx criticized Ludwig Feuerbach for his mechanistic view, in which the individual is transformed into a mere *object*. Feuerbach, Marx argued, had missed the active side of the human being: He had failed to recognize in humanity a creative historical *subject*. Feuerbach and other materialists thus produced a one-sided doctrine in which humans are the mere products of circumstances. What they have overlooked, Marx noted, is that it is none other than human beings who make and change social circumstances.

Hence Marx's theoretical-methodological standpoint is quite complex. And it is more important to understand it than to label it. In his critical mediation between the two major philosophical doctrines of his time, he endeavored to overcome the extreme one-sidedness of each of them in a new synthesis of his own. He stated the basic premise of that new synthesis in one of his earliest attempts to formulate it. In a book entitled *German Ideology,* which he co-authored with his lifelong friend and colleague Frederick Engels, we read: "The premises from which we begin are not arbitrary ones, not dogmas, but real premises from which abstraction can only be made in the imagination. They are the real individuals, their activity and the material conditions under which they live, both those which they find already existing and those produced by their activity."[30]

For Marx, those premises meant that human beings create the sociocultural world by means of their theoretical-practical activity. They create the world in the sense that they produce their tools and other objects with the materials of nature, thus modifying nature, their means of acting upon it, and their relations with their fellow humans. In Marx's view nature is not purely "objective" in the sense that it is entirely independent of the human will or that it would be the same nature in the absence of humanity. Nature is always shaped by human activity.

It is evident that in his conception of humanity and nature, Marx dissociated himself from both idealism and materialism. In one of his earliest writings, the "Paris Manuscripts of 1844," he described his emerging standpoint as "naturalism or humanism [that] is distinguished from both idealism and materialism and at the same time constitutes their unifying truth."[31]

In those manuscripts Marx sketched the rudiments of his approach.   101

He affirmed that while the human individual is a natural being, he is at the same time significantly different from other members of the animal kingdom. The animal "is immediately identical with its life-activity. It does not distinguish itself from it. It is its life-activity. Man makes his life-activity itself the object of his will and consciousness. . . . Conscious life-activity directly distinguishes man from animal life-activity."[32] That was Marx's way of saying that in the course of social life, the human individual develops "mind" and "self"; and, indeed, in a process quite similar to that outlined by George Herbert Mead.

How does a human become a conscious being? By cooperating with other members of his species, by

creating an *objective world,* by his practical activity, in *working-up* inorganic nature. . . . Admittedly animals also produce. They build themselves nests, dwellings, like the bees, beavers, ants, etc. But an animal . . . produces only under the dominion of immediate physical need, whilst man produces even when he is free from physical need and only truly produces in freedom therefrom.[33]

Distinctively human production, then, is creative and free activity, as in the case of the artist whose creativity is not dictated by immediate physical need. On the other hand, when human activity is degraded to a mere means of earning a livelihood, that activity becomes coerced and alienated labor—a dehumanizing process. Marx's main purpose in the Paris manuscripts was to specify the social conditions of *alienation*—a concept to which we shall presently return. In exploring the roots of alienation, he anticipated the central principles of the social psychology of G. H. Mead. "We must bear in mind," wrote Marx, "that man's relation to himself only becomes *objective* and *real* for him through his relation to the other man." Man's "own sensuousness first exists as human sensuousness for himself through the *other* man." Thus Marx recognized that the individual is a "*social being*. His life, even if it may not appear in the direct form of a *communal* life carried out together with others—is an expression and confirmation of *social life*."[34]

But Marx understood that a human is a *natural* being as well. As such, man has a biologic-organic nature with all the needs that such a nature implies. That Marx shares some common ground in that respect with Mead and Freud becomes clear in his statement that the human being is "furnished with natural powers of life—he is an active natural being. These powers exist in him as tendencies and abilities—as *impulses*." But the human individual is a special kind of natural being. He is a *human* natural being. As such, he not only retains his natural needs (hunger and sex) but acquires new ones—*love,* to single out the most important: "If you love without evoking love in return . . . if through a living expression of yourself as a loving person you do not make yourself a loved person, then your love is impotent—a misfortune."[35]

It is clear, then, that Marx's humanist-naturalist conception of the human situation anticipated some of the best and most important ideas of contemporary social thought. That is true even of his understanding of language. "Language," he wrote,

is as old as consciousness, language is practical consciousness, as it exists for other men, and for that reason . . . exists for me personally as well; for language, like consciousness, only arises from the necessity of intercourse with other men. . . . Consciousness is therefore from the very beginning a social product, and remains so as long as men exist at all.[36]

Marx sometimes employed the word *dialectical* to describe his method. In its most rudimentary terms, the dialectical approach means that one must recognize that humans are both natural beings and more than natural beings; that humanity is both author and agent of history and society. To understand the sociohistorical process in which humanity creates the world and itself, one must begin, Marx believed, by studying the most essential human activity: the process of production.

The Productive Process
and the Historical Development of Society

For Marx, the productive process, or the way humans make a living and provide for themselves, was the most fundamental of human activities. The totality of social relations implicated in the productive process he sometimes called the economic *foundation* of society. On the other hand, he occasionally referred to the political, religious, philosophical, and other forms of culture as a society's superstructure. It is often inferred from those terms that Marx believed that economic conditions always determined the other parts of a society and culture. The "foundation" is the cause, the "superstructure" the effect. Thus a rigid, one-way, mechanistic, causal determinism has been ascribed to Marx. However, it cannot be emphasized strongly enough that such an interpretation of Marx's theory is quite wrong.

Marx's real scientific aim was *not* to prove that economics determined everything else. It was rather to show the connections and interrelationships between the economic and other aspects of social organization. It was only in that way, he believed, that one could arrive at an   103 adequate understanding of the structure of a society and the changes it undergoes.

The term Marx used to refer to a society's specific organization of its productive activities is *mode of production*. There have been several such modes of production in history: the primitive communal, the ancient (slave), the feudal, and the capitalist. Marx conceived of modes of production as pure types. He was not enunciating a universal law of history or social evolution, for a specific sequence of modes of production has occurred only in specific Western societies. In other societies, certain modes did not occur at all, or were skipped, or occurred in mixed types, etc. Hence Marx was trying to provide not a suprahistorical, a priori system but a historical-sociological method of inquiry.

The human productive process may be traced to its earliest beginnings hundreds of thousands of years ago, in the Paleolithic or Old Stone Age. In that epoch humans first emerged as *food gatherers*. They lived much like other creatures, catching, hunting, and collecting whatever food nature happened to provide. Although we call it a *gathering* economy, that should not obscure the fact that humans were already toolmakers and tool users. In addition, there existed a division of labor between the sexes in which women collected food and cared for the children while the men wandered farther from home in search of animal prey. As V. Gordon Childe has remarked, the gathering type of economy was the "sole source of livelihood open to any human society during nearly 98 per cent of humanity's sojourn on this planet. . . . It is still practiced by a few backward and isolated societies in the jungles of Malaya or Central Africa, in the deserts of north-western Australia and South Africa, and in the Arctic regions."[37]

But some ten to twelve thousand years ago, humans for the first time learned to increase the food supply by cultivating plants and domesticating animals. That, in contrast to the earlier period, was a *food-producing* economy. In its earliest and simplest form, the new type of economy coincided with the Neolithic or New Stone Age. But *Neolithic* in the economic sense does not merely correspond to a specific epoch, since that way of producing could still be found in the early twentieth century.

The type of social organization to which the Neolithic Revolution gave rise is most familiar to us from history and anthropology. Marx and Engels called it *primitive communal*. Some such form of organization was present among most ancient peoples, and Marx's colleague, Frederick Engels, explored it in detail among the Greeks, Celts, and Germans. But it was also present among certain nineteenth-century

North American Indians, notably the Iroquois. Based on the pioneering studies of the American anthropologist Lewis Henry Morgan and on the historical materials concerning the other peoples, Engels provided a portrait of a primitive communal society. The Iroquois will serve as our illustration.

The Seneca tribe of the Iroquois confederacy was divided into eight clans.[38] At the head of each stood a *sachem,* a peacetime leader chosen from within the clan. The installation of both the peacetime and military leaders required the approval of the common council of the entire Iroquois Confederacy. Men and women voted in those elections. The *sachem*'s authority was strictly moral; he had no means of coercion at his disposal. The war chief's authority was limited to military affairs. The strikingly democratic character of the Seneca organization is also evident in the fact that the *sachem* and the war chief could be removed from office at any time if the tribal members were dissatisfied with them. A deposed *sachem* or chief became a simple brave, like his fellows.

Among the Iroquois there was no such thing as private property. There were, of course, personal belongings—clothing, hunting implements, religious articles, and the like. The belongings of deceased persons remained in the clan and were passed on to the other members. But the land with all of its bounty was conceived of as belonging to the entire confederacy. All clan members were bound by strong ties of brotherhood, owing one another help and protection as well as assistance in avenging injury by outsiders. Prisoners of war could be adopted by a clan, thus imparting to them full clan and tribal rights. In short, the typical Iroquois clan community was highly democratic, egalitarian, and communal. In Morgan's words:

All the members of an Iroquois gens [i.e., clan] were personally free, and they were bound to defend each other's freedom; they were equal in privileges and in personal rights, the sachems and chiefs claiming no superiority; and they were a brotherhood bound together by ties of kin. Liberty, equality and fraternity, though never formulated, were cardinal principles of the gens. These facts are material because the gens was the unit of a social and government system, the foundation upon which Indian society was organized. . . . It serves to explain that sense of independence and dignity universally an attribute of Indian character.[39]

Similar democratic and communal organizations were found among many early peoples on all continents, though Engels focused on Europe —the Greeks, Celts, Germans, and others. As we reflect on later stages in the development of European and other societies, we see that the earlier communal societies may not only be described positively, but   105 also by noting what is *missing* from them. Strikingly absent were the following: the institution of private property; fundamental economic inequalities expressed in the existence of social classes; and a coercive state apparatus. Hence the important question arises as to how, in the Marxian view, those social phenomena came into being.

The key to that question Marx saw in the changing mode of production. To speak generally, it was the growth of the productive forces of the societies in question that brought about basic changes in their social organization. The growth of skill, experience, and knowledge as applied to the cultivation of plants, the domestication of animals, and metallurgy gave rise to marked increases in productivity in all branches of the economy—cattle raising, agriculture, and handicraft industry. The capacity of human labor power, now greatly enhanced, succeeded in producing a larger product than was necessary for its maintenance. A surplus over and above the subsistence requirements of the producers themselves now emerged. At the same time, expanding production and the growing diversification of the division of labor increased the amount of work that needed to be done. It also increased the need for a wide range of handicraft and other skills, and the need for additional laborers became ever more pressing. War, which in the earlier period resulted either in the annihilation of one's enemy or the adoption of the survivors into one's own community, now turned prisoners of war into slaves. Thus the extension and diversification of the division of labor together with the resulting accumulation of wealth brought in its train the first historical social organization based upon socioeconomic classes: *masters and slaves.*

In this way the primitive communal situation was gradually transformed into the earliest forms of ancient class society. Inequalities in land and other resources undermined the communality of the clan community. Rich and poor, freeman and slave now appeared. A new class emerged—merchants—that had nothing to do with production, only with exchange. With time they became wealthy and powerful middlemen.

In the historical transition, the structure of power and authority was also basically transformed. The *state* made its first appearance, emerging together with private property in the means of production, the crystallization of classes, and class conflicts. Private property and inequalities of wealth gave rise to class cleavages, thus disrupting the earlier solidarity of the society as a whole. The state, wrote Engels, "is the admission that this society has involved itself in insoluble self-contradiction and is cleft into irreconcilable antagonisms."[40]

In the earlier communal society, authority was democratic, serving the interests of the collectivity as a whole; in class society the structure of authority became coercive. The means of violence came under the control of a special group serving the slave-owning and other propertied classes. A major function of the state was holding the slaves and disadvantaged masses in check—no mean task when we consider that in ancient Athens, Corinth, and Aegina, the number of slaves was more than ten times the population of free citizens.

The Feudal Mode of Production

In European society chattel slavery eventually gave way to a different type of organization. With the decline and disintegration of the Roman empire, the former seminomadic "barbarian" cultivators were tied down to the soil. The tying down ultimately increased the productivity of the temperate forest zone. The typical cultivator was now emancipated from chattel slavery, but he entered a new form of domination which may be called *serfdom*. Most pre-Marxian discussions of feudalism centered attention on the system of vassalage, upon the juridical relationship between a vassal and his superior. Neglected here was the economic content of the relationship. But Marx focused upon the relationship between the direct producer (whether he be a peasant cultivator on the land or an artisan in a workshop) and his overlord.

In serfdom, definite obligations are laid upon the producer. He is obliged to meet certain economic demands of his overlord, whether they be in the form of services to be performed or dues to be paid in kind or in money. The compliance of the serf may rest on the fear of violence at the hands of the overlord who possesses superior military strength, or it may rest on custom or law, or a combination of those elements. In any event the feudal mode of production contrasts with the slave mode in that the direct producer under serfdom is "in possession of his means of production, of the material labor conditions required for the realization of his labor and the production of his means of subsistence. He carries on his agriculture and the rural house-industries connected with it, as an independent producer," whereas "the slave works with conditions of labor belonging to another." Nevertheless, serfdom implies a "relation between rulers and servants, so that the direct producer is not free."[41]

Historically, feudal serfdom has been associated with instruments of production that are simple and inexpensive; it has also been associated with production for the immediate needs of the household or village 107 community, and not for a market. Finally, feudal serfdom was a highly decentralized political system. In theory, the lord of each domain was supposed to render certain judicial functions in relation to the dependent population.

Since production for a market was nonexistent or minimal, the feudal mode of production, as a pure type, was *not* an exchange economy. That is to say, what was produced here did not yet assume the character of a "commodity." In Marx's terms *commodity* refers strictly to an article produced for exchange, i.e., for a market. Only when the peasant family began to produce more than was enough for its own wants and for the dues payments to the lord did exchange emerge. The surplus products were offered for sale and became commodities.

The artisans of the towns, on the other hand, produced for exchange from the very beginning. But, as Engels observed, even the artisans

supplied the greatest part of their own individual wants. They had gardens and plots of land. They turned their cattle out into the communal forest, which, also yielded them timber and firing. The women spun flax, wool, and so forth. Production for the purpose of exchange, production of commodities, was only in its infancy. Hence, exchange was restricted, the market narrow, the methods of production stable.[42]

One can never date the beginnings of a historical process with precision. Yet many historians are in agreement that from the thirteenth century on, the "natural economy," producing for self-subsistent village communities, was more and more replaced by a developing exchange economy. In Marx's vocabulary, products which earlier had only *use-value* now acquired *exchange-value* as well. Commodity production, or production for the market, became a growing tendency.

Those significant socioeconomic changes gradually brought about other far-reaching social and cultural changes. Although the people who lived through the changes may not have grasped their extent or implications, it is evident in retrospect that the socioeconomic transformation gave rise to new ways of thinking. With the growth of the exchange economy, social relations became more and more *contractual*. In the context of an expanding exchange economy, it was natural enough for people to begin to think of society itself as a contract. And, indeed, we meet that concept in a number of quite diverse thinkers such as Hobbes, Locke, Grotius, Diderot, and, above all, Rousseau in his *Social Contract*. The emergence and prevalence of the social-contract idea enables us better to understand what Marx meant when he said that the socioeconomic "foundation" of a society tends to determine its cultural and ideological "superstructure"; or when he suggested that forms of social existence tend to determine forms of social consciousness. The objective developments that brought the exchange economy to the fore eventually reflected themselves in new categories of thought.

The Capitalist Mode of Production

Although based on commodities and money, the early exchange economy was still a precapitalist one. For Marx, money and commodities are transformed into *capital* under specific historical circumstances. Capitalism requires that "two very different kinds of commodity-possessors must come face to face and into contact; on the one hand, the owners of money, means of production, means of subsistence, who are eager to increase the sum of values they possess, by buying other people's labor power; on the other hand, free laborers, the sellers of their own labor-power and therefore the sellers of labor."[43] Those laborers are free in a double sense. They are neither a part of the means of production, as are slaves and bondsmen; nor do they themselves possess any means of production, as do peasant proprietors. Laborers under capitalism have been "freed" or *separated* from their traditional means of production and are now entirely dependent for their survival on the sale of their labor. In the absence of free labor in that sense, capitalism never could have arisen.

The origins of the new mode of production may therefore be traced to the process separating the producer from his means of production and means of subsistence and placing those means under the exclusive control of the capitalist. Capitalism required the proletarianization of the agricultural population. A "proletarian" possesses nothing but the labor of his hands. How, then, did such a socioeconomic class emerge for the first time?

By the end of the fourteenth century, serfdom had almost disappeared in England. The vast majority of the population consisted of free peasant proprietors who worked with their own instruments of production and provided for their own subsistence. They enjoyed the usufruct of the common lands, which provided them with timber and firewood and their cattle with pasture. However, a century later that state of affairs had begun to change. The feudal lords now had so little power that it was not difficult for the monarchy to disperse their retainers and confiscate their estates. As a result a mass of peasants and yeomen were driven from the soil. At the same time the more powerful feudal lords created an even larger proletariat by the forcible eviction of their peasants and by the usurpation of the common lands.

Those evictions were economically motivated. The continental wool manufacturers were prospering, which led in turn to a marked rise in the price of English wool. Marx described the new nobility as a child of its time, for whom "money was the power of all powers. Transformation of arable land into sheep walks was, therefore, its cry." Humans were displaced by sheep. Describing that tragic spectacle in his *Utopia,* Thomas More wrote: "Your shepe that were wont to be so meke and tame, and so smal eaters, now, as I hear saye, be become so great devourers and so wylde that they eate up, and swallow downe, the very men themselves."[44] Many of the evicted became vagabonds roaming the countryside, and the rest became wage laborers.

Thus what was formerly the peasant's means of production now became *capital* in the hands of the new lords and big farmers; and what was formerly produced by peasants for their own use and consumption now became means of subsistence that the new proletarians could acquire only by selling their labor for wages. Labor had become a commodity, subject to the forces of the market. The expropriation of the self-supporting peasants' property necessarily led to the destruction of their distinctive rural domestic industry. As a consequence, Marx wrote, "The spindles and looms, formerly scattered over the face of the country, are now crowded together in a few great labor-barracks, together with the laborers and the raw material. And spindles, looms, raw material, are now transformed, from means of independent existence for the spinners and weavers, into means for commanding them."[45] Manufacture was thus separated from agriculture, and the new class of proletarians helped create an internal market for the emerging capitalist mode of production.

The Concept of Alienation

For Marx, the proletarianization of the producer and his subjection to capitalist conditions of production resulted in an effect that could be termed *alienation.* That is a condition in which an object produced by the hands of man "stands opposed to [him] as an *alien being,* as a *power independent* of the producer." Humans feel alienated because work has ceased to give them satisfaction and a sense of purpose for which to strive. The worker "does not fulfill himself in his work but denies himself, has a feeling of misery rather than well-being, does not develop freely his mental and physical energies, but is physically exhausted and mentally debased."[46] Under capitalism Marx argued, individuals are controlled and dominated by the creations of their own hands. In fact, the worker's attitude toward the entire system of commodity production may be likened to idolatry. In that regard, Erich Fromm has written:

The essence of what [the ancient Hebrew] prophets call "idolatry" is not that man worships many gods instead of only one. It is that the idols are the work of man's own hands—they are things, and man bows down and worships things. . . . He transfers to the things of his creation the attributes of his own life, and instead of experiencing himself as the creating person, he is in touch with himself only by the worship of the idol.[47]

The idol acquires power over the individual as a result of the latter's failure to recognize the real source of its power. Similarly, the commodities and the instruments of production which the worker himself creates are seen as possessing powers over him. That is the same sort of illusion as seeing a creative force in an idol. To shatter that illusion and to disclose the real social relationships, and hence the real power of the system, became the central objective of Marx's analysis.

Among his other aims, Marx was intent upon showing that the individual worker paid dearly for the growth of productivity facilitated by capitalist organization. The fragmentation of the old crafts and the conversion of craftsmen into detail laborers alienated the worker from his reflective and creative faculties. Knowledge, judgment, and skill, which previously had been exercised by an individual craftsman, now become a function of the productive organization as a whole. The worker is "brought face to face with the intellectual potencies of the material process of production, as the property of another, and as a ruling power."[48]

With machine production, the worker's situation becomes even worse: "The life-long specialty of handling one and the same tool, now becomes the life-long specialty of serving one and the same machine."[49] If earlier the worker used the tool, now the machine uses him. The intellectual powers of the worker become superfluous, and the importance of whatever skills he has vanishes before the gigantic physical forces of the total factory organization and the hidden mind behind it all.

But there is still another consequence of machine industry. The machine becomes a competitor of the worker. The expansion of machine production is directly proportional to the number of workers rendered superfluous by machinery. With the loss of his use-value, the worker also loses his exchange-value. His labor becomes unsalable. The superfluous workers become an "industrial army of unemployed" who swamp the labor market and depress wages even below the subsistence level. For Marx, machine industry increasingly destroyed all previous forms of production, replacing them by the modern capitalist form. But the process which leads to the power of capital, Marx believed, leads also   111 to "the contradictions and antagonisms of the capitalist mode of production, and thereby provides, along with the elements for the formation of a new society, the forces for exploding the old one."[50]

Much more would have to be said in an adequate exposition of Marx's work as a whole. In the foregoing summary we have restricted attention to the Marxian conception of the historical development of whole societies. We can now see some of the dynamics that would account for the emergence of class-structured societies and for the transition from one form of society to another. Marx's theory of classes will receive detailed attention in the following chapter. At present, however, it will be instructive to say a word about the "problem of order" from a Marxian standpoint. That will enable us to compare Marx's theory with those previously considered.

THE MARXIAN VIEW OF THE PROBLEM OF ORDER

At the outset of this chapter, we noted that for Aristotle the stability of a society depended on the existence of middle classes large, prosperous, and independent enough to mediate between the extremes of wealth and poverty. To ensure order, moderate property ownership for most members of the society was essential. The Aristotelian principle had great appeal to later ages. Many subsequent political theorists agreed that moderate property and economic independence are a precondition for a stable polity. That principle made excellent sense so long as the extremes of wealth and poverty were minimal and the vast majority of society's members were in fact propertied. For instance, as poor as some of the English peasant proprietors may have been prior to the Enclosure movements, the vast majority of them had unrestricted usufruct of the land they tilled, and they owned their instruments of production. They did in fact enjoy some economic independence. In those circumstances everyone or almost everyone had a real stake or interest in the existing social order.

But with the emergence and development of capitalism, a form of society came into being which violated Aristotle's principle. For with capitalism one witnessed the appearance and phenomenal growth of a very substantial class of propertyless proletarians: Those who owned nothing. Since they owned nothing, they could have no stake in the system. If such a class had remained minuscule, as it had in ancient and other precapitalist societies, then it would have constituted little or no threat to the existing order; but since the proletarians became a large and rapidly growing class, it was bound to create what Marx called class conflicts and antagonisms.

How was social order maintained in a society containing a very large class of proletarians with no material interests in the existing system? To begin with, we can see that the Hobbesian theory provides a part of the answer. Fear of the state's power doubtless played no small role in the compliance of wage laborers. But there is more to it than that. The fact that wage laborers are so completely dependent for their livelihood on the owners of society's means of production means that fear of losing their jobs also contributed to their compliance. In all likelihood that was no less important a factor than fear of the state. The extreme dependence of wage earners is grasped, as we have seen, in Durkheim's conception of the forced division of labor. Owing to the fundamental inequalities that prevailed between the industrial capitalists and the workers, the contractual relationship between them was objectively a coercive one. That fact was also recognized by another outstanding classical sociologist, Max Weber. He saw that capitalism rests on the existence of a propertyless class, "compelled to sell its labor services to live." Capitalism is possible only where a class of workers offers its services "in the formal sense voluntarily, but actually under the compulsion of the whip of hunger."[51]

But besides the "whip of hunger" and fear of the Leviathan's force which can be brought down upon them, there is still another reason why dominated groups generally tend to obey and comply. For although the organized privileged minority has the vastly superior might of the state at its disposal and can therefore repel challenges to its rule by force, that is done only as a last resort. As a rule it succeeds in stabilizing its dominance by making it acceptable to the ruled masses. That is accomplished by means of what Marx called the *ruling ideology*.

Every ruling class, Marx argued, endeavors to justify its exercise of power by resting it on some higher principle. That important idea was similarly formulated by other classical theorists, notably Max Weber, who spoke of the *legitimation* of power. The ruling ideology or legitimation formula is not mere propaganda. Nor is it simply a great fraud with which to trick the people into obedience. On the other hand, it is a great fiction, "myth," or illusion. It appeals to the common values, beliefs, sentiments, and customs which result from a people's common history and which therefore make that people receptive to the fictions employed by the dominant classes to legitimize their rule.

Nationalism is an obvious example of a dominant ideology in the modern era: "My country right or wrong," "God is on our side," etc.   113

Nationalist ideology has often enabled ruling groups to mobilize the people for military adventures and wars from which they had nothing to gain. In previous eras rule by "divine right" was a prevalent ideology. Ideologies change with the sociohistorical circumstances, but throughout history the consent of the governed has been based on a ruling legitimation formula of some kind. The majority of the people consent to a given political system because it successfully appeals to widely accepted religious and philosophical beliefs. Thus the degree of consent depends upon the extent and ardor with which the dominated groups believe in the ideology with which the dominant groups seek to justify their rule.

A ruling ideology cannot depart too far from the culture of the governed without resulting in conflicts that threaten the foundations of the existing order. The principles underlying the legitimation formula must be rooted in the consciousness of the majority of the people. When the ruling ideology has sunk deeply enough into the consciousness of the people, the rulers, however corrupt and oppressive they may be, achieve remarkable results: the unswerving devotion of the poor, exploited, and oppressed masses.

Such consent, called forth from the working class contrary to its own interests, Marx called *false consciousness*. By this he meant that the ruling ideology gains acceptance among the dominated primarily because they often have so poor an understanding of the social causes of their oppressed condition. Marx envisioned raising the general level of the people's consciousness to the point where they would reject all elements of the ruling ideology which were not in their interest. That would enable them to act rationally and collectively. True consciousness, for Marx, was therefore an essential precondition for the transformation of the existing order and the making of a good society.

In Marx's vision, social order in the good society would have an ever-diminishing need for a Leviathan. Since fundamental inequalities would eventually cease to exist, both the coercive state and ideology would cease to be the basis of social order. A ruling ideology would become superfluous in a classless, stateless society in which the freedom of each would lead to the freedom of all.

Marx's vision, as we all know, has so far failed to materialize. One of his principal errors appears to have been the belief that he was witnessing a mature and dying capitalism, ripe for transformation into something humanly better. But hindsight allows us to see that the capitalism he studied was actually young and virile, the early stages of an economic system that was to reveal remarkable capacities for further

development and expansion.

On the other hand, Marx correctly anticipated certain structural trends under capitalism, even if he was wrong about their political implications. For example, a central thesis of *Capital,* his chief work, was that competitive capitalism would increasingly give way to a capitalism based on oligopoly and monopoly. He correctly foresaw that capital was destined to become concentrated in fewer and fewer hands; that economic power would more and more reside in large, complex industrial and financial organizations such as the modern giant corporation. Yet because he died in 1883, he failed to perceive other master trends of modern society.

BUREAUCRACY

It was Max Weber, the great German sociologist writing in the early twentieth century, who first gave us a systematic analysis of modern bureaucracy. Conceived as a pure type, the modern administrative organization has several distinctive characteristics.[52] A "bureau," or office, is an official jurisdictional area regulated by definite administrative rules. The activities of a typical bureaucrat are regarded as duties for which he has been trained and which he is qualified to carry out thanks to his specialized training. Bureaus are arranged in a *hierarchy,* a system of superordinate and subordinate offices in which the lower have less authority than the higher, and are, accordingly, supervised by them. Each bureau or office contains a body of official records or "files"; and typically an office is a place of business separated from the official's residence or household. The underlying administrative rules of this type of organization are quite *general,* enabling the official to regulate matters abstractly. That is, the people outside the organization are not treated as individuals whose unique situations must be dealt with case by case, but rather as members of categories. The typical bureaucrat is supposed to be impartial and disinterested, an attitude intended to ensure that all clients in a given category will be treated in the same manner.

Weber emphasized that officeholding in a bureaucracy is not just a "job." Rather it is looked upon as a "vocation" or profession requiring specialized training and examinations. The official fulfills his tasks in a dutiful manner and owes his allegiance to the office, not to individuals. He obeys orders and follows the rules not as a personal servant of his superior, but because he is devoted to the organization.

Typically, such an official attains elevated social esteem by virtue of his holding office. Historically, such esteem has been more characteristic   115 of Europe than the United States. In Europe status conventions, the trained expertise of the incumbents, and the fact that they were drawn from the economically privileged strata all contributed to the high esteem associated with officeholding in the state bureaucracy. In the United States such status conventions have been comparatively weak, although since the end of World War II, the rapid growth of governmental bureaucracies has brought with it some "European" characteristics.

There are still other important features of a modern bureaucracy. Normally the official, after a short qualifying period, acquires *tenure,* that is, he holds the position for life. Furthermore, he earns a salary, not a wage, and becomes entitled to an old-age pension. A wage is measured in terms of work done; but a salary is associated with one's status or rank in the organization. Officials aspire to move up from lower to higher positions and thus to earn higher salaries.

Modern bureaucracy presupposes a money economy. Officials are compensated in money, not in kind, which tends to place them in a state of extreme economic dependence. In contrast, officials of agrarian, premodern bureaucracies were paid in kind. Examples are the first bureaucracy in history, in ancient Egypt, as well as the bureaucracies in imperial China, Mogul India, czarist Russia, and the absolutist regime in France. In such preindustrial societies it was scarcely possible to extract from the peasantry a large enough economic surplus to pay the bureaucratic officials a salary. That meant that the crown had to rely on other methods by which to compensate officials. The result was that the official was much less dependent on the center of power.

In imperial China, for instance, the crown employed the method of tax farming, both for the generation of the revenues required and for the compensation of the officials. The Chinese officials were Confucian literati who received a humanist-literary education in schools especially designed for that purpose. Passing their final examinations qualified them for membership in the imperial bureaucracy. Once he had entered the bureaucracy, an official's main duty became tax collecting in the region to which he was assigned. Each official made a salary claim against those collected revenues and then handed over the rest to the crown. That arrangement held out strong temptations for the official to hold on to as much of those resources as he could.

The tax-farming arrangement in itself made the official much less dependent than the modern salaried bureaucrat. But in addition, the Chinese official was rendered even more independent of the crown because of the way he employed his income. The problem that plagued

116

the Chinese peasantry was the parcellization of the land. In the absence of primogeniture, the land of a deceased peasant was bequeathed to all of his surviving sons. As a result the land parcels became smaller and smaller in each succeeding generation. However, some of the peasant families, most often those already better off, found a way within the traditional order of counteracting that process. A son was sent to the Confucian schools in the hope that he would successfully qualify for the imperial bureaucracy. Qualification would enable him to become a tax farmer and to apply his revenues toward the purchasing of land, thus replenishing the family holdings. The larger landholdings enhanced the official's economic independence.

In all preindustrial bureaucracies, tax-farming arrangements imply that the crown's control of the potential resources it requires may be drastically reduced if the tax farmer ruthlessly exploits the crown's subjects. Hence tax farming often entails a conflict of interest between the crown and its tax-collecting agents, for the latter try to squeeze as much as they can out of the peasants while the crown strives to preserve their tax-paying capacity. Modern bureaucracies, in contrast, are assured of continuous revenues by the money economy and the vast surplus generated by the underlying industrial economy.

The increasing expansion of bureaucracy in modern society may be accounted for by both the quantitative and qualitative development of administrative tasks. As Weber noted:

The decisive reason for the advance of bureaucratic organization has always been its purely *technical* superiority over any other form of organization. The fully developed bureaucratic apparatus compares with other organizations exactly as does the machine with the non-mechanical modes of production. Precision, speed, unambiguity, knowledge of the files, continuity, discretion, unity, strict subordination, reduction of friction and of material and personal costs—these are raised to the optimum point in the strictly bureaucratic administration, and especially in its monocratic form.[53]

Speed, precision, and other forms of cost reduction are among the main reasons why we find that the typical modern capitalist enterprise is a large, complex corporation. However, in both the private and the public spheres, it is not merely considerations of efficiency but rather of *power* that have accounted for growing bureaucratization. The bureaucratic tendency has been promoted by power politics, warfare, the creation of large standing armies, and by the immense budgets required for such purposes. At the same time the social-welfare policies of the modern state have also contributed to the enormity, complexity, and costliness of its administrative apparatus.

117

One of Weber's most illuminating observations in that regard was made by elaborating on an aspect of Marx's theory. Marx, it will be recalled, traced the roots of modern capitalism to the *separation* of the producer (peasant proprietor) from his means of production. He then went on to demonstrate that capital was becoming increasingly *concentrated* and *centralized*. His argument, in brief, was that the accumulation of capital in the economy as a whole assumed the form of competition among firms, with some winning and others losing. The latter were either destroyed or absorbed by the victors. The growth of capital in one enterprise was facilitated by the failure of others. Those who remained in the race had successfully reduced their production costs by making larger investments in machinery and the like. As the costs of investment increased, entry into the field of production was restricted to fewer and fewer but larger capitals. Capital became increasingly concentrated in large-scale corporate organizations.

Weber agreed that the concentration of wealth and power was in fact a major tendency of capitalism. But he hastened to add that Marx had explored only one aspect of a much more general historical trend. That is, one could witness parallels to the separation of the producer from the means of production, and the concentration of those means, in several other social spheres. Thus Weber argued that historically the soldier has been separated from the means of violence and the civil servant from the means of administration, while those means have also undergone continual concentration. At one time fighting men owned their own weapons and were economically capable of equipping themselves. That was true of tribal levies, the armed citizens of the ancient city-states, the militias of early cities, and all feudal armies. But modern warfare is a "war of machines" wrote Weber, "and this makes centralized provisioning technically necessary, just as the dominance of the machine in industry promotes the concentration of the means of production and management."[54] Historically, army service has shifted from the shoulders of the propertied to those of the propertyless.

Similar processes have occurred in other spheres as well, notably in scientific research. If we think of scientists and inventors as recently as the turn of the century, someone like Thomas Edison comes to mind. Edison was a "tinkerer" who worked alone in his cellar and produced highly significant inventions. Today's science is a different matter altogether. In order to become a "scientist," one must first pass the examinations of a university or some other large educational organization and obtain the required degrees. One is then permitted to engage in scientific activities by gaining employment in the laboratory of some governmental, corporate, or university organization—all of which are big bureaucratic enterprises. The means of research are large and expensive, and they are controlled by the administrative heads of those organizations, not by scientists. For those reasons Weber believed that "through the concentration of such means in the hands of the privileged head of the institute the mass of researchers and instructors are separated from their 'means of production,' in the same way as the workers are separated from theirs by the capitalist enterprises."[55] In that way Weber observed the concentration of power in several other spheres and not merely in the economy. Everyman, and not just the blue-collar worker, was becoming "proletarianized." Everyman has become a paid laborer, working in a large complex organization and depending upon it for his livelihood.

Once such bureaucratic structures are established, they are practically indestructible. As a power instrument of the first order for those who occupy its command posts, a bureaucracy facilitates the domination and control of large numbers of people. The individual bureaucrat is chained to his specialized activity and is only a small cog in the total operation. His entire mind and body have been trained for obedience, and those who rule organizations expect compliance as a matter of course.[56] Thus Weber made a strong argument for the inevitable growth of bureaucracy. The vested power interests in it, the social control and discipline it facilitates, the specialization of work and the accompanying requirements of expertise—all of those factors make the dismantling of bureaucracy extraordinarily difficult. Weber's argument appears to be further borne out by the fact that much of the state apparatus can be "made to work for anybody who knows how to gain control over it. A rationally ordered officialdom continues to function smoothly after the enemy has occupied the territory; he merely needs to change the top officials."[57] Weber's analysis thus led him to the conclusion that "revolution," in the sense of transcending bureaucracy and creating a new, nonbureaucratic society, was becoming more and more difficult and unlikely.

Prebureaucratic Forms of Organization

For Weber, bureaucracy was, among other things, an expression of *formal* rationality, a mentality dominated by matter-of-factness and a concern with the most efficient means for achieving given ends. Historically, organizations that were nonrational in that formal sense have   119 given way to bureaucracy of the modern rational type. Weber showed the novelty and distinctiveness of modern bureaucracy by contrasting it with patriarchalism, patrimonialism, and feudalism.

In *patriarchalism* one sees several sharp contrasts with bureaucracy. There is nothing impersonal or formal about a patriarch's authority. It is based upon control of a household, not an "office." His subjects owe him personal loyalty and they comply with his requests because his authority has been hallowed by tradition, not by legal rules. The patriarch's power over the household applies not only to his own natural children but to the offspring of his servants as well. The members of the household form a solidary unit despite the distinction between master and servant, for all share common lodging and break bread together. Patriarchal authority is therefore characterized by filial piety and the sanctity of tradition.

With the growth of the household, the patriarch often settles some of his dependents on plots of their own within his extensive holdings. He also provides them with animals and tools. In time, that process leads to the diffusion of the members of the original household and to a decentralized form of organization which Weber called *patrimonialism*. In the new situation the dependents continue to owe the patriarchal lord allegiance. They are expected to deliver him rent in kind while he in turn owes them protection, assistance in times of need, and humane treatment—all of which are dictated by custom and self-interest, and not by laws. The extension of patrimonial organization (by assigning land and equipment to sons and other dependents of the original household) may even give rise to a patrimonial state. In such a state the lord's power extends over a very wide area, which may even become an empire. The rulers and the ruled of a patrimonial state continue to be linked by reciprocal obligations; they constitute a social organization in which the ruler's powers are legitimized by tradition. However, as the lord or prince enlarges his domain by attaching more and more dependencies to it, the growth in the size and complexity of the state gives rise to patrimonial "offices."

Although patrimonialism may thus acquire certain bureaucratic features, it never develops into a formally rational bureaucracy of the modern type. The patrimony never separates the private from the official sphere; and political administration is regarded as an official's personal affair. The jurisdiction of the officials is determined entirely by the ruler's personal discretion. In sharp contrast to the modern bureaucrat, the patrimonial official's allegiance to the ruler is purely personal.

120   The Chinese empire is an example of patrimonial bureaucracy. It

received its greatest impetus from the administrative requirements of large-scale projects such as river regulation, canal construction, and military fortifications. The crown, intent upon retaining its centralized control of the empire, saw to it that no new centers of power should emerge between it and the clan or village communities. Earlier we observed that the Chinese official, as a tax farmer, had more autonomy than the modern salaried bureaucrat. The crown discerned a danger here, for a rich and influential official might aggrandize power to himself and become a feudal lord with a domain of his own. A multiplicity of such domains would threaten the foundations of the centralized state. To forestall such a development, the imperial regime permitted the official only brief terms of service, assigned him to communities other than his own, and kept him under surveillance.

Besides, the Chinese patrimonial bureaucracy, as we have seen, also required schooling and the successful completion of examinations from all candidates. Especially noteworthy is the fact that the education of officials never led to the development of a modern, Western type of bureaucracy. The reason is that the training had nothing whatsoever to do with administrative competence, specialization, or expertise. Instead, the education that officials received was literary and humanistic. It made them into Confucian literati. The

examination really was a test of a person's cultural level and established whether he was a gentleman, not whether he was professionally trained. The Confucian maxim that a refined man was not a tool—the ethical ideal of universal personal self-perfection, so radically opposed to the Occidental notion of a specific vocation—stood in the way of professional schooling and specialized competencies, and time and again prevented their general application.[58]

Feudalism and the Transition to Bureaucracy

A patrimonial state therefore implies that the central regime has successfully monopolized power in its domain. It has prevented the emergence and proliferation of powers so characteristic of genuine feudalism. *Feudalism* is a decentralized system par excellence. It is a "division of powers" in which the numerous power centers check and balance one another and in which mutual obligations of an almost contractual kind exist between lord and vassal. Systems of that kind were known mainly in the West. Eventually such decentralized systems gave way to centralization and bureaucracy. A classical study of that historical process was provided by Alexis de Tocqueville in his *Old Regime and the French Revolution*.

Contrary to the prevailing opinions of the time, Tocqueville demonstrated in his remarkable study that the extraordinary concentration of administrative power in France was the creation neither of the Revolution of 1789 nor of the Napoleonic era, but rather of the old regime. Long before the Revolution the nobility had ceased to rule in the countryside, and the parish had largely come under the control of the central power. The nobles had lost all real authority in the community, yet they retained certain privileges which became a major source of the peasants' resentment. In particular the peasants resented the nobility's exemption from taxes, a privilege now appearing quite unjustified since that class served no useful function in the community. Unlike their English and German counterparts, the French nobles had ceased to have anything to do with the administration of community affairs, except the administration of justice.

Under the old regime, then, the monarchy increasingly aggrandized power to itself by usurping the traditional powers of the nobles. That was accomplished in an alliance with the bourgeoisie. The center of power, the Royal Council, wrote Tocqueville, "was composed not of great seignorial lords but of persons of middle-class or even low extraction." The powerful but inconspicuous council, staffed by the bourgeoisie, controlled "everything that had to do with money, that is almost the entire administration of the country."[59]

In the provinces as well, real authority passed from the nobles to the intendants—men of "humble extraction" appointed by the central government. The intendant was "all-powerful" and yet hardly noticed beside the still lustrous old feudal aristocrats. The crown now controlled either directly or indirectly virtually every aspect of provincial and local life, including public order and even public works. Thus what had been under the feudal arrangement the lord's obligations to the peasant, contributing to a comparatively stable rapport between the two classes, now came under the control of the central government. By increasing its intervention in the social life of the local community, the state usurped the lord's responsibilities. Thus the state not only disrupted the rapport between lord and peasant but immeasurably increased the fiscal burden upon the rural masses. For Tocqueville, the resulting conditions were the major underlying structural cause of the Revolution of 1789.

The absolutist regime in France therefore developed largely as a result of its struggle for power with the lords. But was the bureaucracy efficient and formally rational in Weber's sense? Not in Tocqueville's eyes. "At least a year would elapse before a parish obtained authorization from Paris to repair a church steeple or the priest's house; and more often than not it would take two or three years to honor local requests."[60] By the eighteenth century the immense bureaucratic machine had effectively eliminated all intermediate structures of authority between the centralized state and the individual. The centralized state bureaucracy was therefore not created but rather recreated after the Revolution. Tocqueville looked upon that historical development as the greatest threat to liberty.

Weber also viewed the bureaucratization of modern society with apprehension. The immense concentration of power in fewer and fewer hands was bound to endanger liberal-democratic institutions and to diminish individual freedoms. Increasingly the individual was subjected to an organizational discipline that drastically reduced his initiative; increasingly he was subjected to a *formally* rational regimen that eliminated any opportunities for autonomous and genuinely rational conduct. In Weber's words, bureaucratic "discipline is nothing but the consistently rationalized, methodically prepared and exact execution of the received order, in which all personal criticism is unconditionally suspended and the actor is unswervingly and exclusively set for carrying out the command."[61] What all formally rational, large-scale organizations have in common is regimentation and discipline. A bureaucracy, no less than a factory, tends to mold an individual's psychophysical being in an effort to adapt it to the demands of the organization. In short, bureaucracy "functionalizes" men. It is "horrible to think," wrote Weber, "that the world could one day be filled with nothing but those little cogs, little men clinging to little jobs and striving towards bigger ones."[62]

The challenges that bureaucratization poses for democracy are formidable. Weber correctly foresaw that the powerful bureaucratic tendency is bound to bring with it less individual freedom, not more. Yet in no complex, industrial society has bureaucratization been halted, much less reversed. That is true regardless of whether the societies in question are liberal-capitalist or socialist. Weber's persuasive analysis therefore raises fundamental issues of vital concern to all of us.

6

Social Stratification I:
The Study of
Social Inequality

Sociologists employ the term *social stratification* to describe the study of inequalities based upon wealth, power, prestige, and other social conditions. The unequal distribution of the valued things in life is a prominent feature of almost every society. As we reflect on most societies we know about, we can imagine them as pyramids: hierarchies of strata with the wealthiest, most powerful, and most prestigious at the top, and the impoverished, weak, and despised at the bottom. Such societies may be viewed as social organizations consisting of strata and classes with different and conflicting interests.

We have seen in our discussion of social organization that Durkheim, Marx, and Weber, among others, all employed the concept of class and concerned themselves with the implications of a class-structured society. *Social class* is a key concept in the study of stratification. It is a term that affords us valuable insights into the structure of a society and the changes it undergoes.

THE CONCEPT OF CLASS

As shown in chapter 5, social classes have not always been in existence but emerged under specific historical circumstances. The organization of many early and primitive societies was truly democratic, egalitarian, and communal, and private property and other institutionalized inequalities in wealth and power were unknown. The transformation of primi-

124

tive communal society may be traced to changes in its economic organization. With the growth of productivity resulting from the cultivation of crops, the domestication of animals, and the advance of handicraft techniques, economic surpluses emerged for the first time. Such surpluses, appropriated by groups other than the direct producers themselves, became the foundation for the type of societies best known to us from the study of history. Thus in antiquity there were masters and slaves, patricians and plebeians; in the middle ages, lords and serfs, guildmasters and journeymen; and in the capitalist era following the Industrial Revolution, capitalist entrepreneurs and proletarianized wage earners.

In discussions of social class, Karl Marx's theory has become a natural starting point, for it raises most of the crucial issues that need to be clarified if we wish to apply class concepts in an analysis of present-day society. Yet Marx was certainly not the first social thinker to recognize classes and class conflict in history. Acknowledging his intellectual debts, Marx wrote: "No credit is due me for discovering the existence of classes in modern society, nor yet the struggle between them. Long before me bourgeois historians had described the historical development of this struggle of the classes and bourgeois economists the economic anatomy of the classes." Marx then went on to say that his own contribution was to show that the "existence of classes is bound up with particular historical phases in the development of production" and that the class struggles of the capitalist epoch would eventually lead to a new society without classes.[1]

To grasp the distinctiveness of Marx's conception, it will be helpful to compare it with that of an illustrious predecessor, Saint-Simon. Henri Comte de Saint-Simon (1760–1825) was among the first thinkers following the French Revolution to recognize that a new type of social organization, which he termed *industrial society,* was emerging from the ruins of the old regime in Europe. Saint-Simon regarded "Les Industriels," the productive industrial class, as all those who actively participated in industrial society. He lumped together in one class owners and nonowners of the means of production. Living in the first part of the nineteenth century, prior to the earliest conspicuous struggles between the owners of industry and the proletarians, Saint-Simon failed to discern that capitalists and workers were actually two classes with conflicting interests.

It followed from Saint-Simon's conception that the new industrial order would be relatively peaceful and harmonious. The reorganization of European society on the basis of science and industry would produce a higher organic solidarity, similar to that later envisioned by Durkheim. But for Marx the new industrial order was anything but harmonious. Tension and conflict were endemic to the new system inasmuch as it rested on relationships of domination and exploitation.

Marx's Theory of Classes

The most immediate source of Marx's theory of classes was English classical political economy. He was thoroughly familiar with the works of Adam Smith, David Ricardo, James Mill, and a host of others. His erudition in English economics was vast, and he was profoundly influenced by that body of thought.

Marx's firm grasp of English political economy was acquired in his mature years, in the course of his prolonged stay in Victorian England. He arrived in London in 1849 and made his home there until his death in 1883. During that time, though he also devoted himself to various political activities, he gave his main energy to the writing of his chief work, *Capital* (*Das Kapital* in German). Throughout *Capital* he developed his theoretical ideas in a critical dialogue with Smith, Ricardo, and others.

To take an example of Marx's indebtedness to that tradition, he acknowledged that Ricardo had pointed before him to the inherent conflict of interests existing between the classes of industrial society. Ricardo had assumed that the "general good" was an inevitable outcome of the economic system of his time, but he eventually revised that assumption and concluded that "the opinion entertained by the laboring class, that the employment of machinery is frequently detrimental to their interests, is not founded on prejudice and error, but is conformable to the correct principles of political economy."[2] Thus while Ricardo retained an overall positive view of the industrial system, he was also critical of what Marx later called capitalism.

Marx followed Ricardo in looking upon landlords, capitalists, and wage laborers as the three great classes of English society at the time. There is something which the members within each of the three great social classes share, which surpasses in importance whatever differences of interest they might have. For Marx, it was a common relationship to the means of production: The landowners own (and control) the land; the capitalist entrepreneurs dispose over the industrial means of production; and the workers are nonowners who have nothing to sell but their labor. Marx did not deny that there may be interest conflicts of one kind or another within each class. Far from it. But he wished to underscore that it is the disposition over the vital economic resources of a society and the lack of such disposition that most sharply separate the capitalist and other proprietary classes from the workers.[3] In those terms, although there may be definite conflicts of interest between, say, the landlords and the manufacturers, there is an even sharper conflict of interest between both of those classes on the one side and the wage earners on the other.

Conflicts of Interest
Between the Propertied Classes

In Marx's time there was an evident conflict of interest between the landlords and the manufacturers revolving about the Corn Laws, a series of laws in force prior to 1846 prohibiting the importation of foreign grain to Great Britain. So long as the Corn Laws remained in effect, they served the economic interests of the landowners in a most direct and obvious way, assuring them a steady market at prices considerably higher than they would have been in competition with foreign grain. The manufacturers detested those laws because they artificially raised the price of the workers' subsistence and, hence, the wage bill as a cost of production. The two propertied classes waged their political struggle mainly in the parliamentary arena until the manufacturing class and its allies won out and the Corn Laws were repealed in 1846. The repeal, Marx observed,

merely recognized an already accomplished fact, a change long since enacted in the elements of British civil society, viz. the subordination of the landed interest under the monied interest, of [landed] property under commerce, of agriculture under manufacturing industry, of the country under the city. . . . The price of food was artificially maintained at a high rate by the Corn Laws. . . . [Their repeal] brought down the price of food, which in turn brought down the rent of land, and with sinking rent broke down the real strength upon which the political power of the Tories reposed.[4]

The original enactment of the Corn Laws and their repeal thus reflected the relative economic and political power of the two major propertied classes of the time.

In *Capital* and in other theoretical writings, Marx centered attention on two or three great classes. That fact has sometimes led to the impression that he looked upon classes as homogeneous, undifferentiated entities. But an examination of his historical writings clearly shows that he recognized the complex structure of classes. In his essay on the *Class Struggles in France,* for instance, he wrote: "It was not the French   127 bourgeoisie that ruled under Louis Philippe, but *one* faction of it: bankers, stock-exchange kings, railway magnates, owners of coal and iron mines and of forests, and a part of the landed proprietors associated with them—the so-called finance aristocracy."[5] The finance aristocracy ruled while another faction, *the industrial bourgeoisie,* formed a part of the official opposition. That is only one illustration of the fact that a class, for Marx, consisted of factions and strata which were often in conflict.

What we also learn from Marx's historical writings is that he discerned both an objective and subjective aspect of class. The smallholding peasants made up the majority of the French population at the time. Did the peasants constitute a class? In addressing that question, Marx wrote:

The small-holding peasants form a vast mass, the members of which live in similar conditions but without entering into manifold relations with one another. Their mode of production isolates them from one another instead of bringing them into mutual intercourse. . . . In this way, the great mass of the French nation is formed by a simple addition of homologous magnitudes, like potatoes in a sack. Insofar as millions of families live under economic conditions of existence that separate their mode of life, their interests, and their culture from those of other classes, and put them in hostile opposition to the latter, *they form a class.* Insofar as there is merely a local interconnection among these small-holding peasants, and the identity of their interests begets no community, no nation-wide bond and no political organization among them, *they do not form a class.* (Italics added.)[6]

That well-known passage from Marx's essay *The Eighteenth Brumaire of Louis Bonaparte* is important for the further light it sheds on the complexity of his concept of class. Objectively, Marx was saying, the peasants constitute a class; subjectively, they do not. It is true that they share basic conditions of existence. Their way of life, economic conditions, and culture are different from those of other classes. But since they are fragmented into local residential groups, their similar objective circumstances have failed to engender a nationwide awareness of their community of interests. They have produced no national political organization with which to defend and further their class interests. In a word, the French peasants of the mid-nineteenth century had not acquired what Marx termed *class consciousness.*

Why did Marx consider the industrial proletariat rather than the peasants as the future agent of social change? Though the peasants were poor, they were nonetheless a propertied class, a rural *petite bourgeoisie.* They had a material interest in preserving and, if possible, in expanding

128

their holdings. Then as now, poor peasants were land hungry, and enlarging their land parcels was the chief aim of any political organization they succeeded in forming. Therefore, the nineteenth-century French peasantry had no interest in the radical transformation of the existing society.

The industrial working class, in contrast, bore what Marx called *radical chains*. Having no material interest to preserve and defend, they were more likely to become revolutionary. Moreover, they were a large and growing class, destined to become the majority of the population. For as Marx saw it, small producers and businessmen were increasingly divorced from their means of production as capital became concentrated in ever-larger organizations. Small producers of all kinds would thus fall into the ranks of the wage earners. Handicraftsmen, peasants, small tradespeople, and shopkeepers, Marx said,

sink gradually into the proletariat, partly because their diminutive capital does not suffice for the scale on which modern industry is carried on, and is swamped in the competition with the large capitalists, partly because their specialized skill is rendered worthless by new methods of production. Thus the proletariat is recruited from all classes of the population.[7]

Unlike the peasants, the industrial workers lived and worked in conditions that facilitated communication. Their domination by capital created their common situation and common interests as a class. In time they succeeded in organizing unions and in gaining legislative recognition of their class interests, such as the Ten Hours Bill in England. Eventually, Marx believed, the laborers themselves, the men, women, and children who suffered most from the alienating and oppressive conditions of capitalist industry, would find the political means of changing those conditions.

Soon after Marx died in 1883, critical questions were raised concerning his conception of things. Indeed, a veritable "debate with Marx's ghost" ensued in which intellectuals throughout the world critically examined his theoretical legacy. Among those who participated in that debate, Max Weber stands out as having provided the most fruitful response to the Marxian view.

Weber's Revision of Marx's Class Theory

Weber concurred in many essential respects with Marx's characterization of capitalism. Although capitalistic forms existed in premodern periods of history, Weber agreed that capitalism as described by Marx is a mod- 129 ern phenomenon and that it has become the dominant mode of production since the middle of the nineteenth century. Weber also agreed that modern capitalism presupposes "the appropriation of all physical means of production—land, apparatus, machinery, tools, etc., as disposable property of autonomous, private industrial enterprises." Weber, like Marx, also stressed a free market and "free labor." "Persons must be present," he wrote, "who are not only legally in the position, but are also economically compelled, to sell their labor on the market without restriction." On the face of it, workers hire themselves out voluntarily; but actually it is "under the compulsion of the whip of hunger."[8] Thus "free labor," for Weber as for Marx, was a precondition of modern industrial capitalism. For both thinkers "free labor" had a double meaning: It refers to the fact that workers are free of slavery and other forms of forced servitude; and to the fact that they are free (i.e., nonowners) of any and all means of production.

Weber employed all of Marx's major class concepts: class consciousness, class conflict, class interest, and so on. For Weber the main social classes were:

1. the working class as a whole—the more so, the more automated the work process becomes;
2. the petty bourgeoisie;
3. the propertyless intelligentsia and specialists (technicians, various kinds of white-collar employees, civil servants—possibly with considerable social differences depending on the cost of their training);
4. the classes privileged through property and education.[9]

In this list we can begin to see Weber's departure from Marx, and why he saw the need to revise Marx's theory. Earlier we saw that Marx anticipated the "sinking" of the petty bourgeoisie (small producers and small businessmen) into the working class. But Weber and others, writing early in the present century, noted that that was not in fact happening as dramatically as Marx had supposed it would. At the same time Weber witnessed the phenomenal growth of the "new middle class" —specialists, technicians, and other white-collar employees. Here was a development that Marx never explicitly anticipated. Yet the remarkable growth of the new middle class touched the very heart of Marx's theory, for in his scheme of things, the fact that the members of the new middle class were propertyless meant that they shared with the manual workers a common relationship to the means of production. At least that is the way many Marxists after Marx looked at the matter. It followed that blue- and white-collar workers have common interests and that they would develop a common class consciousness. But it became increasingly clear in the early twentieth century that white-collar employees did not look upon manual workers as class brothers and sisters at all.

Under nineteenth-century conditions Marx may have been justified in ignoring "status" distinctions among various types of workers; but for Weber, the theorist par excellence of growing bureaucratization, it was obvious that differences in education, training, and property other than means of production all played a considerable role in shaping social psychology and hence class identification.

Thus what we find in Weber is a refinement of Marx's categories. Accordingly, Weber stressed that the control of *all* types of wealth—not only the means of production—was a source of power; and that social honor or prestige based upon property, education, or what have you might also be transformed into power. For Weber, then, classes, status groups, and political parties were "phenomena of the distribution of power." "We may speak of a class," he wrote, "when (1) a number of people have in common a specific causal component of their *life chances,* insofar as (2) this component is represented exclusively by economic interests in the possession of goods and opportunities for income, and (3) is represented under the conditions of the commodity or labor markets."[10]

Although Weber was intent upon analytically separating "class" from "status group," his definition is by no means a watering down of the class concept. Class situation, he emphasized, tends to determine life chances, and members of a class tend to share a common fate. In those terms, Weber's view of class situation was not as remote from Marx's as some commentators have suggested. "It is the most elemental fact," Weber observed,

that the way in which the disposition over material property is distributed among a plurality of people, meeting competitively in the market for the purpose of exchange, in itself creates specific life chances. According to the law of marginal utility, this mode of distribution excludes the non-owners from competing for highly valued goods; this favors the owners and, in fact, gives to them a monopoly to acquire such goods. Other things being equal, the mode of distribution monopolizes the opportunities for profitable deals for all those who, provided with goods, do not necessarily have to exchange them. It increases, at least generally, their power in the price struggle with those who, being propertyless, have nothing to offer but their services. . . . This mode of distribution gives to the propertied a monopoly on the possibility of transferring property from the sphere of use as a "fortune," to the sphere of "capital goods," that is, it gives them the entrepreneurial function 131 and all chances to share directly or indirectly in returns on capital. All this holds true within the area in which pure market conditions prevail. *"Property" and "lack of property" are, therefore, the basic categories of all class situations.*[11]

At the same time Weber went on to show that within the broad categories of propertied and propertyless, other important distinctions exist —not only in income, but in prestige or social honor as well. Prestige, for Weber, was associated with the *style of life* of a *status group.* Within any given class one will find several status groups. The relative prestige accorded them may rest on the size and source of their members' incomes, their political positions in the community, and their education, specialized training, or other evaluated social characteristics. Among the wealthy and propertied, we find old and new rich and other status distinctions based on the source of one's wealth; among the propertyless, we find status gradations based upon occupation, education, skill, size of income, expertise, the color of one's collar, and so on. Status differences, Weber maintained, must be taken into account in class analysis because those differences give us an idea of how certain social groups within a class regard themselves and how they are regarded by others.

There is another facet of social structure which Weber brought into relief. Marx had neglected noneconomic forms of power—power not directly derived from wealth and property. But Weber, as we have seen, called attention to the bureaucratization of modern society. Large, formally rational, complex organizations were becoming more and more common. *Power,* for Weber, referred to the ability to realize one's will despite and against the resistance of others. And it was crystal clear that those who occupied the command posts of bureaucratic organizations had little trouble in realizing their wills, whether they were personally wealthy or not.

Thus Weber argued that the concentration of power was not confined to the economic sphere. There were several strategic areas of social life in which one could observe: (1) the concentration of the means of power in the hands of small minorities, and (2) the consequent separation of the majority of the people from those means. That is the inevitable meaning of advancing bureaucratization. For Marx and the Marxists the essential question was: Who controls the means of production? For Weber, it was necessary to ask in addition, Who disposes over the other strategic means of controlling and dominating human beings? Weber did not deny that the control of key economic resources is decisive; but that in itself, he held, is insufficient for an understanding of the general structure of social power. He therefore elaborated Marx's theory, arguing that control of the means of political administration, means of violence, means of scientific research, etc., is also a major means of dominating people.

> Organized domination which calls for continuous administration, requires that human conduct be conditioned to obedience towards those masters who claim to be the bearers of legitimate power. On the other hand, . . . organized domination requires the control of those material goods which in a given case are necessary for the use of physical violence. Thus, organized domination requires control of the personal executive staff and the material implements of administration.[12]

Thus Weber convincingly observed that Marx's "separation" of the worker from the means of production is only one facet of a general social process. If "separation" is one side of the coin, concentration of power is the other. Marx's concentration of the means of production was generalized by Weber to include other means of power, notably the administrative, military, and scientific-technical. Weber's analysis then, critically revised Marx's theory to adapt it to twentieth-century conditions.

CHANGES IN THE STRUCTURE OF CAPITALISM SINCE THE LATE NINETEENTH CENTURY

In the last decades of the nineteenth century, several marked technological changes occurred, resulting in new sources of power and higher productivity in industry. Oil and electricity joined coal. The gas engine and electric motor increasingly replaced steam as a source of industrial power. Steel became a basic industrial material, thanks to the introduction of new processes for hardening it with alloys. Those technological-industrial changes, sometimes referred to as the "second Industrial Revolution," greatly accelerated the concentration and centralization of capital.

One cause of that accelerated concentration was the application of the new technology in highly expensive, large-scale capital equipment. The mounting cost of the initial investment that was required to enter profitable production soon led to a situation in which new industries became the preserve of giant corporations. The immense cost of fixed capital and other investments generally acted as "a sort of 'natural barrier to entry.' "[13]

Another reason for the accelerated concentration was that the main locus of the second Industrial Revolution was not England but rather 133 the United States, Germany, and Japan. These "latecomers" did not have to begin industrialization where the "first arrivals"—Britain, France, and Belgium—had begun it a century or so earlier. Because they entered at the most advanced and modern stage of industrialization, they were less encumbered than the first arrivals by an obsolescent productive plant. Accordingly, the high concentration of industrial production was most marked among the latecomers. Big enterprises in those countries employed a growing proportion of the labor force. In Germany, for instance, enterprises employing 200 or more employees increased from 11.9 percent in 1882 to 45.1 percent in 1961 (i.e., in the German Federal Republic). During the same period of time, the percentage of the work force engaged by enterprises with fewer than 10 employees declined from two thirds to 2 percent.

In the United States, the number of manufacturing establishments with over 1,000 employees increased from 540 in 1909 to 21,106 in 1955. Such establishments engaged 15.3 percent of all employees in 1909; 17.4 percent in 1914; 24.2 percent in 1929; and 33.6 percent in 1955. On the other hand, enterprises with *fewer* than 500 people engaged 72 percent of all employees in 1909 and 54.3 percent in 1955.

Other indexes reveal a similar picture. In fact, incomes and profits appear to be even more concentrated than labor. Thus Department of Internal Revenue data show that companies with a net annual revenue exceeding $5 million increased from 34.17 percent in 1918 to 50.69 percent in 1942. And the data of the Federal Trade Commission indicate that the 200 largest companies in the United States had 35 percent of the turnover of all companies in 1935, 37 percent in 1947, 40.5 percent in 1950, and 47 percent in 1958. Finally, if we look at manufacturing assets, we find that by 1962, 70 percent of those assets were controlled by the 500 largest companies. A comparatively small number of giant manufacturers has acquired control of major industries and a large share of the market; and that has occurred at the expense of small and middle-sized producers.

As a consequence of the concentration and centralization of capital, what we have today in the United States, Western Europe, and Japan is an economic system which economists call *oligopoly*. That refers to a situation in which a few giant firms dominate whole branches of industry, controlling supply and price. Under nineteenth-century competitive capitalism, the enterprise was a "pricetaker." Enterprises in a given industry were too numerous to permit formation of an effective price-controlling organization, and small, new capitals were sufficient to undermine it. Under oligopoly, however, the giant corporation is a "price-maker." With the domination of whole branches of industry by a few giant firms, conditions become much more favorable for the effective organization of oligopolistic and, occasionally, even monopolistic power.

How are prices actually arrived at in an oligopolistic situation? At one time *open* collusion was the rule in the United States. The antitrust laws brought that to an end, and today the process appears to be a form of tacit coordination that has come to be called "price leadership." One firm announces the price at which it plans to sell a product or service, and the others follow suit. As a rule the price leader is the largest and most powerful member of the industry—for example, General Motors and U.S. Steel. Their leadership is accepted by the smaller and less powerful firms because they are the ones most likely to lose by price warfare. Although forms of competition continue to prevail under oligopoly, pricecutting is typically not among them. Instead, large corporations compete in cost reduction, advertising, and promotion.

Of course, even under oligopoly there remain a very large number of small competitive firms in all capitalist economies. Yet since small producers frequently depend almost exclusively on oligopolistic industries for their purchases and sales, their price policies are determined accordingly. The highly competitive automobile-parts industry is a case in point. It must sell its output to the four giant car corporations, so it must follow their lead in pricing its products.[14]

Thus we see that the structure of capitalist organization has undergone very significant changes in the course of the twentieth century. The large corporation has replaced the small family firm as the decisive unit of the economy. How has that process affected the class structure of twentieth-century capitalist society?

THE UPPER CLASS IN PRESENT-DAY CAPITALIST SOCIETY

In the nineteenth century the capitalist entrepreneur was both owner and manager of his firm. Toward the end of the century, however, economists drew attention to something new. With the emergence of the joint-stock company and the dispersion of shares of stock, the roles of owner and manager were not necessarily the same; and sometimes they were distributed over two positions: stockholder and executive.

In 1932 the publication by Adolphe Berle and Gardner C. Means of a book entitled *The Modern Corporation and Private Property* caused quite a stir, for they were among the first to argue systematically that   135 the modern joint-stock company, or corporation, entailed a complete break with earlier capitalism. By separating what has become known as "ownership" and "control," they argued, corporations have given rise to a new group of managers who are altogether different from their capitalist predecessors.

Soon afterward, other writers went even further. Following James Burnham, they began to speak of a "managerial revolution," thereby suggesting that actual economic power had passed out of the hands of the owners and into the hands of the "managers." If that were true, it would mean that the dominance of the corporation has resulted in the drastic reduction and even elimination of the economic power of large property holders. There would be no capitalist class, properly speaking, only a managerial one. But is that theory true?

To begin with, the "managerial revolution" fails to distinguish technicians from managers and the latter, in turn, from other corporate power holders. As G. D. H. Cole has remarked:

If all that is meant [by the term "managerial revolution"] is that the development of modern techniques of mass-production and large-scale administration necessarily requires the existence of a large body of highly trained scientists, technicians, administrators and managers who are bound to claim, for a long time to come, superior rewards and a superior status in society by virtue of their . . . qualities of usefulness, there is nothing to argue about; for no one in his senses is likely to dispute the fact. If, however, it is argued that these occupants of superior positions are bound to become the real masters of society, however it may be nominally governed in its political affairs, I not merely dissent, but see no basis for the contention.[15]

Or, as Maurice Zeitlin has observed, no one can deny the existence of bureaucratic corporate management. But the existence of such management, in and of itself, says nothing about "the bureaucracy's relationship to extra-bureaucratic centers of control at the apex or outside of the bureaucracy proper, such as large shareowners or bankers, to whom it may be responsible."[16]

If the managerial role is no longer played directly by the owners of the corporation, that does not necessarily imply the separation of *control* from ownership. It may simply mean that the managerial *function* has been separated from ownership. The relationship between the propertied and the bureaucracy, between "capitalists" and "managers," has received far too little scholarly attention. It is that relationship that Maurice Zeitlin carefully scrutinizes.

He begins with a thorough examination of *The Modern Corporation and Private Property,* and he finds that Berle and Means carefully qualified their thesis. Although they classified 44 percent of the 200 largest corporations as managerially controlled, they candidly acknowledged that they had "reasonably definite and reliable information" on at most two thirds of the companies. In their detailed tables they noted those corporations about whose locus of control one could only conjecture. Summarizing their qualifications, Zeitlin concludes that the authors

listed 73 corporations under the heading "majority of stock *believed to be widely distributed* and working control held either by a large minority interest or by the management." Of these, 29 were considered *"presumably"* under the control of a minority interest, while 44 were *"presumably"* under management control. Indeed, of a total of 88½ corporations which they classified under management control, they provided *no* information on 44, which they could only consider *"presumably"* management controlled. Among industrials, they classified fully 39 of the 43 management-controlled corporations as only "presumably" under management control. Thus they had information which permitted them to classify as definitely under management control only 22 percent of the 200 largest corporations, and of the 106 industrials, only 3.8 percent. (Italics added.)[17]

Despite those explicit reservations, the Berle-Means thesis has been accepted by virtually all economists and other social scientists.

Sorely lacking have been investigations of the specific circumstances of large corporations. Only such investigations can answer the critical question: What are, in fact, the connections between principal shareholders, officers, and directors? Or, in a word: Who actually controls the modern corporation? A well-known attempt to answer that question was made in the late 1930s by Ferdinand Lundberg. Reexamining the same companies that Berle and Means classified as management controlled, Lundberg found that "in most cases [the largest shareholding] families had themselves installed the management control or were among the directors," while several others were "authoritatively regarded in Wall Street as actually under the rule of J. P. Morgan and Company." Lundberg therefore concluded that "exclusion of stockholders from control, within the context as revealed by Berle-Means, does not mean that large stockholders are excluded from a decisive voice in the management. It means, only, that small stockholders have been [excluded]."[18]

It is not generally known just how scanty is our knowledge concerning the largest corporations. In the United States there is no official listing of such corporations, so most studies have relied on *Fortune* magazine's list of 500. Even the Patman Committee on Banking and Currency of the House of Representatives employed *Fortune*'s list in its analysis of interlocking relationships between large commercial banks   137 and the largest corporations. It is therefore highly instructive to note that in 1966, *Fortune* revealed that in the years that it has been publishing its list, it has been omitting "privately owned or closely held companies that do not publish certified statements of their financial results." The article then went on to name 26 companies which, according to *Fortune*'s sources of information, had a large enough volume of sales in 1965 to qualify for the 500 list. It is therefore evident that

any adequate generalization about the ability of families to maintain control through ownership, indeed private ownership, of very large firms would have to take account of such previously ignored privately owned firms. Were these added to the "list," there would be . . . 31, over six times as many as previously counted among the 500 largest. Whether other such large privately owned firms have still escaped notice is an important question to which there is no presently reliable answer.[19]

We must also consider the fact that the true or "beneficial owners" are often hidden behind the "shareholders of record." Even the presence of large bank holdings may be hidden in that way. As the Patman Committee findings show, 36 corporations earlier classified as "management controlled" are actually under the control of very large banks. Finally, it should be noted that Berle and Means considered the ownership of a 20-percent block of a corporation's shares as the minimum necessary for its control. Recent investigators have used a block of 10 percent on the assumption that stock is now more widely dispersed than it was in the 1930s. However, the Patman Committee concluded that effective control may be assured with even less than a 5-percent holding, "especially in very large corporations whose stock is widely held." If that cutoff point were used, it would significantly raise the total number of U.S. corporations under proprietary rather than management control.

In Europe and Japan there have been very similar structural processes: high economic concentration, the fusion of formerly separate large capitals, the establishment of interlocking directorates, and the development of a large stratum of corporate functionaries. Numerous scholars have therefore generalized the "separation-of-ownership-from-control" thesis to all advanced capitalist-industrial societies. Yet the thesis is no less problematic for those countries than it is for the United States. For while it is clear that the day-to-day administration of the typical giant corporation is left in the hands of the managers, it is equally clear that the largest shareholders hold effective power in it. They stand ready to intervene in the running of the corporation if that proves necessary for the preservation and furtherance of their interests.

138   Therefore, where the "upper class" is concerned, one can make the

following generalization on the basis of available evidence. In all advanced capitalist societies, the large-scale corporations are administered by *strata* of functionaries and managers. A significant proportion of these "propertyless" managers do in fact own shares in their companies. According to Sargent P. Florence, the directors of very large British firms own on an average 1.5 percent of their companies' shares. However, even when the high executives are propertyless, they exhibit a high degree of solidarity with the propertied. There are often direct friendship and marriage connections between them, and where such is not the case, they share a common outlook and sense of common interest. The overlapping of interests between the two groups and their cohesiveness in the face of economic and political challenges indicate that they have a high capacity for common action and policies.[20]

THE NEW MIDDLE CLASS

In early nineteenth-century America, the "old middle class" comprised a large proportion of the working population, perhaps as much as four fifths. In the predominantly agrarian society that the United States was in that era, the old middle class mainly consisted of small farmers, artisans, craftsmen, storekeepers, and the like. As the century advanced, there were added to those occupations small manufacturers and other businessmen who made their living by producing, distributing, and selling manufactured goods.

By 1870 the old middle class as a whole had declined to one third of the employed population, and by 1940 to one fifth. The drastic reduction in the proportion of small, independent producers, small capitalists, and small shopkeepers may be attributed to the advance of large-scale industry. As a result the members of the old middle class

no longer enjoy the social position they once held. They no longer are models of aspiration for the population at large. They no longer fulfill their classic role as integrators of the social structure in which they live and work. These are the indices of their decline. The causes of that decline involve the whole push and shove of modern industrial society.[21]

More recently the trend was summarized this way:

Small businesses are producing less return on their investments than are large businesses. Their total number is showing a relative decline, in that the American [small] business population is growing at a slower rate than either the general population of the country or our gross national product. 139

Many firms continue to disappear because of mergers or other acquisitions by large businesses, increasing the concentration of economic power in the hands of larger concerns.[22]

Although Marx had foreseen a general tendency of that sort accompanying the advance of capitalism, he failed to anticipate the phenomenal growth of the "new middle class."

The growth of the white-collar groups has proceeded further in the United States than in Europe and Japan. As of 1969 the nonmanual and manual categories in the United States each comprised 48 percent of the total work force. In Britain as of 1959, nonmanual workers made up 29 percent of the employed force. That represents a growth of 7 percent since 1921. In Japan in 1963, the proportion of nonmanual workers was 27 percent.[23]

However, the term *white collar* covers a bewildering variety of job categories. The white-collar world is an "occupational salad," in the words of C. Wright Mills, which includes such diverse groups as salaried administrators, managers, technicians, accountants, doctors, lawyers, ministers, professors, teachers, social workers, civil-service employees, clerks, secretaries, typists, nurses, salespeople in and out of stores, mailmen, and many, many more. If we focus attention on the "lower" strata of the white-collar pyramid, we may begin by noting that its members own no means of production. Like manual workers they earn their living by selling their services on the labor market, and they are dependent on large private and public organizations for their livelihood and security. In those terms they appear to be in the same class position as the blue-collar workers. But as soon as we go beyond that fact and take a closer look at other aspects of the white-collar situation, we find that it differs significantly from that of the manual workers.

Historically, white-collar employees in all capitalist societies have enjoyed a higher income than manual workers. Some observers have suggested that that trend has changed in recent years, and that there now exists some overlap between white- and blue-collar workers. But as Anthony Giddens has noted, "The overlap is confined to segments of skilled manual occupations on the one hand, and of clerical and sales occupations on the other." Furthermore, the relative rise in working-class income is in part a function of *the decline in the proportion of men in the clerical and sales force, and a corresponding increase of women.* In Britain, for instance, which may be regarded as typical in that respect, "the proportion of women in white-collar occupations rose from 30 to 45 percent between 1911 and 1961 . . . almost wholly

140

clustered in clerical and sales occupations."[24] Besides income, however, the white-collar worker has enjoyed other advantages.

An admirably thorough study of British clerical workers has shown that the "blackcoated" workers have historically benefited from a much higher degree of *job security* than manual workers. That is perhaps the most significant difference between the two categories of work. For while the manual worker has always been subject to the ups and downs of the labor market, job security has provided the nonmanual workers with considerable immunity to those hazards. Still another advantage of office workers, particularly if they are men, is the greater likelihood of rising in the hierarchy to supervisory and managerial positions. In addition, there are various retirement and other collateral benefits which male office workers have traditionally enjoyed, as well as shorter hours and more frequent paid holidays.[25] Finally, of course, there is the "privileged" cleanliness, quiet, safety, and comfort of office work. Added to those material advantages are others relating to status, which have also tended to divide rather than unite the office and the manual worker.

To understand the distinct situation of the office worker in the United States, we must say a word about the development of the modern office. In the nineteenth century, when industry was conducted on a comparatively small scale, offices were correspondingly small. But with the emergence of large-scale industry, large complex office organizations also became necessary. The "paperwork" mounted steadily. The increase in the size of industry meant that huge personnel and pay records had to be kept, more orders for raw material had to be placed, more correspondence with customers had to be conducted. The volume of paperwork increased with the scale of production but received an additional impetus during both world wars owing to expanded wartime production and government controls. In the 1930s New Deal measures such as unemployment insurance, social security legislation, and corporation tax laws immeasurably added to the already enormous paperwork in industry. The rise of unionism in the same period further contributed to the same effect.

For all those reasons the number of American office workers has grown rapidly in the present century. From 1909 to 1927, the number of salaried employees of all types almost doubled. Between 1940 and 1954, the increase was 67 percent, while the number of blue-collar workers grew by only 26 percent.[26] The expanding volume of the white-collar work force required more space, new facilities, and new equipment. The increased costs that such developments entailed led management to apply formal-rational principles of organization to office work. The common result was the establishment of a central office whose responsibility it was to ensure efficiency and to guard against unnecessary costs. With rational organization came mechanization. During World War I the process began of introducing numerous types of machines into office work, and the process continues unabated today. Many office tasks have become routinized operations performed by relatively unskilled machine operators.

Some observers have suggested that with the mechanization and routinization of office work, the salaried employee experiences a sense of alienation not unlike that of the unskilled factory operative. In both cases the bodies and minds of human individuals have become units in the formal-rational calculations of managers. Creativity, knowledge, judgment, and will are as superfluous in the one case as in the other. Observers have accordingly concluded that the historical differences in the conditions of office and shop workers are becoming less significant.

It is fairly clear, however, that a marked difference in the work and life situations of the two categories of worker continues to exist. Office workers frequently *feel* themselves superior to shop workers, as numerous studies have shown. They believe that their jobs demand greater intelligence, skill, and education, and their claims to higher status have been recognized by management. They are permitted to wear street clothes at work; they are paid in salary rather than in wages; they are assigned different hours of work; and, finally, they are physically separated from the shop.

The divergent material benefits, work situation, and style of life of the white-collar worker no doubt help to account for his resistance to unionization. Union organization in the retail, wholesale, government, financial, insurance, real-estate, and service-employment sectors is quite weak. The unionized portion of the total white-collar force is well below 20 percent; in many job types and in certain geographical areas, unions are practically nonexistent. While approximately 10 million clerical and kindred employees are found throughout the American industrial structure, fewer than one million have been organized.

From the standpoint of the industrial, blue-collar unions, the failure to organize the white-collar force is a major source of weakness and cause for concern. The proportion of nonproduction workers in manufacturing industry is at least 25 percent and in some departments, such as ordnance, 48 percent. In specific industries such as chemicals, the proportion is even higher. As strikes in highly automated plants have revealed in recent years, it is possible for supervisors, technicians, professionals, and other nonproduction workers to maintain the equipment in operation.[27]

The advance of automation, which eliminates many jobs and raises the skill and educational requirements of others, has in recent years reduced the job security of clerical workers in industry. Their displacement as a result of technological change is as real a possibility as it is for production workers. Nevertheless, those white-collar workers have not been persuaded that it is in their interest to organize. In their view, unions are for manual workers. And employers, on their part,

have gone to considerable lengths to harden and widen the antipathy toward unionism. Many have met the [white-collar] workers' economic expectations and provided personnel policies and procedures designed to implant a sense of security, freedom of communications, and individual status that might otherwise be sought through union membership and collective bargaining. Personnel men constantly use the threat of unionism to win management's approval for liberalized practices and policies. Addresses by personnel men at management meetings stress the success achieved in warding off unions by "beating them at their own game."[28]

In sum, it is fairly evident that white-collar workers continue to enjoy a number of material and other advantages: greater income, job security, promotional opportunities, collateral benefits, and a higher status. Such are the significant respects in which the class situations of white- and blue-collar workers continue to diverge.

THE WORKING CLASS

The *working class* may be defined as wage earners engaged in production and other manual occupations in the sphere of manufacturing industry. That class has remained very large in all capitalist countries. In Britain, for example, it comprised 50 percent of the economically active population in 1881 and 49 percent in 1951. In the United States as of 1970, there were over 14 million production workers in manufacturing. In addition there were almost .5 million production workers in mining, 2.8 million in construction, and 3.9 million in transportation and utilities—hence just over 21 million all told.[29] Such production workers roughly correspond to what Marx had in mind when he spoke of the industrial working class.

The next largest employment sector is in distribution and trade and related business activities such as finance, insurance, and real estate. Here we find a total of 18.7 million employees. Government is third   143 with 12.6 million on its payroll in 1970. Finally, there is the private-service sector, which in 1970 employed 11.6 million persons. Aside from a minority of professionals, small businessmen (restaurateurs, shopkeepers, etc.), and the like, the service sector is mainly composed of unorganized, poorly paid laborers. It is in that category that we find a disproportionately large number of nonwhites. As compared with the organized working class in manufacturing industry, the service laborers are a poor and powerless *underclass*—a category to which we shall return.

We have said that there are over 14 million production workers in manufacturing industry. But there are also some 6 million nonproduction workers in that sector. The latter are professional and technical workers, white-collar personnel employed in production-related scientific and technical jobs. Are they to be counted among the industrial working class? Are production and nonproduction workers to be regarded as one class? That question has been frequently raised in recent years.

Marxian scholars in particular have argued that salaried professional, technical, and managerial employees now constitute a "new working class." Like their blue-collar brothers and sisters, they own no means of production and are only to be distinguished by their higher education and training. Here, then, is an issue very similar to the one discussed earlier concerning the relationship of clerical and manual workers. There we saw that because the situations of the two categories of workers are distinct in several material and nonmaterial respects, their common position of nonownership of means of production has not sufficed to produce a sense of common interest or a relationship of solidarity between them.

The same logic may be applied to the new working class. It is true that salaried professional, technical, and managerial personnel own no means of production, but it is evident that those white-collar workers, no less than others, occupy a position in the industrial and social structure that tends to distance them from production workers. Compared with production workers, the new working class enjoys a number of significant advantages: more paid holidays, more frequent and longer paid vacations, medical insurance and sick leave with full pay, retirement pensions and other fringe benefits, and greater job security. Furthermore, the work situation is comparatively privileged: members of this class work in clean and comfortable offices, unregimented by machines, at tasks that are not inherently hazardous to health, life, and limb. Finally, their status situation is, of course, a potent consciousness-shaping factor.

144

Their higher formal education and training together with the fact that they typically emanate from middle-class families tend to set them apart.

It is worth noting that even among the production workers themselves, their common objective position has not led to the results envisioned by Marx. In the United States the traditional working class has so far not developed organizational forms of class consciousness beyond trade unionism. It has yet to create a political party of its own. And although Socialist and Communist parties have existed in Western Europe for some time, they have in recent years disavowed the kind of insurrectionist politics Marxists have traditionally anticipated. The fact is that the present-day industrial working class has become something other than a proletariat.

The Transformation of the Working Class

Marx called the industrial workers of his time a "proletariat" because they had no material interest in the existing system. He expected that that class would become more and more homogeneous owing to the advance of machine industry, which rendered skill superfluous among machine tenders. With the growth in the scale of industry, the workers would increase numerically by absorbing the failures of the old middle class. At the same time they would develop solidarity due to their common life chances and work situations. Finally, they would become a revolutionary force because of the stark contrast between their material conditions and those of the bourgeoisie. In time the workers would come to realize that it was only through the construction of a society based on different principles that they might attain a decent life for themselves and for the vast majority of the people.

Marx's critics have argued, however, that he was wrong on all those counts. Contrary to his expectations, the working class has become more highly differentiated in respect of skill levels and divided into occupational and other sectional interests; the growth of the new middle class has diminished the proportion of industrial workers in the total employed population; social mobility has weakened working-class solidarity; and the rise in the standards of living of the working class has led to its *embourgeoisement*—i.e., to an outlook and way of life that is "middle class" or bourgeois.

As one reflects on those critical observations, it becomes clear that they are, by and large, upheld by the evidence. One exception is the assertion that industrial workers have diminished as a proportion of the total employed population. We have already seen that that proportion, 145 even if it has not grown, has remained comparatively stable in most capitalist societies since the end of the nineteenth century.

However, the most disputed of the critical observations is the *embourgeoisement* thesis, of which the following statement is representative: "Working-class life finds itself on the move towards new middle-class values and middle-class existence. . . . The change can only be described as a deep transformation of values, as the development of new ways of thinking and feeling, a new ethos, new aspirations and cravings."[30] Some commentators maintain that the worker's main concerns are with home ownership, upward mobility for himself and his children, savings and insurance, prestige and respect, and other so-called middle-class values. That thesis has been examined in light of a careful and extensive study of the British working class by John Goldthorpe and David Lockwood. They found that on a wide range of material, social, political, cultural, and psychological variables, there were significant differences between production workers and the middle class. So far as material conditions are concerned, the researchers found that relative to that of salaried white-collar groups, the economic well being of the working class has been exaggerated. Production workers still lag behind in job security, opportunities for promotion, and a wide range of fringe benefits.[31]

Similar conclusions were reached in a study of the French working class by Serge Mallet. However, he makes the additional point that in studying working-class conduct and attitudes, one must distinguish between the spheres of consumption and production. In the first, "the working class has ceased to live apart. Its level of living and its aspirations for material comfort have led it out of the ghetto in which it was confined at the beginning of industrialization." In the area of production itself, on the other hand, "the fundamental characteristics which distinguish the working class from other social strata seem to have remained unchanged."[32]

How does the *embourgeoisement* thesis hold up as applied to the United States? We have already seen that there are substantial material, situational, and status differences between blue- and white-collar workers. Those who contend that present-day production workers have become affluent tend to overlook those differences. They also exaggerate both the actual upward mobility that has taken place and the effects it has had on the economic system. Although the available studies of mobility have so far failed to yield firmly established conclusions, a few tentative generalizations may be made on the basis of the evidence.

146   To take the internal structure of the working class first, it has in fact

changed significantly since the turn of the century. There are today more workers in semiskilled jobs than earlier and fewer in unskilled jobs. The proportion of skilled workers has remained about the same. As for interclass and intergenerational mobility, it seems clear that a very substantial proportion of the working class is not mobile at all. Approximately two thirds of all production workers tend to remain within their class of origin. However, a significant but much smaller proportion (perhaps as much as one third) has moved into nonmanual occupations. Yet, at the same time a roughly equal proportion of persons whose fathers were in nonmanual occupations moved "down" into manual ones.[33] Equally salient is the fact that most social mobility occurs between social levels which are very close together. Children of blue-collar workers, for example, enter into the lowest levels of the white-collar pyramid or, less often, into the small proprietor class. Movement from the ranks of the industrial working class into the upper class is very rare indeed.

Finally, it should again be emphasized that the proportional size of the industrial working class has remained relatively stable since the turn of the century. The numerical size is increasing, of course, but not faster than the general population. That seems to be equally true of other capitalist–industrial societies as well. Thus Anthony Giddens informs us that in Britain "in 1881, 50 percent of the economically active population was in manufacturing; in 1951 the figure was 49 percent, with only minor fluctuations in between."[34] The evidence on mobility therefore indicates that individuals move freely between skill levels within the working class and they freely enter the bottom layers of the white-collar pyramid. But the American industrial working class is remarkably stable in relation to the class structure of the society as a whole.

CLASS CONSCIOUSNESS

It remains for us to address the question of how American workers think of themselves. In 1940 *Fortune* magazine published a poll which seemed to show that most Americans looked upon themselves as middle class. Thus when a sample of individuals was asked whether they belonged to the upper class, middle class, or lower class, 7.6 percent replied upper class and 79.2 percent replied middle class. 5.3 percent were unable to place themselves. The respondents also provided data concerning their "actual" positions in the social structure: prosperous, 147 upper middle, lower middle, and poor. It was especially interesting to find that 74.7 percent of the prosperous and 70.3 percent of the poor said that they were middle class.[35]

Subsequent polls asking the same type of questions tended to confirm the accuracy of the *Fortune* poll. A study conducted in Minneapolis in the early 1950s found that 76 percent of the sample referred to themselves as middle class, 5 percent as upper class, and 10 percent as lower, while 9 percent identified themselves with no class.[36]

However, given the inherent vagueness of such terms as "middle class," "lower class," etc., some social scientists seriously questioned these and other studies yielding like results. For it seemed highly improbable that the majority of working people should not be aware of the position they occupied in the social structure. For that reason an additional study was designed by Richard Centers and administered to a national sample. His questionnaire differed in one significant respect: It substituted "working class" for "lower class." The results were highly illuminating. Now 75 percent of the manual laborers identified themselves with the working class.[37] Other studies have corroborated Centers's findings.

What we learn from those studies is that the self-respect of many American working people makes them reluctant to identify themselves as lower class. The word *lower* carries a connotation which they prefer not to apply to themselves. But when they are offered the option of "working class," they readily apply the term to themselves because there is nothing offensive or stigmatic about it. We can also go a step further in the interpretation of Centers's findings. They strongly indicate that working people recognize the existing differences between themselves and other groups in economic situation, style of life, and culture.

Another facet of class consciousness is how classes regard each other. Do the members of a given class see a conflict of interest between themselves and another class? Do they fear or resent the power of another class and desire more power for themselves? Although Centers did not pursue such questions in depth, his findings afford us a glimpse of interclass attitudes. When he asked whether the working class should have more power than it presently has or less, nearly three quarters of the executives of large enterprises in the sample opposed an increase in working-class power. The white-collar workers were about equally divided on the question, but 60 percent of the urban workers desired more power for themselves.

148   Where working-class consciousness is concerned, we may safely con-

clude that American production workers have an awareness of their common class situation. That awareness not only expresses itself in the attitudes reviewed here but in the trade-union movement and in the long history in the United States of bitter strife between labor and management. Yet there is some indication that trade-union organization has slowed down in recent decades and has even suffered setbacks. The proportion of the nonagricultural work force in unions declined from 33 percent in 1955 to about 27 percent in 1969, in large part due to the substantial shift in industrial location from the East and Midwest, traditional union strongholds, to the South, where unions have been virtually nonexistent. As one student of the problem has explained, "Bargaining rights do not move with the plant. Unions have to start organizing drives at the sites of the new plants and frequently find their task most difficult because of unfriendly local attitudes."[38]

Another reason for the relative decline in union strength is the fact that the vast white-collar and service sectors remain for the most part unorganized. Our comparative analysis of the manual and nonmanual situations suggests that white-collar workers may continue to resist entering blue-collar unions. On the other hand, there has been a significant measure of independent white-collar organization, prompted by the desire to defend and further their general interests. In some economic circumstances the likelihood of such independent unions affiliating with the large trade-union organizations may increase.

THE UNDERCLASS

The vast sector of service employees (11.6 million persons as of 1970) presents a quite different picture. That sector mainly consists of blacks, Puerto Ricans, and Mexicans who are not only poorly paid and nonunionized but who suffer from chronic unemployment as well. Though they are manual laborers in the fullest sense of the term, their general situation differs markedly from that of the workers in the corporate industrial sector. For the latter work in semiskilled and skilled occupations; they are more highly unionized; and they are employed in the high-paying manufacturing and other industries.

We can readily understand, then, why the huge mass of service employees has been called an *underclass*. They comprise the most disadvantaged section of the American work force. A substantial majority are employed in very small establishments, a condition which places special obstacles in the way of those service workers who wish to organize. 149

Small proprietors can ill afford the increased labor costs that unionization would entail. Small producers and other small businessmen, operating in those sectors of the economy which are still price competitive, cannot pass those costs on to the consumer (as do giant corporations) without putting themselves out of business.

The condition of the present-day underclass can best be understood by comparing it with that of the American poor of a previous era. America is a country of immigrants, as we all know. The most massive influx occurred in the late nineteenth and early twentieth centuries, when millions of poor peasants, craftsmen, and workers left Europe and poured into the United States, looking to make a better life for themselves. Although the conditions they found upon arrival were largely an improvement over those they left behind, poverty, even terrible poverty, remained the lot of the vast majority. Men and women labored long hours each day under highly oppressive circumstances, living out their lives in squalid slums and ghettos. That situation was typical of the "old poor" in America right up to the 1930s. Yet if we compare their situation with that of the "new poor" of today, we see both objective and psychological differences.

Yesterday's poor participated in the stupendous growth of American capitalism. At least until the Great Depression of the 1930s, economic opportunity was expanding. This trend continued during and after the Second World War. With all of its cyclical ups and downs, the economy nonetheless absorbed large masses of unskilled and semiskilled workers. Well-paying factory jobs were available to grade-school dropouts and to European immigrants who hardly spoke English. The availability of jobs together with the conspicuous success of some families in moving out of the slums after a generation or two contributed to the hopeful mood of the poor.

The hopeful mood was further sustained by other circumstances. The old poor lived in ethnic residential communities, sharing a common language, religion, and cultural background. Among members of each ethnic group considerable solidarity prevailed, expressing itself in a wide range of self-help institutions. Through the creation of the big-city political machines, the ethnic minorities also acquired a measure of political influence and one of the first welfare systems in the United States. With the depression of the 1930s, the ethnic minorities became a base for something more than benevolent associations. Millions of the poor lent a hand in the organization of the new industrial unions, notably the CIO. The poor became an important part of the political coalition of the New Deal years, which produced significant reform legislation such as the Wagner Act, Social Security, a minimum wage law, and more. In short, though the living standards of the old poor may often have been lower than those of today's underclass, they had grounds for the expectation that either they or their children or their children's children would have a better life.

The situation of today's underclass is a point-for-point contrast with that of the old poor. While the old poor were a majority, the new poor are a minority. The former consisted of a wide variety of ethnic groups; the latter largely comprise blacks, Puerto Ricans, and Mexicans. The former used their ethnic solidarity as a means of gaining influence; the latter have remained comparatively powerless. Finally, whereas the corporate economy of the earlier era had an insatiable need for unskilled and uneducated laborers, the corporate economy of today has no such need. Automation and cybernation have increasingly eliminated the need for unskilled labor in industry while raising the skill requirements of the remaining jobs. The result is an impoverished, unemployed, or underemployed underclass, concentrated in the service sector of the economy and laboring in the most menial and least desirable tasks.

Blacks are the largest single constituent of the American underclass. Prior to World War I they were a predominantly agrarian people concentrated in the southern rural areas of the United States. Since then they have continued to migrate to northern and other urban areas. Today a majority of black Americans reside in the ten or twelve largest cities of the United States.

What prompted that massive immigration? A major cause was the extraordinary growth of farm productivity in the United States. In the period following World War II, there were greater technological changes in rural America than in industry. Farm output per man hour expanded twice as fast as nonfarm. At the same time federal policies tended to favor corporate farming, with the result that small producers, who received no substantial support, were driven out of business. In 1940 there were over six million farms in the United States, half of them consisting of units smaller than 100 acres; in 1967 the number of farms was cut in half, with most of the units now being larger than 100 acres. Consequently, great masses of poor Southerners, mainly blacks, were forced to abandon the countryside. The U.S. Economic Development Administration reported in 1967 that during the 1950s ten million people were driven from the land into the metropolitan areas. Between 1960 and 1965, the black population of American cities increased by 20 percent.

As of 1967 about one third of America's poor were nonwhite while two thirds were white. Most of these whites are Puerto Ricans, Southwest Mexican Americans, and old American stock from Appalachia. Taken as a whole then, the American poor of today include both urban and rural workers barely eking out a living as well as the jobless and indigent. Although all of these groups form an underclass sharing similar life chances or a common social fate, their similar objective circumstances have not as yet given rise to class-conscious organizations. This is easy to understand if we compare the situation of the underclass with that of the organized production workers in the corporate sector. The latter are concentrated in large industrial enterprises, and they have a common class adversary in management. Besides, they have real potential power which can be expressed by withholding their services and disrupting the economy.

Members of the underclass, in contrast, are scattered in innumerable small firms. They have no easy means of communicating with one another and no obvious class adversary. Accordingly, they have not been successful in creating organizations for the defense and furtherance of their interests. It appears that since they cannot create such organizations by themselves, the assistance and cooperation of others will be needed. The industrial workers would seem to be the most obvious ally of the underclass, but the socioeconomic chasm separating the two categories of worker suggests that the creation of an alliance between them would be enormously difficult. A fuller discussion of black Americans and their position in the underclass will be found in chapter 8.

In sum, the concept of social class continues to be a very valuable analytical tool. It enables us to discern the major socioeconomic groupings in a society and their unequal life chances. *Life chances,* as we have seen, refers to the opportunities members of a given class have of obtaining the valued things in life. The concept comprises everything that is essential for human life, happiness, and dignity. Whether one will survive the first year of birth; whether one will remain healthy and, if sick, recover; whether one will live in spacious or crammed quarters, receive a good education, and be able to earn a living at an occupation that provides self-fulfillment and commands respect—such are only a few of the questions implied by the all-important and all-encompassing term *life chances.* For it is clear that all those aspects of human life are in a large measure determined by the class position one is born into.

Are basic social inequalities necessary and unavoidable? Or can they be reduced and even eliminated? Those are the questions we need to address next.

7

Social Stratification II: Other Approaches to the Study of Social Inequality

A central question in the study of social inequality is how it comes into being and how it is perpetuated. Social-class theory explains fundamental inequalities such as class divisions by studying their historical origins. If, for example, we wish to know how the industrial working class originated, we would have to explore the Enclosure movements which began in the late fifteenth century. There we would discover the process by which small independent peasant proprietors were separated from their land and other means of production and turned into a proletariat.

As we have seen in chapter 5, Enclosure was initiated by the powerful lords and the crown. It was they who forcibly expelled the yeomen and turned arable land into sheepwalks. They did so in response to a definite material incentive, the rising price of wool on the continent. In time that process resulted in a radical transformation of the social structure of the countryside. In place of lords and peasants there now emerged three classes: (1) the lords of giant estates; (2) capitalist farmers who leased a portion of the estate and provided capital for agricultural production; and (3) rural proletarians forced to sell their labor in order to survive. With the Industrial Revolution, the descendants of those who were expelled from the land became the industrial working class.[1]

The same sort of approach would be taken to the question of the origins of slavery, serfdom, and any other system of domination. Our historical analysis would be guided by such pertinent sociological con-   153 cepts as mode of production, class, material interests, power, force, and ideology. Hence both history and sociology would be employed to explain how a specific social system emerged and how it was maintained. Yet some highly influential sociological theorists have taken a quite different approach.

THE FUNCTIONALIST THEORY OF STRATIFICATION

Sociological functionalism, we will recall, starts from the assumption that all social structures—or at least the major ones—contribute to the maintenance of the system in which they operate. The existence of a given structure (or pattern) is explained by means of its *consequences,* which presumably are necessary and beneficial to the society in question. Generally functionalists have tried to show that a given pattern meets some vital "system need," the fulfillment of which accounts for the existence of the pattern.

Functionalists have argued, for example, that the function (i.e., consequence) of religion is group solidarity and the function of the family is the socialization of the child. The most telling logical criticisms of that type of analysis have been presented in chapter 5 and will not be repeated here. Instead, we will center attention on the functionalist view of stratification, a view which holds that social inequality is *necessary* because it fulfills system needs vital to the functioning of the entire society.

The Davis-Moore Thesis

In an article published in the *American Sociological Review* [ASR] in 1945, two distinguished sociologists, Kingsley Davis and Wilbert E. Moore, first presented their theory of stratification.[2] They began by noting that no society is unstratified. That being the case, the universality of inequality suggested to the authors that it is also necessary and unavoidable. The main task they set themselves was to explain why some *positions* in society are regarded as higher than others and why the higher positions carry greater rewards.

Every society, the authors argue, must distribute its members among the various positions of the social structure. Members must be induced to fill certain positions and, once in them, to perform the necessary duties. That is a requirement of any system whether it be competitive or not.

154   Not all social positions, the argument continues, are equally pleasant

to human beings, equally essential to society's survival, or equally in need of the same ability and talent. It follows for Davis and Moore that society must have rewards of some kind with which to induce individuals to fill positions; and that those rewards must be distributed unequally, corresponding with the functional importance of the positions. The rewards which society has at its disposal for distributing personnel and ensuring the performance of essential services are of both a material and nonmaterial kind. Society

has, first of all, the things that contribute to sustenance and comfort. It has, second, the things that contribute to humor and diversion. And it has, finally, the things that contribute to self-respect and ego expansion. . . . *In any social system all three kinds of rewards must be dispensed differentially according to positions.* (Italics added.)[3]

Thus the authors conclude the first part of their argument by maintaining that a society's social positions must be stratified and that the rights and rewards associated with positions must be unequal. They write:

Social inequality is thus an unconsciously evolved device by which societies insure that the most important positions are conscientiously filled by the most qualified persons. Hence every society, no matter how simple or complex, must differentiate persons in terms of both prestige and esteem, and must therefore possess a certain amount of institutionalized inequality.[4]

The authors emphasize that there are two factors which determine the ranking of positions: (1) their functional importance, and (2) the amount of training and/or talent required. In other words, given the functional importance of a position, its rewards will be higher the more scarce are the personnel trained to fill it. As the authors note, the talent required for the practice of medicine may be fairly abundant in the population at large; but the long and costly training program enables only a few to qualify. "Modern medicine," they write, "is within the mental capacity of most individuals, but a medical education is so burdensome and expensive that virtually none would undertake it if the position of M.D. did not carry a reward commensurate with the sacrifice."[5] In general, then, Davis and Moore maintain that the higher income, power, and prestige of a position are due to its functional importance and to the scarcity of trained personnel.

Qualifications of the Thesis

Critics of Davis and Moore point out that the authors introduced a sizable modification of their thesis in later publications. But the authors' 155 qualification of their thesis in the original article has apparently gone unnoticed. That qualification, as we shall see, is a significant concession to class theory and therefore deserves to be quoted in full:

> In a system of private property in productive enterprise, an income above what an individual spends can give rise to possession of capital wealth. Presumably such possession is a reward for the proper management of one's finances originally and of the productive enterprise later. But as social differentiation becomes highly advanced and yet the institution of inheritance persists, the phenomenon of pure ownership, and reward for pure ownership, emerges. In such a case it is difficult to prove that the position is functionally important or that the scarcity involved is anything other than extrinsic and accidental.[6]

The authors then observe that "ownership of production goods consists in rights over the labor of others" and that "property in capital goods inevitably introduces a compulsive [coercive] element even into the nominally free contractual relationship." Finally, "the control of the avenues of training may inhere as a sort of property right in certain families or classes, giving them power and prestige in consequence. Such a situation adds an artificial scarcity to the natural scarcity of skills and talents."[7]

Clearly, Davis and Moore recognize just how problematic the functionalist thesis is in the context of a class-structured society. They acknowledge that some rewards are not functionally determined at all but rather the result of ownership of wealth and the institution of inheritance. They also avow that the control of access to training by powerful and privileged groups creates an artificial scarcity of talent. All in all, the authors concede that under such circumstances the presumed functional importance of a social position is difficult to prove.

Indeed, in his textbook *Human Society,* also published in 1945, Kingsley Davis adds a major modification to the original thesis. In the *ASR* article the authors contend that the functionalist theory would fit any society, whether competitive or noncompetitive. In the textbook, however, Davis states that one may object to the functionalist explanation of stratification

> on the ground that it fits a competitive order but does not fit a non-competitive one. For instance, in a caste system it seems that people do not get their positions because of talent or training but rather because of birth. This criticism raises a crucial problem and forces an addition to the theory. . . . The necessity of having a social organization—the family—for the reproduction and socialization of children requires that stratification be somehow accommodated to this organization. Such accommodation takes the form of *status ascription.* (Italics added.)[8]

The addition of "status ascription" is a significant change of the original *ASR* statement. For Davis is now saying that the distribution of social positions in a class-structured society is largely based on *ascription* and not merely on *achievement*.

Those terms require a word of explanation. Sociologists speak of *achieved status* when a society enables individuals to move from lower to higher positions on the basis of merit; and they speak of *ascribed status* when a society prevents such mobility. Those concepts are pure types, for no actual society is perfectly based on either principle. A good example of a social structure based largely on ascription is a caste society. Those born into a certain caste remain there throughout their lives, and their life chances are rigidly determined by their position in the caste hierarchy. All class societies are also highly ascriptive. So we see that in acknowledging the powerful influence of ascription, Davis and Moore have substantially modified the original functionalist argument.

Yet the authors have remained committed to the view that (1) social positions have varying degrees of functional importance (i.e., make different contributions to society's preservation and survival); (2) adequate fulfillment of position duties requires talent and training which are scarce; and (3) stratification, or unequal rewards, ensures that the most talented and trained individuals will fill the social roles of greatest functional importance. In sum, the general tenor of the Davis-Moore argument is that the rich, powerful, and prestigious are at the top because they are the most talented and the best trained and also because they make the greatest contribution to society's preservation. It is such implications that aroused criticism and opposition.

The critical controversy began in 1953, with Melvin M. Tumin's "Some Principles of Stratification: A Critical Analysis."[9] Tumin begins by questioning a key concept in the Davis-Moore theory, namely, "functional importance." He argues that no one has thus far been able to demonstrate the varying functional importance of social positions. Are engineers functionally more important to a factory than unskilled workers? To that question Tumin replies that "at some point along the line one must face the problem of adequate motivation for *all* workers at all levels of skill in the factory. In the long run, *some* labor force of unskilled workmen is as important and as indispensable to the factory as some labor force of engineers."[10]

Next Tumin questions the scarcity of talent as an adequate determinant of stratification. He contends, first, that we have no sound knowledge of the actual range of talent in a society; and, second, that stratifi- 157 cation systems artificially limit the development of whatever potential skills there are in a population. For instance, where "access to education depends upon the wealth of one's parents, and where wealth is differentially distributed, large segments of the population are likely to be deprived of the chance even to *discover* what are their talents." Besides, those who find themselves in privileged positions and professions have a tendency to restrict further access to those positions. "This is especially true in a culture where it is possible for an elite to contrive a high demand and a . . . higher reward for its work by restricting the numbers of the elite available to do the work. The recruitment and training of doctors in modern United States is at least partly a case in point."[11]

Tumin then asks, Should the loss of earnings or the cost of training be looked upon as a "sacrifice" made by the trainee? No, he replies, because those costs tend to be paid out of income which the parents of the trainees "were able to earn generally by virtue of *their* privileged positions in the hierarchy of stratification." Further, the notion of "sacrifices" completely overlooks "the psychic and spiritual rewards which are available to the elite trainees by comparison with their age peers in the labor force."[12] Such rewards would include greater opportunity for self-development and greater leisure and freedom. Finally, Tumin argues that the motivating of individuals to fulfill social tasks conscientiously does not require unequal rewards. There are, he suggests, at least two possible alternatives to unequal rewards: intrinsic job satisfaction and social service. Tumin concludes that social stratification, far from being positively functional for society, is just the opposite.

Kingsley Davis, replying to Tumin soon afterward, agreed that the "functional importance" of a social position is difficult to establish. He also agreed that stratification systems often restrict the discovery and development of talent in the population, but he protested that he had already met that objection by acknowledging the role of status ascription. On the issue of "functional alternatives" to unequal rewards, Davis insisted that intrinsic job satisfaction and social service "are supplementary rather than alternative."[13]

In his "Reply to Kingsley Davis,"[14] Tumin continued to challenge the view that the universality of stratification is proof of its necessity. He maintained that alternatives to unequal rewards should not be ruled out, and that if Davis wishes to argue that such alternatives are impossible, he should provide some evidence to that effect.

To summarize, Tumin has effectively challenged major elements of the functionalist theory. First, he successfully questioned the scientific validity of the view that social positions vary in their functional importance; and second, he called into question the alleged necessity of "unequal rewards" by convincingly arguing the possibility of alternatives. Tumin's critique has cast doubt on the functionalist theory that "social inequality is . . . an unconsciously evolved device by which societies insure that the most important positions are conscientiously filled by the most qualified persons."[15]

CONFLICT-COERCION THEORY

Functionalism, as we have seen, looks upon society as a highly integrated social system. A society is held together in a stable condition because it rests primarily on the common values of its members. Social order results from consensus, which maintains itself rather easily and automatically. Typically, functionalists have not considered the role of power, force, and indoctrination in maintaining the consent of the governed.

Conflict theorists, in contrast, view society as consisting of a large number of groups and organizations with different, competing, and even conflicting interests. The basic social process taking place among those groups is a struggle over their material and ideal interests. Stable relations between two or more groups are not the result of consensus pure and simple but rather a result of the recognition by each group of the others' relative power. The social structure of a society is therefore largely maintained by power and constraint, while changes in the social structure are the consequence of struggles and conflicts.

Elements of that approach may be found in numerous classical social theories, notably those of Marx and Weber. But it is only in recent decades that contemporary sociologists have attempted to develop and apply so-called *conflict theory*. One of the best-known contemporary applications of that theory to advanced industrial society is found in the work of Ralf Dahrendorf. As we shall see, however, his version of conflict theory results in a truncated concept of class.

Dahrendorf's Views of Modern Society

Following the lead of several classical theorists, Dahrendorf views every social organization as one in which power and authority are unequally distributed. "Power," if we accept Weber's definition, refers to the probability that one will realize his will against the resistance of others.   159

Bullies and armed thugs, for example, frequently get their way. Yet power does not always take the form of naked force. Most often it is *legitimized* in one way or another so that it becomes *authority*. The control a parent has over a child or a military officer has over his men or a higher civil-service officer has over his subordinates is authority.

Authority relations are found in all social organizations. The most powerful organizations in modern society are large, complex, formal associations. In such associations, Dahrendorf maintains, one may not only observe a line of authority but a *dichotomy* as well. That is, one may distinguish the rulers from the ruled, those who command from those who obey. Authority, in those terms, is a "zero-sum" concept. That concept, borrowed from game theory, means that if some win others necessarily lose. Similarly, we find in formal associations that there are those who possess authority and others who are subject to it. It is the dichotomous division of authority which generates two basic "conflict groups" in every association, the rulers and the subjects. Thus every association is based on domination. Those who rule ultimately have at their disposal effective means of coercing their subjects. For Dahrendorf, rulers and subjects are "conflict groups" because their respective interests are opposed and contradictory.

Although we have been calling rulers and subjects *groups,* that is not quite accurate. For they are only potential groups or what Dahrendorf calls *quasi-groups.* It is from the latter that actual groups are formed. In every association, two quasi-groups, each with its common latent interests, may be distinguished. "Their orientations of interest," writes Dahrendorf, "are determined by possession of or exclusion from authority. From these quasi-groups, interest groups are recruited, the articulate programs of which defend or attack the legitimacy of existing authority structures. In any given association, two such groups are in conflict."[16]

It is Dahrendorf's central purpose here to argue that the present-day industrial societies of Western Europe, the United States, and Japan are *post-capitalist.* For Dahrendorf, that term means, among other things, that modern society is no longer class-structured. There are no longer any classes or class conflicts in the society as a whole. There are only conflict groups, which are strictly confined to *particular associations.* If Marx, Weber, and other classical theorists viewed society as a whole as divided into classes, Dahrendorf looks upon Western-type industrial societies quite differently. In place of capitalists, workers, and other classes, he sees a multitude of conflicting groups. His two-class model

160 of rulers and subjects "applies not to total societies but only to specific

associations within societies. . . . If in a given society, there are fifty associations, we should expect to find a hundred classes, or conflict groups in the sense of the present study."[17]

In capitalist society, Dahrendorf argues, the dominant groups of industry were also the dominant groups of the state, either personally or through their agents. Subjected groups, such as industrial workers, were excluded from political authority. Classes were for the most part closed units. Upward mobility was rare. Class fronts were hardened, and violent class conflicts not unusual. But all of that has changed, Dahrendorf maintains. If asked, "Do we still have a class society?" Dahrendorf replies that there are no classes stretching beyond specific associations. In his view, therefore, there are either no social classes at all or innumerable ones since "classes and class conflict are present wherever authority is distributed unequally over social positions." Social conflict, he insists time and again, is as universal as the relations of authority in associations, "for it is the distribution of authority that provides the basis and cause of its occurrence."[18]

Dahrendorf goes even further and asserts that in "post-capitalist" society, industry and society have been separated and dissociated. "Industry and industrial conflict," he writes, "are, in post-capitalist society, institutionally isolated, i.e., confined within the borders of their proper realm and robbed of their influence on other spheres of society." That is not all. Since society-wide classes no longer exist, only a multiplicity of conflict groups in a multitude of associations, it is quite possible for those at the top of some associations to be at the bottom of others and vice versa. It follows "that in post-capitalist society the ruling and the subjected classes of industry and of the political society are no longer identical; that there are, . . . in principle two independent conflict fronts. Outside the enterprise, the manager may be a mere citizen, the worker a member of parliament; their industrial class position no longer determines their authority position in the political society."[19]

Critical Observations

Dahrendorf's "post-capitalist" society consists of a plurality of associations, each composed of two opposing groups—one in authority and the other out. It is the exercise of authority and the subjection to it, in Dahrendorf's view, that cause conflict between groups. Wherever there is authority, Dahrendorf asserts, men will struggle for it. But that sounds more like a psychological law than a sociological proposition. For it implies that every individual is inherently authority-seeking, and that conflicts are generated by *authority differences themselves,* rather than by oppression, exploitation, abuse of authority, or other substantive issues.

Dahrendorf dissolves all social classes, including the dominant ones, into a multitude of groups confined to single associations. There are as many ruling groups in society as there are associations. Yet Dahrendorf himself finds his extreme pluralistic view less than satisfactory. Reflecting on industry, the state, and the church, he acknowledges that it "is more probable that the workers of industry are at the same time mere members of the church and mere citizens of the state. One might expect that the dignitaries of the church are in some ways connected with the rulers of the state and possibly even with the owners or managers of industry."[20] But he says repeatedly that whether or not the above pattern is typical is a matter for investigation.

One can only agree that such questions should be answered by appropriate research. But does Dahrendorf mean to imply that an investigation of "post-capitalist" society would reveal that manual workers occupy the command posts of associations? Would such an investigation disclose that workers exercise authority in economic, political, educational, or other associations? Or that while corporate executives control industry, workers control the state? One doubts that Dahrendorf would seriously reply to those questions in the affirmative. Yet he sticks to his bizarre notion of a plurality of associations, with those at the top of some being at the bottom of others.

Equally problematic is Dahrendorf's assertion that today's industrial conflicts are well insulated—that conflicts in the sphere of industry do not spill over and affect other spheres of society. The available evidence tends to contradict that view. For as T. B. Bottomore has noted, "Numerous studies have shown that in the European industrial countries, and to a lesser extent in the USA, the major political conflicts are closely and continuously associated with industrial conflicts, and express the divergent interests of the principal social classes."[21]

The fact is that Dahrendorf himself has misgivings about his notion of insulated industrial conflict. Indeed, his own reservations amount to a refutation of his own theory. "First," he writes,

it would seem that the life of people in modern societies is increasingly rationalized, mechanized, in this sense "industrialized." The products of industry dominate consumption in all social strata and mold the standard and style of living of all. Secondly, industrial and political problems are more closely connected than ever. If industrial production is stopped (e.g., by a strike) every member of society feels the consequences. In this sense, industry is by no means isolated. Thirdly we can observe an increase rather than a decrease of government influence on industry. From this point of view, industry and society are more closely connected than ever. Finally, there is even now a wealth of evidence of the existence of interrelations, often of a personal kind, between the ruling classes of industry and society. The composition of recent or current governments in the United States, Britain, France, Germany, and other countries would seem to sustain the thesis of C. Wright Mills that the "power elite" of post-capitalist societies is relatively uniform and dominated by the carriers of industrial authority.[22]

Yet, Dahrendorf, in the face of his own cogent objections, nevertheless tries to save his theory by arguing that it describes a "tendency," not the actual state of affairs. That is the context in which he asserts that "outside the enterprise, the manager may be a mere citizen, the worker a member of parliament."[23] To that assertion one can only reply that if such a tendency exists, its manifestations have yet to be observed.

Finally, one must note that Dahrendorf looks upon the modern political state as just another association. It is no more inclusive for Dahrendorf than any other organization, and it is a "unit of social analysis strictly equivalent to an industrial enterprise, a church or an army."[24] That view implies that the state has no greater authority than any other association; it denies the wide range of circumstances in which state authority supersedes the authority of all other associations.

As the foregoing criticisms suggest, Dahrendorf's conception of things fails to reflect the social and political reality of any modern industrial society.

ELITES AND RULING CLASSES

It remains for us to consider one more body of ideas which has figured prominently in the study of stratification. The term *elite* is often used to refer to superior and powerful groups in society. This term was introduced into sociological discourse by a well-known Italian economist named Vilfredo Pareto (1848–1923).

Pareto believed that elites possessed superiority of one kind or another—intelligence, character, skill, capacity, power, and so on. He sometimes allowed for the possibility that persons and groups may acquire the label of elite without actually possessing such qualities. On balance, however, he thought that those who possess elite qualities are the ones who become social elites. One can assess the degree of excellence of every human endeavor, of prostitution and theft as well as law and medicine, and assign to the individuals in each endeavor scores ranging from zero to ten. A grade of ten would be assigned to the very best in each field, reserving zero for the totally incompetent. Thus the elite of a society consists of those with the highest scores in their respective branches of activity.

Pareto divided the general elite into two parts: A *governing elite,* i.e., those "who directly or indirectly play some considerable part in government, and a *nongoverning elite,* comprising the rest."[25] Together they constitute the upper stratum of society. The lower stratum or nonelite comprises all those whose political power is practically nil from Pareto's standpoint. The governing elite is made up of the effective rulers of society; it dominates the nongoverning elite and the masses. As we shall see, Pareto's theory has nothing in common with class theory and is not even very sociological. To understand the essential difference between elite theory and class theory, we must consider another basic idea of Pareto's.

Pareto firmly believed that human beings are impelled into action not by rational considerations but by internal biopsychic drives. His view of human motivation was very similar to that of the instinctualist psychology of his day. For Pareto human rationality, or what he called "logico-experimental" reasoning, was confined to a very few areas of social life, notably the scientific and the economic. Ideally, the scientist in his laboratory is rigorously logical. So is the entrepreneur in a competitive economy, because he must choose the most economical means of producing and marketing his goods. There are a few other situations in which individuals exercise their rational faculties, but the masses are motivated by nonrational considerations, though they themselves fail to recognize that fact.

The three basic elements in Pareto's theory of human conduct are: (a) "sentiments," (b) "residues," and (c) "derivations." The term *sentiment* may be taken as a synonym for *drive* or *instinct.* It is the internal, unconscious, biopsychic force that impels individuals into action. Since sentiments are not directly observable, Pareto could only postulate them. He therefore preferred to focus attention on the residues, which are observable nonverbal and verbal actions. Like the behaviorists, Pareto argued that residues provide the real data for the study of human conduct. Finally, there are the derivations, verbal attempts to justify behavior by placing pseudological constructions upon it. Most people believe that the derivations are the true cause of their behavior. They imagine that the causal sequence of their behavior is this: (c) derivations → (b) residues → (a) sentiments. But, said Pareto, that is an illusion. For the true and actual sequence is: (a) → (b) → (c).

Although Pareto introduced six types of residues, he actually employed only the first two in his analysis. He called the first type *Instinct for Combinations*. That is a synonym for "ability to think," "inventiveness," "imagination," "ingenuity," "originality," and other such qualities. The Type-one residue is characteristic of the various elite elements in society. He called the second type of residue *Instinct for Persistence of Aggregates*. That cumbersome term refers to tendencies of the mind which create certain persistent, nonrational belief systems. Examples of "persistent aggregates" are the belief in the "devil," in "Santa Claus," in "democracy," etc. For Pareto all such beliefs were substantially the same: tenaciously held, nonlogical notions ultimately rooted in human nature. Pareto linked the rulers and the ruled to the two types of residue. Type-one residues predominate among the rulers and type-two among the ruled. It is among the "masses" at large that type-two residues are most active.

The social "equilibrium" as well as the decline of one governing elite and the rise of another depend on how successfully the elite can invent formulas that appeal to the sentiment-rooted beliefs of the masses. While the masses are passive and their sentiments unchanging, the elite is actively exploiting those sentiments by means of the ingenious political formulas. Just as their sentiments remain unchanged, so do the conditions of the masses, regardless of how often elites change positions. The masses always remain blindly nonrational becaue they are controlled and moved by unconscious "forces," "instincts," or "sentiments."

That is the way Pareto tried to explain both the structure of society and the changes it undergoes. In the governing elite type-two residues gradually lose in strength until they are now and again reinforced by tides upwelling from the masses. Revolutions are such great tides, the upward thrusts of the lower strata strong in type-two residues. Pareto also invoked residues to explain why "history is the graveyard of aristocracies." Elites decay as they lose those residues that "enabled them to win their power and hold it. The governing class is restored not only in numbers, but—and that is the more important thing—in quality, by families rising from the lower classes and bringing with them the vigor and the proportions of residues necessary for keeping themselves in power. It is also restored by the loss of its more degenerate members." If such circulation ceases, the governing elite collapses, sweeping "the whole of a nation along with it. A potent cause of disturbance in the equilibrium is the accumulation of superior elements in the lower classes and, conversely, of inferior elements in the higher classes."[26] A governing elite, then, is invigorated by transfusions from below.

But governing also requires something else—*force.* "Superior elements" are not only those "fit to rule" but those willing to use force. Inferior and decadent elements fear its use and thus become unfit. The "decaying" elite, shying away from the use of force, tries to buy off its adversaries; it becomes less the lion and more the fox and thus increasingly vulnerable to new lions. When the rule of the governing elite is threatened, and when out of humanitarian or similar sentiments it declines to meet force with force, even a small group can impose its will upon it. And when the governing elite resorts instead to fraud and deceit in an effort to outwit its adversaries, that eventually brings about a change in its composition—power passes "from the lions to the foxes." Foxiness, resting on type-one residues, becomes preponderant in the elite, while type-two residues decline. But now it is precisely that foxy resourcefulness employed to outsmart one's opponents which makes the ruling elite more and more vulnerable to those willing and able to use force. The latter are the "lions," either from within the governing circles or from the leaders of the masses.

The leaders of the subject masses (who are themselves an elite), employing force, topple the existing rulers. That is accomplished all the more easily if the rulers are moved by humanitarian sentiments, and if they have found no way of assimilating into their midst the elite of the subject masses. A closed aristocracy is the most vulnerable and insecure. On the other hand, the more adept the governing elite is in absorbing those subject elements that are talented and skilled, the more secure is its rule. For the absorption of such elements undercuts the possibility that they will "become the leaders of such plebians as are disposed to use violence. Thus left without leadership, without talent, disorganized, the subject class is almost always powerless to set up any lasting regime."[27] Pareto thus wished to prove that there will always be a subject class. A subject class is inevitable because it has no real leadership of its own; its elite elements are continually coopted by the governing circles.

The governing elite, being small, is greatly strengthened by the influx of individuals with type-one residues. However, the subject class is enfeebled not only by the loss of those elements, but also by the fact that though it is still left with many talented individuals, they do not apply themselves to politics. "That circumstance," wrote Pareto,

lends stability to societies, for the governing class is required to absorb only a small number of new individuals in order to keep the subject class deprived of leadership. However, in the long run the differences in temperament between the governing class and the subject class bcome gradually accentuated. . . . When that difference becomes sufficiently great, revolution occurs.[28]

Unfortunately, Pareto never systematically considered any causal conditions outside of his sentiments-residues, the prime mover to which he returned again and again. There is no attempt anywhere to relate the contrasting character of the "elite" and the "mass" to their respective sociocultural and class situations. The alleged incompetence of the mass is an eternal trait because it is a result of those constant "sentiments." For Pareto, there will always be a ruling elite and always a subject mass —for they unavoidably follow from his method.

Pareto's dichotomy of rational/nonrational allows for no gradations in between. Actions are always either one or the other. The effectiveness of his polemic rests on proving that it is the nonrational which dominates social life and must continue to do so. Hence he conveniently ignored the subtle mixtures of rational and nonrational encountered in all humans and in all societies.

By arbitrarily restricting rationality to a few narrow spheres and then defining all other spheres as nonlogical, Pareto unswervingly led himself to the very conclusion he wished to find: Human conduct is so thoroughly nonrational as to preclude a conscious and rational altering of the social order along more democratic lines. An individual acts in response to "sentimental" causes which are so powerful that they cannot be counteracted by his weak and insignificant efforts at rationality. One seeks in vain for any kind of sociological analysis designed to illuminate the interdependence of the social situations of the various groupings in society and their consciousness and behavior. So although Pareto called his major work a treatise on "sociology," it is actually something quite different. For it virtually ignores the *social* context of human conduct.

THE RULING CLASS

A theory similar to Pareto's was put forward by his distinguished contemporary Gaetano Mosca (1858–1941). Throughout history, argued Mosca, from the dawn of civilization to the present, two classes of people have been evident: One that rules and another that is ruled. Such has been the case regardless of whether the political system has been democratic, aristocratic, despotic, or what have you. The division of society into rulers and ruled is inevitable and will continue as long as there are human societies on earth.

167

To rule effectively, the ruling class of any political system must take the mood and sentiments of the masses into account. Pressures arising from the discontent of the masses can influence the rulers' policies, and massive discontent might even lead to the overthrow of a ruling class. But that overthrow would never result in a lasting and genuine democracy, for a new class of rulers would soon emerge to take the place of the old.

The power of the ruling class rests on the fact that it is an *organized* minority confronting an *unorganized* majority. Mosca, like Pareto, looked upon the majority of the economically disadvantaged populace as an atomized, unorganized mass. Precisely because the ruling class is a comparatively small group, it can achieve what the majority cannot: communication, mutual understanding, and concerted action. "It follows," wrote Mosca, "that the larger the political community, the smaller will the proportion of the governing minority to the governed majority be, and the more difficult will it be for the majority to organize for reaction against the minority."[29] An inflexible social law rooted in human nature makes it inevitable that the representatives of the people—whether elected or appointed—will transform themselves from servants into masters. Even in a political democracy, those elected to represent and defend the interests of their constituencies soon develop interests of their own; and in their zealous pursuit of those interests they become a well-organized, powerful, and dominant minority. The ruling group is strengthened not only by its organization but by the superior qualities—material, intellectual, moral—which distinguish it from the mass.

For Mosca there existed a basic psychological law which impels individuals to *struggle for preeminence,* his term for social competition and conflict over wealth, power, and prestige. This struggle for "control of the means and instruments that enable a person to direct many human activities, many human wills as he sees fit"[30] always results in the victory of a minority which, by virtue of its organization and other superior qualities, gains decisive control over certain "social forces." Control of any one social force—e.g., military, economic, political, administrative, religious, moral—may lead to the control of others.

The military power of warrior lords, for instance, once enabled them to demand and receive "the community's whole produce minus what was absolutely necessary for subsistence on the part of the cultivators; and when the latter tried to escape such abuses, they were constrained by force to stay bound to the soil, their situation taking on all the characteristics of serfdom pure and simple." This has been generally true where land was the chief source of wealth: Military power led to wealth just as later wealth in the form of money led to political and military power. When "fighting with the mailed fist is prohibited whereas fighting with pounds and pence is sanctioned, the better posts are inevitably won by those who are better supplied with pounds and pence."[31]

Mosca believed that the various advantages of the ruling minority—organization, superior qualities, and control of social forces—lead to a condition in which "all ruling classes tend to become hereditary in fact if not in law."[32] There is no eliminating all those special advantages which individuals acquire as a birthright simply by being born into privileged social positions. Although the organized ruling minority has superior might and can therefore repel challenges to its rule by force, it does so only as a last resort. Usually it consolidates its dominance by making it acceptable to the masses. That is done by means of a *political formula,* a term roughly equivalent to Marx's "ruling ideology," Weber's "legitimation" of power, and Pareto's political "derivations." Every governing class, said Mosca, seeks "to justify its actual exercise of power by resting it on some universal moral principle."[33]

A careful reading of Mosca's work reveals that like Pareto, he was intent upon showing that genuine democracy and equality are unrealizable. Both thinkers wished to discredit socialism and, in particular, Marx's vision of a classless society. Mosca's point was that there is no realistic basis for such a vision since the struggle for preeminence is determined by human nature and not by social conditions. Hence what has been true of past societies will continue to be true of all future ones. A ruling class is a permanent attribute of every society's structure.

Mosca's convictions led him to conclude that the best one can hope for is a society based on the principles long ago enunciated by Aristotle and Montesquieu. Aristotle was right, according to Mosca, when he asserted that a good society is one in which there exists a large middle class, with the extremes of rich and poor being negligible. And Montesquieu was right to insist that a good society requires the division or separation of powers. Mosca strongly agreed with Montesquieu that only power can check power, but he stressed that the separation of powers cannot be truly effective so long as it exists on the governmental level alone. A really effective system of checks and balances must be founded on a plurality of well-balanced *social* forces, for in the absence of such a balance, the system will tend toward tyranny. "When no other social forces exist apart from those which represent the principle on which sovereignty over the nation is based, then there can be no resistance, no effective control, to restrain a natural tendency in those who stand at the head of the social order to abuse their powers." A good   169 society is one in which the many different groups within it can check one another "effectively enough to prevent absolute control by the individual, or individuals, who stand at the head of the social order."[34]

So we see that although Mosca argued the necessity of a ruling class, he nevertheless qualified his thesis rather severely. He recognized that a good society requires (1) the elimination of extreme inequalities, and (2) the creation of a balance of social forces in which none could aggrandize sufficient power to itself to tyrannize over all the rest. But those qualifications tend to undermine Mosca's central thesis. What really emerges from his analysis is not the necessity of a ruling elite nor even the necessity of a plurality of competing elites. Rather, it is that a good polity presupposes many different social groups (not elites) which can check and limit one another's powers and especially the power of the state. Mosca understood that democracy requires the diffusion of power in a wide range of social and political organizations which can set limits on the powers of the leadership. A society which fails to institutionalize a division of powers along those lines will unavoidably become totalitarian.

Thus interpreted, Mosca's theory appears in a new light; and his work can be read quite differently from the way it has been in the past. For his analysis, far from proving the necessity of a ruling class, effectively shows how the traditional gap between rulers and ruled can be significantly reduced, if not eliminated.[35]

Ethnic Minorities
in the United States

THE CONCEPTS OF RACE AND ETHNIC GROUP

The word *race* is firmly a part of the popular vocabulary. People speak of the white race, the yellow race, and the black race, and the term is often applied to religious and cultural groups as well. But what is a "race"? How many "races" are there? How do we distinguish one from the other?

Throughout the nineteenth century and well into the twentieth, scientists were practically unanimous in subscribing to the concept of race. As recently as the 1950s standard anthropological textbooks described humankind as divided into three major races, Caucasoid, Mongoloid, and Negroid, or, in plain English, white, yellow, and black. At about the same time, however, anthropologists began to modify this view. W. Boyd, for example, proposed a fivefold race classification: (1) *European* (Caucasoid), (2) *African* (Negroid), (3) *Asiatic* (Mongoloid), (4) *American Indian,* and (5) *Australoid.*[1]

Coon, Garn, and Birdsell went further and distinguished nine major "geographical races":

1. *Amerindian*—the pre-Columbian populations of the Americas.

2. *Polynesian*—islands of the Eastern Pacific, from New Zealand to Hawaii and Easter Island.

3. *Micronesian*—islands of the Western Pacific, from Guam to the Marshall and Gilbert islands.

171

4. *Melanesian*—Papuan—islands of the Western Pacific, from New Guinea to New Caledonia and Fiji.

5. *Australian*—Australian aboriginal populations.

6. *Asiatic*—populations extending from Indonesia and Southeast Asia to Tibet, China, Japan, Mongolia, and the native tribes of Siberia.

7. *Indian*—populations of the subcontinent of India.

8. *European*—populations of Europe, the Middle East, and Africa north of the Sahara; now worldwide.

9. *African*—populations south of the Sahara.[2]

The same scholars also introduced a finer subdivision of *Homo sapiens* that yielded more than 30 races:

1. *Northwest European*—Scandinavia, northern Germany, northern France, the Low Countries, the United Kingdom, and Ireland.

2. *Northeast European*—Poland, Russia, most of the present population of Siberia.

3. *Alpine*—from central France, south Germany, Switzerland, and northern Italy eastward to the shores of the Black Sea.

4. *Mediterranean*—peoples from both sides of the Mediterranean, from Tangier to the Dardanelles, Arabia, Turkey, Iran, and Turkomania.

5. *Hindu*—India and Pakistan.

6. *Turkic*—Turkestan, western China.

7. *Tibetan*—Tibet.

8. *North Chinese*—northern and central China and Manchuria.

9. *Classic Mongoloid*—Siberia, Mongolia, Korea, and Japan.

10. *Eskimo*—arctic America.

11. *Southeast Asiatic*—South China to Thailand, Burma, Malaya, and Indonesia.

12. *Ainu*—aboriginal population of northern Japan.

13. *Lapp*—arctic Scandinavia and Finland.

14. *North American Indian*—indigenous populations of Canada and the United States.

15. *Central American Indian*—from southwestern United States through Central America to Bolivia.

16. *South American Indian*—primarily the agricultural peoples of Peru, Bolivia, and Chile.

17. *Fuegian*—nonagricultural inhabitants of southern South America.

18. *East African*—East Africa, Ethiopia, a part of Sudan.

19. *Sudanese*—most of Sudan.

20. *Forest Negro*—West Africa and much of the Congo.

21. *Bantu*—South Africa and part of East Africa.

22. *Bushman and Hottentot*—the aboriginal inhabitants of South Africa.

23. *African Pygmy*—a small-statured population living in the rain forests of equatorial Africa.

24. *Dravidian*—aboriginal populations of southern India and Ceylon (Sri Lanka).

25. *Negrito*—small-statured and frizzy-haired populations scattered from the Philippines to the Andamans, Malaya, and New Guinea.

26. *Melanesian-Papuan*—New Guinea to Fiji.

27. *Murrayian*—aboriginal population of southeastern Australia.

28. *Carpentarian*—aboriginal population of northern and central Australia.

29. *Micronesian*—islands of the western Pacific.

30. *Polynesian*—islands of the central and eastern Pacific.

31. *Neo-Hawaiian*—an emerging population of Hawaii.

32. *Ladino*—emerging population of Central and South America.

33. *North American Colored*—the so-called black population of North America.

34. *South African Colored*—the so-called black population of South Africa.

This classification was innovative inasmuch as it defined races as Mendelian populations that change in time. Some of the races listed above were formed within the last 400 years. The North American black people (33) emerged from a mixture of 20, 21, 1, 3, 4, and probably others. The South African blacks (34) arose from 21, 22, 1, and 3; the Ladinos (32) came from at least 15, 16, 4, 20, and 21; and the Neo-Hawaiians (31) emerged from 30, 1, 9, and some 4, 8, and 11. The fact that these races are recent hybrids makes them no less real and natural than others in the list. Most or all of the populations that existed in the more remote past also had mixed origins. The so-called Neanderthal Man, for instance, was differentiated into local races (varieties), some of which exchanged genes with the varieties labeled *Homo sapiens*.[3] In this perspective it is clear that some races were more prominent in the past than they are at present; others are new and they continue to be assimilated by intermarriage with their neighbors.

We see, then, that scientific anthropology has given us classifications yielding 3, 5, 9, and 34 races. Does this mean that some of these classi-

173

fications are necessarily wrong? The answer is no, they all may be right. The races distinguished in all of these lists are based on observable physical traits; to that extent the racial differences are objective and real. Mendelian populations of all kinds, from small groups to inhabitants of countries and continents, may or may not differ in the frequency of some genetic variants. If they so differ, they may be said to be racially distinct. That is the reason why some scientific anthropologists continue to defend the use of the term *race* as applied to human beings. Although human races are difficult to study, they argue, they are neither more nor less real than races in other species.

All scientific anthropologists recognize that most human populations isolated for study would be found to differ at least slightly in the frequencies of some genes. They also acknowledge that the *number* of races we choose to recognize is a matter of convenience. One could, if one wished, multiply and subdivide races indefinitely. That being the case, some anthropologists have questioned the scientific usefulness of the term *race* as applied to humanity. What is the point, they ask, of distinguishing 3, 5, 9, 34, 100, or 1,000 races? In differentiating humankind into races, whatever the number, are we doing anything more than saying that the human species consists of individuals varying in size, color, and other physical features? Then why not drop the term *race,* which has done so much harm and, evidently, no good at all?

In order to assess the scientific usefulness of the concept of race, we have to review the available evidence concerning the origins and biological significance of the so-called racial traits. Let us begin with skin color, a most conspicuous characteristic. If we assume that differences in skin color arose through natural selection as adaptations to the environments in which people lived, then the geographic distribution of dark- and light-skinned peoples makes sense. Dark skins appear to be adaptive in climates with strong sunshine and clear skies and light skins in lands with cold and cloudy climates. The darkest peoples live in the savannas of Africa, south of the Sahara Desert but north of the equatorial rain forests. The forest dwellers are lighter in color and the aboriginal populations of South Africa (Bushmen) have yellow-brown skin. The center of "blondism" or "leucodermia," in contrast, is northern Europe, which was covered by the Pleistocene ice sheet and where still today there is precious little sunshine.

A similar correlation is discernible among other members of the animal kingdom—compare the habitat of the white bear with that of the black or brown bear. As a rule, melanin or black pigmentation in mammals and birds increases in hot and arid regions. But the pattern is by no means consistent either among animals or humans. Among American Indians, for instance, we find that those living in Central and South America are on the average only slightly darker, if at all, than the Indians in the temperate and cold regions of North America. On the other hand, the Mongoloid peoples of northern Siberia and the Eskimos of arctic America have more pigment than one would expect in those climates. With these qualifications in mind, we can say that dark skins appear to be a character of adaptive value. For there is some evidence that dark skins are better able to withstand the effects of prolonged exposure to sunlight. The principal pigment responsible for a dark skin is melanin, which is produced in cells known as melanocytes by a reaction between tyrosine and oxygen.[4] Exposure to the ultraviolet rays of sunlight activates the process of converting tyrosine into melanin. Racial groups do not differ in the number of melanocytes they possess. Differences in pigmentation are discernible among individual members of a racial group, and even in different parts of one person's body. Such differences are due to the unequal distribution of melanin particles in the melanocytes, a process partly under genetic and partly under environmental control. We all know that the exposure to intense sunlight brings about increased pigmentation in a white person's skin, turning it dark and even black—though, of course, this has no effect on the genes for white skin.

If the biological significance of differences in skin color is poorly understood, the significance of other so-called racial features is even more elusive. What difference does it make whether one's hair is straight, wavy, or curly? It is worth noting that Darwin had refused to ascribe human racial differences either to natural selection or to the Lamarckian theory that acquired characteristics are inherited. But Darwin also advanced a theory of "sexual selection," which appears to be the most plausible one by which to account for such differences as the quantity and texture of body hair. Applying the principle of sexual selection to humans, a distinguished geneticist writes: "If tastes were unlike in different countries and if those whose figures and features were popular with the opposite sex produced on that account more surviving progeny, then races would have become different in appearance."[5] This would have occurred even though the racial characteristics were not in themselves either useful or harmful. Finally, race differences—that is, differences in appearance among populations—may be accounted for by genetic drift: accidents of sampling from the gene pool.

In sum, modern science has no solid and conclusive evidence regarding the adaptive significance of the so-called racial traits in human beings. All that science has been able to provide so far are speculations and surmises, more or less plausible. And even those scientists who insist upon retaining the term *race* acknowledge that "the differential adaptations of the races of man are most probably concerned with environments of a remote past, largely superseded by the environments created by civilization, to which all races may be equally adapted or unadapted."[6]

That point needs to be underscored: Whatever the observed differences may be among so-called racial groups, there is no evidence of an associated superiority or inferiority, mental or physical. There are no superior and inferior races. The achievements of all human groups depend upon their historical experiences—upon the geographical, social, economic, and political conditions of their existence. Even a moment's reflection will make it quite clear that biological races have not been the major actors in the history of humanity. Whatever classification of race we might wish to accept, whether of 5, 9, 34, or more races, it is evident that it is not races that have made and continue to make history but rather clans, tribes, and nations—not to mention other collectivities such as socioeconomic groups.

If the concept of race is vague, and the number of races one chooses to distinguish is arbitrary; and if races, however delineated, are not the real actors in history, then perhaps the term *race* ought to be abandoned in intelligent and rational discourse. This suggestion was first made by the British biologist Julian Huxley. It "would be highly desirable," he wrote, "if we could banish the question—begging term 'race' from all discussion of human affairs and substitute the noncommittal phrase 'ethnic group.' That would be a first step toward rational consideration of the problem at issue."[7]

The term *ethnic,* derived from the Greek *ethnos,* refers to a number of people living together who share a common culture and language. In Homer and Pindar *ethnos* is used to indicate a clan, tribe, people, or nation. Substituting *ethnic group* for *race* may not accomplish much by way of combating racial prejudice. After all one may hate an ethnic group as virulently as one does a race. But a change of terms will help us to perceive more clearly the real phenomena with which we are dealing. In the present chapter, for example, we shall be surveying the conditions of the various groups that immigrated to the United States. What distinguishes these groups from one another, as we shall see, is not their racial but their social and cultural characteristics.

The misleading nature of the term *race* may be further illustrated by noting a parallel between antifeminism and race prejudice. As Ashley Montagu has observed, "Almost every one of the arguments used by the racists to 'prove' the inferiority of one or another so-called 'race' was not so long ago used by the antifeminists to 'prove' the inferiority of the female as compared with the male."[8] In the nineteenth century, Montagu continues,

It was fairly generally believed that women were inferior creatures. Was it not a fact that women had smaller brains than men? Was it not apparent to everyone that their intelligence was lower, that they were essentially creatures of emotion rather than reason—volatile swooning natures whose powers of concentration were severely limited and whose creative abilities were restricted almost entirely to knitting and childbirth? For hundreds of years women had played musical instruments and painted, but to how many great female musicians and painters could one point? Where were the great women poets and novelists? Women had practically no executive ability, were quite unable to manage the domestic finances, and, as for competing with men in the business or professional world, such an idea was utterly preposterous, for women were held to possess neither the necessary intelligence nor the equally unattainable stamina. Man's place was out in the world earning a living: women's place was definitely in the home.[9]

The antifeminist argument is identical with the racist one: you deny a particular group equality of opportunity and then look upon that group as biologically inferior because it has achieved less than yours. The parallel between antifeminism and race prejudice brings out rather clearly the misleading character of race. For in the case of both women and other disadvantaged groups, it is their *socially* inherited disabilities and not their biological makeup that accounts for their relative lack of achievement.

Today there are few people who any longer doubt that women can produce great writers and that given the opportunity, they can achieve equally with men in any field. The same truth is increasingly recognized with respect to ethnic minorities. The decisive point is that nobody can know what cultural capacities human individuals and groups possess until they have been given an equal opportunity to demonstrate their capacities.

The term *minorities* has been employed by sociologists to refer to various types of disadvantaged and oppressed groups. Often such groups are in fact numerical minorities, but that is not always the case. We have seen that dominant groups such as governing classes are likely to be comparatively small, and that in many societies the disadvantaged and oppressed constitute the majority. So the term *minority* as employed in this chapter will have a specific meaning: A social category of persons the members of which experience disabilities of various sorts at the hands of advantaged and more powerful groups. Minorities, by that definition, are subject to any of the following disabilities: prejudice, discrimination, segregation, exploitation, persecution, and oppression.

In the history of the United States *ethnic* groups have been the most conspicuous minorities. The term *ethnic* is derived from a Greek word meaning "nation" or "people." An ethnic group is a "collectivity based on presumed common origin."[10] It is a collectivity with a distinct cultural heritage. In the American context most ethnic minorities are the descendants of immigrants who freely left their lands of origin to settle in this country. Irish, Germans, Poles, Jews, Italians, Chinese, and Japanese are examples of such ethnic groups. There are however other minorities in the United States who, though they share a presumed common ethnic origin, must be placed in a different category. Black Americans are a case in point, for their ancestors, far from having been free immigrants, were brought to this country in chains.

American Indians, on the other hand, are natives of this land, a small remnant of a conquered and nearly annihilated people. Both blacks and Indians therefore have unique histories which distinguish them from all other American ethnic groups. Mexican Americans too, though descended from immigrants, are native to this continent. We will now briefly examine the history of ethnic immigration to the United States and then turn to the main subject of this chapter: the disadvantaged minorities of the present era.

ETHNIC IMMIGRATION IN EARLIER ERAS OF AMERICAN HISTORY

The colonial period was of course one of English Protestant dominance. The majority of immigrants during that period were also Protestant, though non-English.[11] The largest single non-English group comprised about 250,000 Scotch-Irish, Presbyterian Scots who had settled in Ulster County, Ireland, in the early seventeenth century and whose descendants began immigrating to the American colonies in the early eighteenth century. The second most numerous European minority consisted of some 200,000 Germans. In addition, there arrived 15,000 Huguenots—Protestants expelled from France when their privileges of worship there were withdrawn following the revocation of the Edict of Nantes in 1685. To those may be added the much smaller numbers of Scots, Dutch, and Swedes; a few small and scattered enclaves of Roman Catholics; and minute communities of Jews in the port cities of Savannah, Charleston, Philadelphia, New York, and Newport.

As is well known, the English majority showed no great tolerance toward the newcomers. In 1698 South Carolina passed an act awarding bounties to new arrivals but excluding the Scotch-Irish (who were referred to simply as "the Irish") and Roman Catholics. Soon afterward Maryland temporarily halted the importation of Scotch-Irish servants, and Virginia imposed similar restrictions. In 1729 Pennsylvania placed a 20-shilling duty on each imported servant. A common fear prevailed that if the Scotch-Irish continued to pour in, they would make themselves proprietors of the colonies.

Many of the newcomers had lived in extreme poverty in their countries of origin and could not even afford to pay the passage fees. Hence they signed an "indenture," or contract, agreeing to work in the New World for a period of three to seven years in order to pay for their passage. In that way immigrants "sold" themselves aboard ship, and not infrequently members of the same family found themselves in the hands of different "masters." If a parent or spouse died, the surviving members of the family had to pay for the deceased person's passage by serving additional time. Children orphaned during the journey remained indentured until the age of 21. Typically, an indentured servant who completed his term of service received tools, money, new garments, and, occasionally, land. The indenturing of servants continued until about 1820, and it is estimated that two thirds of the white immigrants arrived in the colonies in this manner. The colonies, as noted, were thus predominantly settled by two major European minorities, the Scotch-Irish and the Germans, and by other smaller Protestant groups. Those minorities eventually blended with the dominant English settlers whose descendants became increasingly tolerant of the differences within Protestantism.

LATER IMMIGRATION PATTERNS

From 1820 to 1930 more than 37 million immigrants, mostly Europeans, arrived in the United States. While northern and western Euro-   179 peans predominated until the end of the 1880s, it was southern and eastern Europeans who did so after that date. The major factor impelling that great mass of people to uproot themselves and to emigrate was poverty.

The first of the impoverished Europeans to leave their homes in the nineteenth century were the Irish. The Poor Law, the enclosure of the land, and the potato famine of the late 1840s caused untold misery and starvation to millions. Alexis de Tocqueville, who visited Dublin in the 1830s, recorded the most appalling manifestations of poverty even before the Great Hunger caused by the potato blight of the late 1840s. Between 1,800 and 2,000 paupers were received in the poorhouse each day. There he saw

a very long room filled with women and children whose infirmities or age prevent them from working. On the floor paupers lying pell-mell. . . . One finds it difficult not to step on one of the half-nude bodies. In the left wing a smaller room, filled with old and infirm men. They sit on wooden benches, all turned in the same direction and huddled together. . . . They neither speak, nor move, nor look at anything.

As he left the poorhouse, Tocqueville saw two paupers pushing a small closed wheelbarrow. They were on their way to the homes of the rich to collect their garbage and bring it to the poorhouse, so that broth could be made from it. "There is no doubt," he concluded, "that the most miserable of English paupers is better fed and clothed than the most prosperous of Irish laborers."

Such were the conditions which led to a massive outflow of Irish in the 1830s and 1840s. More than a million came to the United States. The vast majority settled just where they landed, in the port cities of New York and Boston, because they were too poor to move elsewhere. With the improvement of conditions in Ireland in the mid-1850s, emigration temporarily diminished. But the new potato blights of 1863 and the 1880s again caused great swells in Irish emigration, so that by the close of the nineteenth century, nearly four million had arrived.

During the same period millions of Germans left their homes for similar reasons. More than five million reached the United States by the end of the nineteenth century, and another two million arrived in the early decades of the twentieth. Though most of those millions were poor peasants, the immigrants also included well-to-do farmers and a variety of skilled artisans and craftsmen. Along with the Irish and Germans came the Scandinavians, with well over two million settling in the United States between 1820 and 1930. The Germans, the Irish, and the Scandinavians were the largest non-English immigrant groups of that period. But there were other, smaller groups as well: Dutch, French Canadians, and Chinese. It was those great masses of newcomers who made the astounding economic growth of the United States possible. They provided the cheap labor for the plowing of the American fields, the building of canals and railroads, the digging of the mines, and the tending of the industrial machines.

The Germans, as we have noted, were the most numerous of the nineteenth-century immigrants. By 1900 they constituted the largest single ethnic minority in 27 states. Wherever they settled they perpetuated their cultural institutions, spoke their mother tongue, and published German-language newspapers. They lent a distinctive cultural flavor to such cities as St. Louis, Cincinnati, and Milwaukee. In New York, where they outnumbered all other foreign-born groups, they created a *Kleindeutschland* (Little Germany) in which life was almost the same as in the old country, and where one did not even need to speak English in order to make a living.

Although all of the new ethnic minorities generally improved their living standards in the United States, they also experienced hardships in the new land. They lived in highly congested urban slums and suffered from unemployment, poverty, and disease. They also met with ethnic discrimination. The best jobs were most often closed to them, and employers frequently posted signs like this one: "Irish need not apply."

After 1890 the immigration patterns changed significantly. Now it was no longer the northern and western but the southern and eastern Europeans who dominated the immigration statistics. The "new immigrants" were Italians, Jews, Poles and other Slavic groups, Magyars (Hungarians), Greeks, Portuguese, Armenians, Syrians, and others. In the period between 1881 and 1910, more than three million Italians and two million Jews entered the United States. The Slavic groups— Russians, Poles, Ruthenians (Ukrainians), Slovaks, Slovenes, Croatians, Serbs, and Bulgarians—accounted for about one quarter of the new arrivals. The Poles, the largest of the Slavic groups, ranked third in numbers behind the Italians and the Jews. More than a million Poles arrived in the United States prior to World War I. To those must be added one million Magyars, 300,000 Greeks, 146,000 Portuguese, 80,000 Armenians, and thousands of Syrians. From the Orient and Hawaii came about 90,000 Japanese. Although the massive influx of immigrants was temporarily halted by World War I, it was resumed in the 1920s, bringing another 350,000 Italians, 300,000 Scots, and 200,000 Germans.

With respect to all the ethnic minorities that entered the United

States from colonial times until the 1930s, a sound generalization may be made: *They eventually improved their economic circumstances considerably.* To cite just one example, Irish Catholics, who were so severely discriminated against in an earlier era, have today joined the educational, occupational, and income elites of the American population.[13] However, the lot of those groups that have immigrated *since* the 1930s has been quite different.

Beginning in the 1930s a new immigration pattern emerged, a major influx of Spanish-speaking minorities from Central and Latin America. As of 1973 there were more than ten million people of Mexican, Puerto Rican, Cuban, and other Spanish-speaking origins in the United States. Of these groups the Mexicans make up 6.3 million, 80 percent of whom live in the Southwest, the majority in urban areas. The Mexican Americans are the most numerous Spanish-speaking group, and they also constitute the largest ethnic minority in the United States after black Americans. With the exception of the Cubans, the vast majority of the Spanish-speaking people in the United States find themselves at the bottom of the socioeconomic pyramid—or in what we have called the "underclass." They therefore share the fate or life chances of the majority of black Americans. American Indians, though native to this continent, are also among the most disadvantaged. The particularly disadvantaged ethnic minorities in the United States require special attention.

BLACK AMERICANS

The historical experiences of black Americans are unique when compared with the experiences of all immigrant ethnic groups. In the first place, black Americans never left their homeland voluntarily. They left chained two by two, left leg to right leg, and that is also the way they arrived upon this continent. Some of them were captured in native wars and sold to black slave merchants who in turn sold them to Europeans. Others were kidnapped by Europeans. Many made forced marches to the African coast where they were examined and branded like cattle. Then they were herded into the holds of slave ships in preparation for the dreadful Middle Passage across the Atlantic. The holds were often no higher than 18 inches, and as one captain remarked, "They had not so much room as a man in his coffin, either in length or breadth. It was impossible for them to turn or shift with any degree of ease."[14] The voyage, lasting many weeks in those conditions, was unbearable. Epidemics of dysentery and smallpox were common. Wedged together al-

182

most immovably and chained by the neck and legs, some captives went mad before suffocating. Many took their own lives. Once they arrived at their destination, they were sold into bondage.

The enslavement of blacks evolved gradually in the colonies. Indeed, a few blacks lived in the South many years before chattel slavery emerged as a legal institution. The first black settlers found a niche for themselves which carried with it no racial prejudice. Some owned land, voted, testified in court, and interacted with whites on an equal basis, and others became servants. At first their status was not unlike that of white servants: They gained their freedom after several years of service or with conversion to Christianity. It is true that even at that early stage their servitude was often more severe, with longer and even indefinite terms. But the greater severity of black servitude merely reflected the fact that as compared with the white indentured servants their coming to America was not a matter of free choice. Nevertheless, as Kenneth Stampp has observed,

The Negro and white servants of the seventeenth century seemed to be remarkably unconcerned about their visible physical differences. They toiled together in the fields, fraternized during leisure hours, and, in and out of wedlock, collaborated in siring a numerous progeny. Though the first southern white settlers were quite familiar with rigid class lines, they were as unfamiliar with a caste system [based on color] as they were with chattel slavery.[15]

With time, however, clear distinctions between white and black bondsmen did appear. Blacks lacked the protection of the English government; and they had no written indentures defining their rights and limiting their terms of service. Relatively unprotected, they were increasingly subjected to special treatment. The enslavement of blacks emerged only after efforts to enslave whites and Indians had failed. White men and women not only enjoyed the protection of the English king; they could escape and blend into the general population. Indians could also escape easily; the terrain was familiar to them, and their tribal brothers were only a forest or a hill away. What is more, the small and weak early white settlements valued the friendship of the neighboring Indians more than their labor. Accordingly, they soon passed acts prohibiting the enslavement of Indians.

In the 1660s the first legal distinctions between white and black servants were made. New laws were passed stipulating that blacks were to be slaves for life. Children automatically inherited the servile status of their parents, and baptism no longer spared one from that status. In the following decades additional laws defined the slaves as property and conferred upon the masters considerable disciplinary power, reinforced by the white governing structure. Interracial marriages were now prohibited. In that manner the African's skin color became by the eighteenth century a badge of slavery and degradation. Chattel slavery had become firmly a part of American custom and law.

In time "Jim Crow" became a synonym for Negro. A great social barrier was erected between whites and blacks. Its purpose was to prevent interracial mixing of any kind and especially intermarriage. Separate drinking fountains, separate privies, separate schools, etc., became the order of the day. White nurses were forbidden to treat black men; white teachers were not allowed to teach black students. In Florida even "negro" and "white" textbooks were segregated in warehouses. In Oklahoma there were separate telephone booths; and in Atlanta courtrooms, Jim Crow Bibles were provided for black witnesses and regular Bibles for white.

Especially interesting is the relatively recent origin of racial segregation. As C. Vann Woodward has shown in his superb book *The Strange Career of Jim Crow,* such segregation never existed in the slavery era. It emerged considerably later. So long as blacks were slaves, their social status was clear and unambiguous. Blacks and whites living and working together entailed no threat to the status and power of the whites. But following the Civil War and the Emancipation Proclamation, when the black man became a citizen, a threat to white status and power was recognized. Something had to be done to keep the blacks in their traditionally subordinate place.

The new social mechanisms for accomplishing that aim were worked out gradually. For several years after Reconstruction, blacks and whites were served in the same inns and buried in the same cemeteries. In 1885 T. McCants Stewart, a black lawyer and reporter, made the first freedom ride through the South. In a railway-station dining room in Virginia he was served at a table with whites. He traveled through Delaware, Maryland, Virginia, the Carolinas, Georgia, and into Florida, all old slave states with large black populations, and found that "a first-class ticket is good in a first-class coach." From South Carolina he wrote: "I feel about as safe here as in Providence, R.I. I can ride on first-class cars on the railroads and in the streets. I can go into saloons and get refreshments even as in New York. I can stop in and drink a glass of soda and be more politely waited on than in some parts of New England."[16]

But that state of affairs had already begun to change. Beginning with Florida in 1887, every Southern state passed Jim Crow railroad laws. In

184

1896, in the case of *Plessy* v. *Ferguson,* the Supreme Court effectively legalized racial segregation. State laws requiring "separate but equal" facilities for blacks, the Court ruled, were a "reasonable" use of state power. The Fourteenth Amendment, the Court added, "could not have been intended to abolish distinctions based on color, or to enforce social, as distinguished from political equality, or a commingling of the two races upon terms unsatisfactory to either." In his dissenting statement John Marshall Harlan wrote that "the judgment this day rendered will, in time, prove to be quite as pernicious as the decision made by this tribunal in the Dred Scott case." The Plessy decision, as Justice Harlan foresaw, did indeed open wide the door to an encroachment on the rights of black Americans. Prior to 1899, only three states required Jim Crow waiting rooms. But "in the next three decades," writes Lerone Bennett, Jr.,

other Southern states fell into line. Only Georgia, before 1900, had required Jim Crow seating on streetcars. North Carolina and Virginia fell into line in 1901; Louisiana in 1902; Arkansas, South Carolina, and Tennessee in 1903; Mississippi and Maryland in 1904 and Florida in 1905.

A Jim Crow mania seized men. . . . The laws came in spurts and waves. Each year brought some new twist or elaboration. Negroes and whites were forcibly separated in public transportation, sports, hospitals, orphanages, prisons, asylums, funeral homes, morgues, cemeteries. Mobile, Alabama, required negroes to be off the streets by 10 p.m. Birmingham, Alabama, forbade negroes and whites to play checkers together.[17]

It was also in the decade of the nineties that the infamous lynchings began on a large scale. Blacks were lynched on the slightest pretext. Worse still, they were burned to death over slow fires and their bodies mutilated.

It is clear that the Fourteenth and Fifteenth Amendments and the Civil Rights Bill of 1875 had failed to ensure the emancipated blacks the most elementary human rights. Furthermore, in the several decades following the Civil War, blacks drifted back into a relationship of dependency on their former masters. Most Southern blacks became share-croppers, that is, tenant farmers. They worked the land owned by the whites for a small share of the total yield, and they were maintained in a subordinate caste position by means of intimidation and physical violence.

Those were the conditions under which the vast majority of black Americans lived and worked in the rural South until recently. They were a predominantly rural people well into the second half of the twentieth century. But increasing urbanization has been a major trend among them. Beginning soon after the turn of the century, they moved from rural areas in the South to urban areas of the North and West; and from the Southern countryside to Southern urban centers. The north-ward migration gained impetus with the entry of the United States into World War I and with the restrictive immigration laws against Euro-peans in the 1920s. This movement temporarily subsided during the de-pression of the 1930s, but it continued and accelerated again when the United States entered World War II. From the end of that war the rural-urban migration has continued so that today black Americans are a preponderantly urban people. It is primarily the urban context, there-fore, that has to be examined if the distinctive disabilities of black Americans today are to be understood.

Urban Blacks and the American Economy

Prior to World War I, black Americans were denied even the most menial jobs in the manufacturing sector. But with the labor shortages of the war and the expanding automobile and steel industries of the 1920s, many blacks obtained unskilled jobs. World War II created even greater opportunities. Acute shortages of labor together with government-subsi-dized training programs enabled blacks to move into a wide range of semiskilled positions. In fact, the stream of black migrants moving from the rural South to the cities throughout the 1940s and early 1950s was easily absorbed by the urban labor market. After 1954, however, certain economic and demographic changes occurred, causing severe unemploy-ment problems for a large section of the black population.

In the 20-year period from 1954 to 1974, the unemployment rate for blacks was twice that for whites. And for young blacks, 16 to 19 years of age, the unemployment rate has reached disastrous proportions. As William J. Wilson has observed:

In 1954, the black teenage unemployment percentage was only slightly greater than the white rate. However, each year since 1966, a greater than two-to-one black-white unemployment ratio has been officially recorded. From 1970 to 1974, black teenagers' unemployment has averaged 32 per-cent, and the 1974 rate of 32.9 percent was close to two-and-a-half times greater than the recorded white teenagers' unemployment.[18]

One reason for the high and steady rate of joblessness among black teenagers has been their rapid growth as a proportion of the population. The number of black youths aged 16 to 19 in American cities increased almost 75 percent from 1960 to 1969, while the number of white youths of the same ages increased by only 14 percent. During the same period, the number of black young adults (aged 20–24) increased by 66 percent, which was three times the growth rate for white young adults. But the explosive growth of black youths is only one side of the problem, since that growth coincided with structural changes in the economy.

Following World War II technological and economic changes led to the wide dispersion of industry and other businesses. Improved transportation and communication facilities made it economical to use large and inexpensive tracts of land for manufacturing, wholesale, retail, and residential development. Firms previously located in the central city owing to their reliance on port, freight, and passenger facilities now turned to truck transportation, thus enabling them to locate in outlying districts. The phenomenal expansion of automobile travel and the metropolitan expressway system meant that firms no longer needed to locate themselves near public-transportation facilities in order to ensure themselves a work force. As a consequence of those developments, we find that in the nation's 12 largest metropolitan areas, "the central city's proportion of all manufacturing employment dropped from 66.1 percent in 1947 to less than 40 percent in 1970." [19]

That shift in the *proportion* of industrial employment does not mean that a large number of central-city business moved to the suburbs. The problem is not so much that the central city has lost industry, but that most economic growth has occurred in the suburbs, resulting in shrinking opportunities for better-paying jobs in the central city.

Changes in Skill and Education Requirements

At the same time other significant changes have taken place. In an earlier era the low skill requirements of American mass-production manufacturing made it possible to absorb large numbers of immigrant and other workers in industry. However, the demand for unskilled and poorly educated workers has steadily declined, while skill and educational requirements have continually risen. That has led to an imbalance between skill requirements and the skill levels of the work force. Among potential employees, it is the relatively unskilled blacks (and other minorities) who are hardest hit by this imbalance. The result is that young blacks who are willing and able to work become discouraged and abandon an active search for employment. Thus the Department of Labor reported that "among persons not in the labor force in early 1975, 4.4 percent of the black and other minorities and 1.6 percent of whites were discouraged workers." [20]

The central cities do not have an absolute shortage of jobs but rather a deficiency of stable, well-paying, and preferred types of employment. A study of the labor market in Boston, for instance, "showed that employment in hospitals, hotels, warehouses, building maintenance services, industrial sweatshops, and so forth, is readily available to the disadvantaged. But attractive high wage employment—jobs which constitute, in the current language of manpower, 'meaningful employment opportunities'—were much less accessible." [21]

The Dual Labor Market

Some economists and sociologists have argued that today's American labor market is divided into two sectors, a primary and a secondary one. Wilson has summed up the difference between them this way: "Jobs in the primary labor market are distinguished by stable employment, high wages, good working conditions, advancement opportunities, and due process and equity in the administration of work rules. On the other hand, jobs in the secondary sector tend to be dead-end and to have high labor turnover, low wages, inadequate fringe benefits and capricious and arbitrary supervision." [22] The "dual labor-market" theory has clear implications for black workers as well as those from other central-city minorities who suffer from such social disabilities as poor education and training. It is they who are most likely to be imprisoned in the secondary labor market.

Evidence also suggests that there are many black workers qualified for the more desirable blue-collar jobs of the primary sector. However, studies of discrimination in employment in 30 of the nation's largest cities have shown that blacks encounter discriminatory barriers to employment in this sector. Another important finding was this: The greater the role of the private business sector in city employment, the higher is the unemployment rate in the slums. Conversely, the greater the role of government in city employment, the lower is the slum unemployment rate. The difference may be accounted for by the fact that blacks have found it easier to obtain desirable jobs in federal, state, and local government than in the private sector, which is dominated by a unionized white labor force that has consolidated its hold on the best blue-collar jobs. [23]

What we find, then, is that the primary sector has become the preserve of privileged white workers while the secondary sector has become predominantly nonwhite. The fact that the mainly menial and least desirable jobs are available to nonwhite workers has had a profound impact on their attitudes. It has lowered their morale and self-confidence, often prompting them to abandon the search for employment. Indeed, inner-city black men have the same contemptuous attitude toward menial, dirty, and low-paying jobs as do most members of American society. When the only jobs available are those commanding no respect and carrying no prestige or opportunities for advancement, it is no wonder that such jobs are shunned. Doubtless that is a new attitude on the part of young, poor, black Americans, whose fathers and grandfathers were more inclined to submit to a life of menial labor. But the civil-rights movement of the 1960s brought about a revolution in black consciousness. New attitudes of self-respect and dignity now prevail, attitudes which are difficult to maintain when low-status tasks such as sweeping, dishwashing, and garbage collecting are the only ones available. In that light it is not difficult to understand why so many young blacks have rejected low-wage, menial employment in favor of lucrative illegal activities. It was estimated that in 1966, 20 percent of the adult residents of Harlem derived their income from numbers, gambling, narcotic sales, prostitution, and the like.[24]

The foregoing analysis enables us to see why the term *underclass* is an apt description of the socioeconomic condition of the majority of black Americans. Poor and ghettoized, they are ensnared in what appear to be hopeless circumstances. To change those circumstances, a deliberate and concerted reorganization of the inner-city economies will be necessary.

PUERTO RICANS

At the close of the Spanish-American war in 1898, Spain ceded Puerto Rico, a small Caribbean island 35 miles wide and 100 miles long, to the United States. In 1917 Puerto Ricans were granted American citizenship, and throughout the twentieth century they have been migrating to the mainland. By 1930 the census recorded 53,000 who had moved stateside. The Puerto Ricans, like so many other groups, were coming to the United States to escape conditions of misery.[25]

In the 1930s, even after 35 years of American administration, Puerto Rico was a scene of abject poverty. In the words of Rexford Tugwell, the American governor of the island in the early 1940s, it was a "stricken land." The tiny island held a population of 1.75 million. Though the death rate had been reduced to 20 per thousand, the birth rate remained among the highest in the world. The island's economy was based on sugar. With the Great Depression, the market for that cash crop collapsed, causing widespread unemployment and adding substantially to the already extreme suffering of the people. The vast majority of the population consisted of agricultural workers who even before the depression lived in conditions no better than those in India and China. "Sanitary facilities," writes Nathan Glazer,

were primitive; shoes were rarely worn in the country districts; the ground was infested with sewage and parasites and so, too, was the population; and a prevalent malnutrition produced a stunting of growth and susceptibility to a wide range of diseases. The details of the infant mortality rate, death rates from various causes, all showed the effects of a grinding poverty that is scarcely imaginable in contemporary industrial countries. Most of the population was unemployed and underemployed and suffered from hunger.[26]

The flow from the island to the mainland was cut off during the depression and World War II, but it resumed and swelled in 1945. The introduction of cheap air service between San Juan and New York made it possible for thousands upon thousands to leave the island in search of better opportunities. They were attracted by the postwar abundance of unskilled and semiskilled jobs available in New York City. In 1940 New York had about 60,000 Puerto Ricans. By 1950 there were a quarter of a million, and the 1960 census recorded about 613,000 of Puerto Rican birth or parentage in the city. In the early 1970s there were some 1.5 million Puerto Ricans in the United States, with perhaps as many as two thirds of them concentrated in the New York area. Other major Puerto Rican centers were Chicago with about 80,000 and Philadelphia with at least 27,000. But there are also Puerto Rican communities in several cities of New Jersey; in Rochester, New York; Bridgeport, Connecticut; Dayton, Ohio; Boston, Massachusetts; Miami, Florida; and Milwaukee, Wisconsin.

It is clear that in the United States the Puerto Ricans have become an urban people. As such, they share the conditions of America's urban poor. In New York and other cities they have moved into the lowest-level jobs and into the worst slums. Like the European immigrants of a previous era, they speak a foreign language and share a common ethnic culture. But unlike European immigrants, Puerto Ricans face a color problem. Reflecting centuries of mixing between the black and white populations of the island, the Puerto Ricans in the United States are black, intermediate, and white. Although a dark skin in Puerto Rico does not evoke the prejudice and discrimination which it has historically in the United States, darker complexions are nonetheless associated with lower status. On the mainland, however, the Puerto Ricans discovered that a dark skin can be a severe handicap. Hence for all their similarities with the European immigrants, the Puerto Ricans are in a plight like that of inner-city black Americans and indeed appear in several respects to be even worse off.

Data were gathered in the 1970 Census Employment Survey for seven poverty areas in New York City. The survey revealed that the unemployment rate for Puerto Rican males was 1.3 times higher than that of all poverty-area males. Of all Puerto Rican men employed, only 7 percent were in professional, technical, and managerial jobs, substantially below the percentage for all poverty-area residents. The salaries of full-time working Puerto Rican men were 88 percent of the poverty-area average. About a fourth of the two-parent families and 82 percent of the families headed by women were on welfare; the corresponding figures for all poverty-area families were 13 and 65 percent, respectively. Undoubtedly the Puerto Ricans are among the most disadvantaged and impoverished of American groups.

Labor-force participation has declined among young Puerto Ricans much as it has among young blacks. Among men aged 25 to 34, participation rates fell in the 1960s from 92 to 87 percent. And while unemployment rates for all age/sex cohorts dropped in the 1960s, the rate rose for Puerto Rican males from 5.6 to 11.1 percent between 1970 and 1973. In that period the proportion employed in labor and service occupations grew by several points while that in professional, technical, managerial, and craft jobs correspondingly fell. [27]

On the other hand, the Puerto Ricans appear to have a well-developed small-business tradition. As of 1964 they owned more than 6,000 small businesses in New York City, including groceries, barber shops, and dry-cleaning stores. That figure is considerably higher than the number established by blacks, though the latter are a much larger group whose major influx to New York preceded that of the Puerto Ricans by some 30 years. For the most part the Puerto Rican small merchant services the needs of his own ethnic community. He provides for distinctive needs in food, books, records, and other consumer goods. But the small stores also service a wider clientele. They have shown remarkable vitality and growth in the face of fierce supermarket and chain-store competition. In the last two decades a conspicuous number of Puerto Rican–owned "superettes" have appeared on New York's West Side. [28]

Yet the entrepreneurial tradition, strong as it may be, does not change the general picture. The annual median income of New York's Puerto Ricans was $4,969 in 1970, lower than that of any other urban minority. Their unemployment rate was roughly equal to that of blacks.

191

One third of Puerto Rican families live in poverty, and one out of every two families is on relief. Educational deficiencies likewise parallel those of blacks. It therefore appears clear that the Puerto Ricans in New York and other American cities are subject to some of the same economic forces as black Americans. Largely closed off from the corporate industrial labor market, they are confined to the least desirable sources of employment in the secondary labor market. To that extent they share the fate of the huge black underclass.

MEXICAN AMERICANS

Like the native peoples of the continental United States—the so-called American Indians—Mexican Americans or Chicanos are also descended from peoples who were indigenous to the North American continent. They lived in the American Southwest long before it came under United States rule. The homeland of Mexican Americans is not only the present country of Mexico but rather that vast stretch of territory which today consists of Mexico, Arizona, Colorado, Texas, California, and New Mexico.

In 1848 there were about 75,000 Mexicans in the area referred to today as the American Southwest. Since the vast majority of today's 7 million Mexican Americans settled the area after that date, they are not for the most part the physical descendants of that region's early settlers. They are however the cultural heirs of the Southwest. The experience of Mexican Americans has been quite unlike that of immigrants to the United States. They have not uprooted themselves from their native cultures in quite the same way as the immigrants. From a cultural standpoint, the border between Mexico and the Southwest is of doubtful significance. Until recently the border was unguarded, and Mexicans had easy access to El Paso, Tucson, San Antonio, Los Angeles, and other Southwestern cities. But the migrants have settled and continue to settle in established Chicano communities, and this enables them to preserve their traditions, language, and values.[29]

The Southwest was originally coveted by both Mexico and the United States.[30] The first Anglo-American colony was established in Texas in 1821. In 1835, ostensibly resisting Mexican domination, the Texans rebelled. Though they sustained initial defeats at the Alamo in San Antonio and in Goliad, the Texans eventually overwhelmed the Mexican forces at the Battle of San Jacinto in 1836. With that victory the Texans declared their independence from Mexico, a status which the Mexican government refused to recognize. In 1845 statehood was granted to Texas, whereupon Mexico broke off diplomatic relations with the United States. Hostilities between the two countries intensified, soon leading to a declaration of war by the United States on May 13, 1846. The United States decisively defeated the Mexican forces, and with the signing of the Treaty of Guadalupe Hidalgo on February 2, 1848, the war officially ended.

Mexico now recognized the American annexation of Texas and at the same time ceded most of what is present-day Colorado, New Mexico, Nevada, Arizona, and California. The treaty guaranteed that all of the Mexican citizens in the ceded territories would be permitted to retain their property and traditions; those who chose not to leave within a year would be accorded American citizenship with all of its rights and privileges. Ever since that time the history of Mexicans in the United States has been tied to the Southwest; and despite the treaty guarantees, the experiences of Mexican Americans have been those of a subordinate, persecuted, and oppressed people. Vigilante groups such as the Texas Rangers were founded to "keep the Mexicans in their place," and there were more lynchings of Mexican Americans in the Southwest from 1865 to 1920 than of blacks in the Southeast during the same period.[31]

As the Southwest developed economically, the Mexican influx to the region mounted steadily. The construction of the Southwestern railroads, the expansion of cotton planting in Texas, Arizona, and California, and the irrigation of farmlands in the Imperial and San Joaquin valleys in California all required abundant cheap labor, which the Mexican workers provided. In the early twentieth century Mexicans made up the majority of the common laborers laying the railroad tracks, working the mines of Arizona and New Mexico, and picking and packing the fruit crops of Texas and California. They also cultivated sugar-beet fields in such faraway states as Montana, Michigan, and Ohio.

Prior to 1910 most of the Mexican migrants were temporary laborers in the United States. But with the revolutionary upheavals in Mexico of that year, many permanent settlers arrived. Although the overwhelming majority consisted of poor agrarian workers, the new migrants also included artisans, professionals, and businessmen whose property had been lost during the revolutionary violence.

The migration was spurred by other factors besides the revolution. From 1877 to 1910, Mexico's population increased from 9.4 million to 15 million without a commensurate increase in the production of means of subsistence. A handful of *hacendados* (rich and powerful landlords) controlled most of the country's land, which was tilled by landless agricultural laborers. A patron-peon relationship existed between the hacienda owners and their laborers, with the latter held in a state of servitude. Even when the Mexican economy expanded and prices rose, laborers' wages remained constant and even declined to a level insufficient for the support of their families. However, at the beginning of the twentieth century, the construction of the Mexican Central and Mexican National railroads and the opening of mines in northern Mexico attracted large numbers of laborers. Moving from central and eastern Mexico, many saw no reason to stop at the border since wages in the United States were at least five times as high as in Mexico. About 10 percent of Mexico's population thus moved into the American Southwest.

Mexican Migrants in the Twentieth Century

The earliest Mexican migrants in the twentieth century were for the most part men, transient laborers working in railroad gangs and living in boxcars. They moved from place to place laying tracks for the Southern Pacific, Santa Fe, Chicago, Rock Island, and Pacific railroads. Their daily wage, $1 to $1.25, was less than that of their Greek, Italian, and Japanese predecessors. The major Southwestern Railroad employed more than 50,000 Mexicans. Many of the Mexican-American *colonias* (settlements) of today originated as railroad labor camps.

With the outbreak of World War I, European immigration almost ceased and Americans went off to war. To meet the labor needs of American farms, more Mexicans, hitherto ineligible for immigration visas, were brought into the country. The economic depression of 1921–1922 threw many of them out of work, but the recovery created a further demand for Mexican labor. The large agricultural growers of the Southwest successfully pressed Congress to exempt Mexicans from the existing immigration restrictions. But legal immigration entailed visa and medical-examination costs, proof of literacy, and other formal hurdles. Such formalities tended to deter many who, knowing that the border was open, preferred to enter the United States illegally. (It was not until 1924 that Congress appropriated funds for border patrol.) It is estimated that during the 1920s there were at least as many illegal immigrants as there were legal ones—about 500,000.

The depression of the 1930s severely curtailed Mexican immigration. Many Mexicans, together with their American-born children, were encouraged and often forced to return to Mexico. Between 1929 and 1940 nearly a half-million persons, or more than a third of the Mexican-

American population in 1930, were compelled to return. Those who remained suffered extreme deprivations. Agricultural wages throughout the Southwest fell from 35 to 15 cents an hour. Mexican farmhands earned as little as 60 cents a day; in California in the late 1930s, the average annual income of Mexican families was $254. Worse still, they were denied work altogether whenever Anglo-Americans were available. In fact by 1939, 90 percent of California's field workers were dust-bowl refugees who had replaced Mexican Americans.

That state of affairs changed only with the coming of World War II, and the vast employment opportunities it created. Large numbers of Mexican laborers moved to the cities where they found jobs in airplane-manufacturing plants, shipyards, and other war-related industries. They also moved to the urban centers of the Midwest, filling vacancies caused by the massive conscription of American men. No less starved for workers was the agricultural Southwest. The general labor shortage prompted the United States to inaugurate a new program, the importation from Mexico of contract laborers, known as *braceros.* In accordance with the agreement between the two countries, Mexican laborers came into the United States for temporary, seasonal work and then returned to Mexico at the close of the season. From 1942 to 1947 there were about 220,000 *braceros* in the United States. They worked in 21 states, more than half laboring in California. However, because Texas practiced extreme discrimination against Mexicans, the Mexican government refused to allow any of its nationals to work there.

The first *bracero* program ended in 1947 but was renewed in 1951 in response to the pressure of Southwestern growers. The renewed program continued until 1964. Thus cheap labor from south of the border helped build up the multi-billion-dollar agricultural corporations of the Southwest. The program was terminated for several reasons. The Southwestern growers had introduced mechanization, which considerably reduced the need for manual labor; labor costs rose with the dollar-an-hour minimum wage of 1962; and the Mexican government was anxious to terminate the agreement because Mexican growers had their own expanding need for cheap labor. By the time the *bracero* program ended in 1964, almost five million *braceros* had worked in the United States.

Besides the *braceros,* or legal migrant laborers, the Southwestern growers also employed illegal Mexican immigrants—the *wetbacks*—a term derived from the fact that many Mexicans waded across the Rio Grande River, separating Mexico from Texas, during the dry seasons when the water was shallow. Though they had not been selected for the *bracero* program, they were equally impoverished and equally in need. 195

The conditions they met in the United States were appalling: wages of 20 to 30 cents an hour, shacks without plumbing or electricity, washing in irrigation ditches. From the growers' standpoint the wetbacks were ideal laborers. They cost next to nothing, and since they feared disclosure of their illegal status, they remained compliant and obedient. More than four million wetbacks were apprehended in the United States between 1947 and 1954. How many escaped detection by the authorities is anyone's guess.

Chicanos and Anglo-American Society

The Mexican laborers who settled in the United States joined the existing Southwestern Mexican communities, in which they felt ethnically at home. Newcomers had no need to change their faith, language, or culture. The continual influx of newcomers reinvigorated those Mexican enclaves in the United States so that assimilation into American culture was considerably retarded. It was further retarded by Anglo-American discriminatory practices.

Discrimination against Mexicans has existed throughout the Southwest. Chicanos have been confined to their *barrios* (ghettos) even when they have preferred to live elsewhere, excluded from many public recreational facilities, and in general kept in a subordinate position. For the most part, only the most menial and least desirable jobs have been available to them. Although the forms and intensity of discrimination have varied from one Southwestern state to another, Mexicans have borne the brunt of extreme bigotry in California and even more so in Texas.

In the Lone Star State, Mexicans were routinely refused service in food shops and restaurants; their children were referred to as "greasers"; and many churches held separate services "for Colored and Mexicans." Anti-Mexican violence was also common in Texas. The Texas Rangers "often killed Mexicans who had nothing to do with the criminals they were after. Some actually were shot by mistake. . . . But perhaps the majority of innocent Mexicans who died at Ranger hands were killed much more deliberately than that."[32]

During World War II, as was earlier noted, Mexican Americans enjoyed new opportunities for employment and mobility and were hired for jobs previously denied them. Although the new opportunities momentarily strengthened their faith in the American system, several incidents soon demonstrated that attitudes toward Chicanos had not in fact changed. In 1942 the members of a Mexican teenage gang were arrested and convicted of murder even though the prosecution presented no evidence at the trial to justify their conviction. Existing community prejudices, however, combined with the unkempt and disheveled appearances of the youths (the prosecuting attorney refused to allow them to bathe or change their clothes during the first week of the trial) sufficed to bring forth a guilty verdict. . . . Inability to raise bail forced the defendants to spend two years in San Quentin prison before an appeals court unanimously reversed the lower court's decision "for lack of evidence" and reprimanded the trial judge for his injudicious behavior during the proceedings.[33]

A second incident was California's "Zoot Suit Riots" of 1943, involving a number of Mexican-American youths who sported the then fashionable "zoot suit," distinguished by a long coat with wide shoulders and a loose back, baggy trousers with a high waist and tight cuffs, and a wide-brimmed flat hat to match. On the evening of June 3, 1943, a group of sailors was attacked in a slum area enclosed by the Mexican *barrio*. The sailors reported the incident to the police, claiming that their assailants had been Mexicans. The police investigated the matter but found no one to arrest. The next night 200 sailors went into the Mexican district and began beating up every zoot suiter in sight. The beatings were repeated for several nights while the police did nothing to stop them. *Time* magazine described those events as "the ugliest brand of mob action since the coolie riots of the 1870's.[34] And in 1948 public officials in Three Rivers, Texas, refused to allow a Mexican-American veteran to be buried in the local cemetery. Those events aside, the recently hired Mexican Americans found soon after the war that they were being steadily replaced by returning Anglo-American veterans, while Mexican-American veterans had no jobs awaiting them.

During the 1960s Mexican Americans made modest income gains. From 1959 to 1969 the median income for Chicano males in the Southwest rose to about 75 percent of the national male median. For Chicano females it rose to almost 90 percent of the national female median. But in the ensuing years (1969–1973), the income of Chicano males remained at 74–75 percent of the white male median, while for Chicano females income dropped to 85 percent of the national female median. In 1972 the median annual income for Chicano families was $8,759; that was higher than the $6,864 for black Americans but substantially lower than the $11,549 for white or Anglo-Americans. In San Antonio a study indicated that life expectancy for Chicano males was 2.6 years less than that for Anglos, and that the differential between Chicano and white females was 6.6 years.[35] In Colorado the average age at death of Chicano males was 57, as compared with 67 for Anglos. Infant mor-

197

tality, perhaps the most sensitive index of living standards, was twice as high among Chicanos.

Mexican Americans have been less dependent on welfare than other disadvantaged minorities. In 1970 one in nine Chicano families received public assistance, as compared with one in four Puerto Rican, one in five American Indian, and one in six black families. The main problem for Chicanos is the same as for blacks: They end up in the most menial and lowest-paying jobs. As Levitan et al. have observed:

Overrepresented in the lowest rungs in each occupational category as well as in the lowest-paying establishments within industries, Chicanos earn less. In 1970, earnings of Spanish-surname males were 74 percent of the Anglo median in California and 79 percent in Texas. Chicano male professionals earned 17 percent less than Anglos in California and 30 percent less in Texas; operatives earned 7 and 28 percent, respectively; similar differentials occur in all broad occupational categories except laborers.[36]

Chicanos also suffer from educational deficiencies more serious than those of the other minorities. In 1970 their median-average number of school years completed was 8.1, compared with 9.8 for blacks and Indians and 8.7 for Puerto Ricans.

In the 1950s Chicanos were a preponderantly rural group, but by 1970, 84 percent of Mexican Americans lived in metropolitan areas and 47 percent in central cities. Urbanization has brought with it certain improvements. In 1969 the mean-average income of urban Chicano males was a third higher than for rural nonfarm males and three-fifths higher than for farm workers. The urban mean of females was also substantially higher: 40 percent above rural nonfarm workers and 52 percent above farm workers. As for professionals, managers, and craftsmen, the proportion for Chicano males rose from 21 percent in nonfarm rural areas to 32 percent in the cities. The corresponding figures for women were 25 and 35 percent respectively. Finally, we find that urban Chicanos aged 25–34 have completed 3 to 4 more years of school than their rural counterparts.

But Mexican immigration continues to play an important role in determining the income levels of the Chicano community. In 1971 there were 50,000 immigrants compared with the annual average of 40,000 from 1965 to 1968. The vast majority (80 percent) planned to reside either in California or in Texas and were only qualified for unskilled employment as laborers or service workers. Along with constant immigration, a form of the *bracero* program continues. In 1972 about 735,000 "green carders" were legally admitted to the United States, presumably to fill labor shortages. Green carders may move back and forth between Mexico and the United States at will. A very large proportion do so daily, living in Mexico but working in the States, thus benefiting from the lower living costs in Mexico and paying no United States income taxes. The net result for Mexican-American workers is a marked depression of wage levels.

In addition, illegal immigration, the true magnitude of which nobody knows, continues to be a powerful wage-depressing factor. In each of the years from 1965 to 1972, an average of 200,000 illegals were returned to Mexico, the number reaching as high as 400,000 in 1972. It is estimated that for every illegal returned, another remains in the States undetected. If that is true, the consequences are far-reaching. As Levitan et al. have remarked: "If there are, indeed, more than 400,000 illegals in the Southwest, and if most are workers, this has a massive impact, since there were only 683,000 employed persons of Spanish [speaking] origin in the area in 1970."[37]

In the 1960s the social and economic disabilities of Mexican Americans gave rise to an activist Chicano movement whose aim was the general advancement of the entire ethnic minority. César Chavez, probably the best-known leader of the movement, led the California grapepickers in a five-year strike which in 1970 ended with higher wages and better working conditions. There are other less-well-known but equally determined leaders who continue to work for the interests of Chicanos and other oppressed and highly exploited working people. These leaders believe that the way in which American society will reply to Chicano and other demands for equity will be a good gauge of its "responsiveness and openness."[38]

AMERICAN INDIANS

Archeological evidence indicates that the American Indians are descended from the peoples of northeastern Asia who discovered the continent some 20,000 years ago.[39] Anthropologists estimate that in 1492 there were between 700,000 and 1,000,000 Indians in the area now comprising the United States.[40] Yet this most native of American peoples was treated as a minority from the moment that the first European settlers arrived. The Indians, though romanticized as "nature's noblemen," were regarded as inferiors by the overwhelming majority of the European newcomers of the colonial era. Though they attempted in their own way to deal justly with the Indians, they never acknowledged their legal rights to the land upon which they had dwelt for so many centuries.

Nor did the English colonists ever doubt that the New World belonged to the crown, and that the king alone had the authority to bestow land upon his subjects. That attitude, together with the firm conviction that the Christian white man was superior to the pagan red man, led to social conflicts from the start. The first experiences between the Pilgrims and the Indians set the stage for future relations between the native peoples and the settlers.[41] As the European settlements grew, they increasingly appropriated the tribal territories, driving the Indians farther and farther inland.

Soon after the American Revolution, the new government inaugurated a policy of negotiating treaties with the Indians, who were required to cede their tribal lands and move west of the Mississippi. When the Indians refused to move, military force was used against them and warfare ensued. The Seminole War in Florida and the Black Hawk War in the Illinois territory are among the best-known struggles carried on east of the Mississippi.

The Seminole resisted from 1835 to 1842, costing the United States some 1,500 troops and $20 million. Many tribes of the Iroquois confederacy sought sanctuary in Canada while the Oneida and the Seneca were moved westward, though fragments of Iroquois tribes remained in western New York State. In Illinois the Sac and the Fox made a desperate stand, but the survivors eventually succumbed and moved. Expulsion from their ancestral lands was also the fate of the Ottawa, Potawatomi, Wyandot, Shawnee, Kickapoo, Winnebago, Delaware, Peoria, Miami, and numerous others who are today remembered only in the name of some town, lake, county, or state.

The history of the Cherokee is particularly interesting for the light it sheds on government policy. More than any other Indian group, they had attempted to adopt the ways of the conqueror. In about 1790 they began to modify their traditional culture, establishing churches and schools and practicing agriculture and other European crafts. By 1826 the Cherokee lived in European-type houses and raised thousands upon thousands of cattle, swine, and sheep. They also had their own looms, spinning wheels, plows, sawmills, gristmills, and blacksmith shops. "In one of the Cherokee districts alone," writes Peter Farb,

there were some 1,000 volumes "of good books." In 1821 . . . a Cherokee named Sequoyah . . . perfected a method of syllabary notation in which English letters stood for Cherokee syllables; by 1828 the Cherokee were already publishing their own newspaper. At about the same time they adopted a written constitution providing for an executive, a bicameral legislature, a supreme court, and a code of laws. [42]

Yet all of those efforts availed them nothing. With mounting white pressure for Indian lands, the Cherokee were treated like all the others. On December 19, 1829, the Georgia state legislature passed an act appropriating the bulk of Cherokee lands. The Cherokee appealed to the federal government, but President Jackson was totally unsympathetic. Indeed, the idea that the Indians were independent nations with a right to their lands he regarded as absurd. The treaties, he believed, were a farce. He enthusiastically supported moving the Indians to new lands west of the Mississippi; and he persuaded Congress to pass the Removal Act of 1830, giving the president the authority to expel and, in effect, to extirpate all those Indians who had managed to survive east of the Mississippi River.

With the failure of their appeal, the Cherokee were uprooted and forced west over the 1,000-mile "trail of tears." It was estimated at the time that 4,000 died en route. The fate suffered by the Cherokee and numerous other tribes was the direct result of government policy to push the Indians to the western plains and to confine them there. But the western migration of whites created ever more pressure for the expropriation of the new reservation lands, and so the terrible tragedy of expropriation and annihilation continued west of the Mississippi.

By the 1870s it became widely recognized among the Indians that there would be no end to the white man's expansion, and that his treaties concerning the newly allotted reservation areas in the West were worthless pieces of paper. That is the era in which they made their last stand, an era celebrated in countless cowboy movies. A series of uprisings and rebellions now broke out—in the southern plains in 1874, of the Sioux in 1876, the Nez Percé in 1877, the Cheyenne and the Bannock in 1878, the Ute in 1879, the Apache numerous times in the 1870s and 1880s, and the Modack, in northeastern California, in 1872–73. The wars against the Indians were waged with the intention of getting rid of them. In 1864, at Sand Creek in Colorado, militiamen attacked an encampment of Cheyenne who had been guaranteed safe conduct and massacred most of them. And in the winter of 1890, United States forces armed with Hotchkiss machine guns slaughtered nearly 300 Sioux at Wounded Knee, South Dakota.

The costliness of those wars, both in human life and resources (in 1870 it was estimated that it cost the federal government $1 million for every dead Indian), gave rise to a new policy, the Dawes Act of   201

1887. Those who were sympathetic to the plight of the Indians, including Senator Henry L. Dawes, the act's sponsor, hoped that the policy would provide the remnant of Indian populations with the dignity and security of land as private property. The plan was widely supported in the halls of Congress, in the press, and in meetings of religious associations. Unfortunately, however, the act had loopholes which also made it attractive to those whose hunger for Indian land was unabated. Under the Dawes Act the president was authorized to break up tribal lands and to parcel them out to individual Indians. Each adult was to receive 160 acres and each minor 80 acres, the aim being to make the recipients into small industrious freehold farmers. The tribal lands remaining after distribution were declared "surplus" and put up for sale, with the proceeds presumably to be spent by the Department of the Interior for the benefit of the Indians. From the Indian standpoint the breaking up of tribal lands was foreign and shocking. The Creek, Choctaw, Chicasaw, Cherokee, Seminole and several others resisted the policy and were exempted from it. The act failed in both of its aims: splitting up the tribes, and converting them into small farmers who would then more easily assimilate into the surrounding society. Instead, the act's net effect was once again to expropriate their lands and to impoverish Indians to an extreme degree. Between 1887 and 1932, approximately 90 million acres out of the 138 million initially held by the Indians passed into the hands of white owners. Ninety thousand Indians were now landless.

In the 1930s a new policy emerged, reversing the previous one and encouraging the Indians to retain their tribal cultures. The allotment of the remaining Indian lands ceased, and effort was made to restore a land base for Indian communities. The new policy also encouraged tribal government and native arts and crafts; it fostered social services on tribal lands and restored freedom of religion for Indians.

In the 1950s, however, that policy was again reversed as Congress resolved to terminate federal relations with the tribes as speedily as possible. "Termination" proved to be a disaster for all those who felt its impact. Indians now lost any special standing they had under federal law. The tax-exempt status of their land was discontinued, and federal responsibility for Indian economic and social services was ended. Wherever it was applied, termination led to the undermining of Indian communities and to great suffering and bewilderment.

Four bands of Paiute were the first to have their relations with the Bureau of Indian Affairs (BIA) terminated. The Shivwit, Koosharem, Indian Peaks, and Kanosh bands of the Paiute held some 46,000 acres of land. They were terminated in 1957. A year or so later their situation had become desperate, and they were asked why they had accepted the new policy in the first place. A Kanosh spokesman replied that they had not understood what was happening at the time of the hearings. Upon termination, custody over Paiute land was transferred from the BIA to trustees of the Walker Bank and Trust Company in Salt Lake City, 160 miles away. "The Paiutes," write Aberle and Brophy,

> had difficulty getting transportation to the bank and then communicating with the trust officer. An Indian Peaks man said that they finally collected enough money for gasoline to go to Salt Lake City, where they saw the trustee, but they could not understand his remarks and after a few minutes were shown out of his office. They also tried unsuccessfully to get advice from an attorney appointed by the bank and paid with Paiute money.[43]

Termination for the Paiute as for the other tribes meant that they now had somehow to survive in the larger society, though they had neither the resources nor the experience which such survival required. Not surprisingly, the termination policy was widely opposed by the Indians, who were now confronted with the destruction of what little they had been able to preserve of their old way of life. In response to Indian protests, the federal government first slowed the program during the Kennedy-Johnson administrations and then abandoned it during Nixon's presidency.

American Indians Today

In 1970, 827,000 people identified themselves as Indians, Eskimos, or Aleuts. Of those, it was estimated by the BIA in 1973, 543,000 lived on or near reservations and were eligible for its services. More than 75 percent of all American Indians are located west of the Mississippi, with the highest concentrations in Arizona, New Mexico, Oklahoma, California, and Alaska. Although 481 tribal groups have been identified in the United States, the vast majority of Indians belong to ten tribes, the largest of which are the Navajo, Cherokee, Sioux, and Chippewa.

The social, economic, and cultural conditions of those tribes vary greatly. The Navajo, the largest tribe, residing on or near their reservation, have preserved their tribal traditions while participating little in the surrounding white society. In contrast, only a few of the Cherokee, the second-largest tribe, live on their reservation in North Carolina. With family incomes and education levels almost twice as high as those of the Navajo, the Cherokee are far more assimilated into the larger society and culture.

Following World War II the conversion to peacetime production and the general growth of employment opportunities lured increasing numbers of Indians to the cities. Between 1960 and 1970 the urban Indian population more than doubled, from 165,000 to 340,000. By 1970 Los Angeles, Tulsa, Oklahoma City, and other cities had larger Indian populations than any reservation except that of the Navajo. A comparison of those who have left the reservations with those who have remained behind presents a significant contrast. Urban Indians have more formal education, far lower unemployment rates, and much larger average family income. Indeed, as Levitan et al. have noted, "By such standards as family income, male labor-force participation, percent of high school graduates, dependents per breadwinner, and poverty status, Indians in metropolitan areas are better off than metropolitan Blacks."[44]

In contrast, reservation Indians are economically worse off than any other minority group in the United States. In 1969 the median family income for rural Indians was only $4,088, and 55 percent lived in poverty. Most reservations are "open-air slums." In 1973 a BIA study disclosed that out of 107,000 Indian families, 66,000 needed housing assistance. And according to the 1970 census, 44 percent of all rural Indian families lived in housing with more than one person per room. Fifty percent had no bathroom and a third no indoor running water. Two thirds of all rural Indian dwellings were valued at less than $7,500. The lot of reservation Indians would be even worse were it not for government-created employment. Forty-six percent of all jobs on Indian reservations are provided by state, local, and federal agencies. That percentage is more than three times the national average for government employment. In the private sector, employment rates for Indians are far below the average for other groups.

In education, on the other hand, both rural and urban Indians have fared better than other minorities—though they still lag considerably behind whites. In 1970, among those 25 years of age and older, 33 percent of Indians had graduated from high school. That compares with 31 percent for blacks and 55 percent for whites. However, though enrollment rates for Indians aged 5 through 17 have approached national averages, completion rates have not improved correspondingly. In 1970, 57 percent of urban Indian males aged 20 to 24 years had graduated from high school, thus lagging behind blacks (63 percent) and far behind whites (85 percent). The rural dropout rates for Indian men were even higher, with fewer than 50 percent finishing high school. As for college education, Levitan et al. report that "only 12 percent of Indians aged 18 to 24 years were in college in 1970, compared to 15 percent of Blacks and 27 percent of whites."[45]

American Indians have been conquered and uprooted and then isolated in reservations. The prolonged suffering they have endured has left many of them in a state of demoralization which expresses itself in many ways, notably in extraordinarily high rates of suicide, homicide, and alcoholism. Today the majority remain confined to their reservations, where the severest difficulties are encountered. By force of historical circumstances they have been placed in a state of dependency. Toward all those Indians who choose to remain in their tribal communities, American society has an urgent responsibility: To provide them with the means of achieving economic independence so that they may sustain themselves and their way of life in dignity.

9

The Family in Flux

Is the family changing? No doubt it is. But in what ways? Is the family in a state of decline? Is it becoming an obsolete institution? Numerous commentators have maintained for some time now that the family is a mere shadow of its former self. Some say it is falling apart; others describe the transformation it is undergoing as "decay." Still others argue that *all* of the family's functions can now be better performed by other institutions.

Whether they welcome the presumed decline of the family or decry it, many observers agree that the family is bound to be profoundly affected by a whole series of new developments. The women's-liberation movement, the introduction of nonadversary (no-fault) divorce laws, the liberalization of divorce laws in general, the movement for a four-day work week, the spread of contraceptive information to unmarried women and minors, the movement for zero-population growth, the various experiments in communal living, the marked influence of the youth culture, and the greater tolerance of homosexuality are only a few of the most striking of such developments. Can the family survive the absolute sexual equality demanded by the radical feminists? Will equal-employment opportunities remove the incentives men have for marriage and the care of families? Will the increasing equality and success of women in the world of work lead to their rejection of the bearing and raising of children? Will changes in task division within marriage undermine its affective and erotic foundations? Those are among the most salient questions which have been raised following the events of recent decades. To answer those questions somewhat adequately, we must have a clear idea of the basic forms and features of the family, both historically and transculturally. Only then can we assess the nature and extent of family change in our time and anticipate some of the changes to come.

THE NUCLEAR FAMILY

Social scientists employ the term *nuclear family* to refer to the most basic form of the family, consisting of a married man and woman and at least one offspring. It is characterized by a socially approved sexual relationship between a man and a woman who typically share a common residence and cooperate economically. The basic family form is called *nuclear* because it is out of such nuclei that larger kinship aggregates are formed. In many societies the polygamous family is such an aggregate, formed from two or more nuclear families affiliated by having one married parent in common. *Polygamy* is the general term for plural marriage, of which there are two types: (1) *polygyny,* the marriage of one man to two or more women; and (2) *polyandry,* the marriage of one woman to two or more men. Under polygyny, by far the more common form, one man unites several wives and their children into a larger familial group by playing the role of husband and father to all of them. Another larger aggregate made up of several nuclear families is the *extended family,* which consists of three generations, all living under one roof or, occasionally, in closely clustered dwellings.

The nuclear family appears to be a cultural universal. In the words of anthropologist G. P. Murdock, it "is a universal human social grouping. Either as the sole prevailing form of the family or as the basic unit from which more complex familial forms are compounded, it exists as a distinct and strongly functional group in every known society."[1] Murdock based his study of kinship forms on 250 nonliterate ("primitive") societies described in the Cross Cultural Survey files of the Institute of Human Relations at Yale University. He found no exception to his generalization, either in those societies or in any other.

Let us be quite clear about what Murdock is asserting. It is *not* that everyone everywhere lives in a nuclear family. Rather, he maintains that the nuclear unit exists in every human society and that it is *either* the "sole prevailing form" *or* a form enmeshed in larger aggregates such as polygamous and extended families.

To account for the universality of the nuclear family, Murdock posits four basic functions that it fulfills everywhere: sexual regulation, eco- 207 nomic cooperation, reproduction, and education (socialization). However, some anthropologists have challenged Murdock's thesis. They argue on the basis of ethnographic evidence that the nuclear family is not as universal as Murdock's widely accepted theory suggests. Two of the best-known critiques have been put forth by Kathleen Gough in her study of the Nayar and by Melford E. Spiro in his study of the Israeli *kibbutzim.* A brief examination of those two cases will help us assess the validity of the universality thesis.

The Nayar

The present-day kinship structure of the Nayar, Kathleen Gough acknowledges, is no exception to the thesis that the nuclear family is found in all societies.[2] At the time of the study, at least 30 percent of all Nayar households were nuclear-family units, and another 50 percent were extended families. The remaining 20 percent were not easily classifiable since they contained a miscellaneous assortment of relatives. It is presumed that there were no nuclear families among the Nayar prior to British rule in 1792. The Nayar men were a military caste in Indian society, and as such they spent much of their time away from home training for combat. The major agency of socialization among the Nayar was a matrilinear group called the *taravad.* Typically, a well-to-do *taravad* consisted of several hundred persons divided into numerous residential groups, each headed by the eldest living ancestress.

The Nayar practiced two forms of marriage. The first was a ritual ceremony, a *tali* rite, in which a Nayar girl was married to a group of men in her caste, though the group was represented by a ritual husband. With that rite the girl was considered a sexually mature woman ready to bear children. The relation between the newlyweds lasted three to five days. Following that period the woman was free to receive men in a second marriage ceremony, the *sambandham,* as visiting husbands. The men were either of the same or of a higher caste than the woman, and all children born to her of such unions were regarded as legitimate.

Since the men were so often absent from the community, child care was the responsibility of the mother, her female relatives, maids, and the older girls. In each dwelling a postnuptial female had a room of her own on the lower floor. She shared that room with her children until they reached the age of six. Then the children were placed in common rooms where they remained until puberty, when they were separated by sex. Unmarried women lived on the upper floor. The Nayar children were therefore almost totally in the hands of the mother and the other female members of the lineage, and fathers were little more than occasional visitors. Ordinarily men were not in residence with the women who bore their children. Women were permitted to change lovers frequently, and husbands played little or no role in childrearing. Finally, economic cooperation was the responsibility not of the husband and wife but of the entire *taravad*. If the description of the Nayar *sambandham* is accurate, it may be viewed as a form of group marriage in which the mother's lineage and not the nuclear family was the central kinship form. Hence the case of the Nayar appears to contradict the universality thesis.

But one must seriously question whether the Nayar were a genuine exception. For we must bear in mind that Murdock and others who have held the universality thesis claim that the nuclear family, by itself or embedded in larger kinship forms, exists in every *society*. The Nayar, however, do not in and of themselves constitute a society. They are a caste within the larger Indian society, and in that society the prevalence of the nuclear family is quite evident. On those grounds we would have to say that the universality thesis has not been shaken.

The Kibbutz

The claim that the nuclear family is found in every society has been challenged from another quarter. The Israeli *kibbutzim* have been frequently cited as an exception to the rule. They are small, cooperative agricultural communities in Israel which today comprise over 90,000 persons. In the *kibbutz*, Melford E. Spiro argues, the nuclear family is *not* a functioning unit.[3] Husband and wife generally do not live together with their children in one residence, nor do the mother and father cooperate economically as a couple. Children are reared from infancy by specially trained nurses of the community, and at each stage of their development, the children share a residence with their age peers. The loyalty of all concerned is to the community as a whole and not primarily to the family. Since most *kibbutzim* are comparatively small, about 250 persons, everyone knows everyone else. Thus many adults besides a child's own parents contribute to his socialization, just as his peers and older youths do. For those reasons Spiro and others maintain that the nuclear family in the *kibbutz* operates minimally or not at all.

But again as in the case of the Nayar, one must question whether the *kibbutz* is a genuine exception. In the first place the *kibbutzim*, like the Nayar communities, do not constitute a society in the largest sense. They are a social movement within the larger Israeli society. Second, and

209

more important, it is simply wrong to suggest that the four basic social processes associated with the nuclear family are absent from the *kibbutz* family. Sexual regulation, reproduction, education, and economic co-operation are all operative. Let us look briefly at each in turn.

Before a man and a woman in the *kibbutz* may live together as a couple and share a room, they must either marry or receive official *kibbutz* approval. In those terms their sexual relations are no less regulated than those of other husbands and wives. It is within the context of such a socially approved man-wife relation that reproduction occurs. It is true that children typically do not share their parents' residence; but there can be no doubt that parents play a key role in the child's education. Indeed, *kibbutz* parents spend a good deal of time with their own children, several hours a day after work. Since mothers do not have to prepare family meals (all meals are prepared and served by a rotating staff in the communal dining room), the time spent with their children is free of the pressures and distractions that mothers and housewives usually face in the late afternoon and evening. The father is likewise free of chores during those hours. As a rule *kibbutz* children visit with their parents two to three hours daily, playing, walking, and talking together. Afterward the children return to their dwellings for supper with their peers. After supper, however, parents in many *kibbutzim* accompany their children to their sleeping quarters and again spend time with them before tucking them in. So we see that sexual regulation, reproduction, and education are no less present in the *kibbutz* family than in others.

As for economic cooperation, that process too is fully operative. It is true that a husband and wife work only indirectly for the support of their own family since all their material needs are met by the *kibbutz*. Yet a definite division of labor exists between men and women not only in the *kibbutz* as a whole, but between husbands and wives in their own apartments. Keeping the apartment clean and preparing snacks for the evening hours which the couple spends either alone or with guests— such are tasks in which husbands and wives cooperate. Finally, one should note that the feeling of kinship that is commonly associated with families is at least as strong in the *kibbutz* as outside of it. That is evident not only in the everyday life of the *kibbutz* family but also when *kibbutz* members decide for some reason to leave the collective. In the vast majority of cases, they leave as a family. There is no solid ground, then, for the assertion that the nuclear family has ceased to exist in the *kibbutz*. Though the nucleus is enmeshed in a larger communal context, it is no less vital on that account.

210   It is therefore highly probable that the basic family unit which has

been called nuclear is in fact universal. That unit has varying degrees of autonomy and is most commonly embedded in a larger kinship system. In "primitive" societies the basic unit is so thoroughly intertwined with a larger kinship network that anthropologists find it difficult in practice to assess the degree of its autonomy. But in Western societies, as we shall see, a high degree of nuclear-family autonomy is unmistakable.

MONOGAMY AND POLYGAMY

Though the sex ratio varies from one society to another, it is approximately 1:1 the world over. Nowhere do deviations approach 1:2. That means that the ratio of men to women militates against bigamy and even more so against polygamy. "No man," Dr. Johnson once remarked to Boswell, "can have two wives, but by preventing somebody else from having one." In point of fact monogamy is the most prevalent form of marriage in all societies, though it is obligatory only in a few. A single mate at a time is the actual state of affairs even when a society poses no objection to plural marriage. Among Moslems, for instance, a common believer may espouse four wives, and a ruler may acquire a large harem. But the typical Moslem man everywhere practices monogamy simply because he cannot afford anything else.

Historically, very few societies have objected to plural marriages in principle; but the majority of people in all societies have lived monogamously in practice. Among so-called primitive peoples only a few have positively prescribed monogamy, but that does not mean that sexual relations are restricted to the married couple. Anthropologist Robert H. Lowie tells us that the Canella men, who have been described as uncompromisingly monogamous, show their friendship to their tribal brothers by the occasional exchange of wives, with the prior consent of the women. "Such exchanges," writes Lowie, "are invariably temporary and in no way affect the permanence of the marital tie."[4]

The two basic forms of plural marriage, as noted earlier, are polyandry and polygyny. The first, the marriage of one woman to two or more men, is quite rare. In an earlier era anthropologists compiled a rather long list of allegedly polyandrous peoples. But after a careful scrutiny of those cases, it was found that in almost all of them the prevailing marriage practices were misunderstood. Today it is generally agreed that southern India is the center of institutional polyandry. The best-known case is that of the Toda, where it was preferred that one woman marry two or more men, typically brothers, and where that marriage form was the dominant one. Polyandry became possible owing to an artificial tampering with the sex ratio. Prior to British rule and even afterward for a time, the Toda practiced female infanticide with the result that in some generations men outnumbered women 2:1. According to the ancient custom that continued for a while under the British, a married man's brothers automatically shared in his conjugal rights, all forming one household. Less common was an agreement between unrelated men to take the same wife as joint husbands.

If polyandry is made possible by a surplus of marriageable males, polygyny rests on a surplus of nubile females. That surplus nowhere results from male infanticide but rather from warfare, hunting, and other life-endangering masculine occupations. Thus Eskimo men, hunting with their native implements, frequently lost their lives. In one Greenland settlement the ratio of men to women was 5:16, an exceptional disparity. Less extreme was the imbalance among the Igulik, of whom 146 were men and 161 women. The imbalance permitted a minimal indulgence in polygyny. The various Caribou Eskimo bands uniformly revealed a preponderance of adult women. Lowie informs us that the ratios of women to men were: "30:25; 28:22; 18:13; 31:25; 21:17. This is all the more interesting because just as regularly boys exceed the girls, the figures being 24:11; 31:26; 13:10; 28:20; 23:15. In these groups about 25 percent of the men were bigamists."[5]

Lowie goes on to observe that the situation was quite different among the Copper Eskimo, where one settlement had a man-woman ratio of 61:60 and another settlement a ratio of 46:42. Those ratios, probably a consequence of the frequency in those groups of female infanticide, gave rise in turn to polyandry, which the Eskimo allowed in such circumstances just as they permitted polygyny in others. Hence the Eskimo culture permitted plural marriage, but its form and incidence were determined by local demographic and economic conditions.

Among the Murngin of Australia, the middle-aged and older men had at least three wives. A sufficient number of young men were killed in the incessant feuds of that people to enable the elders to indulge in polygyny. In Tikopia (Melanesia) before the coming of Christianity, all men of higher rank had more than one wife; their polygyny was made possible by the fact that many young men perished in dangerous sea voyages. Throughout Africa in the late nineteenth and early twentieth centuries, polygyny was associated with a preponderance of women.

212   But it is not that imbalance alone which explains the prevalence of

polygyny in Africa. Wealth and power are crucial factors. In Africa as elsewhere, poor men were monogamous, the better-off had two or three wives, rich men from six to ten, and high chiefs as many as twenty or thirty.

Where polygyny is an institution it is not masculine lust that primarily accounts for it. The man's motives are varied, but most often he desires several wives for the economic utility, prestige, and additional children he gains from them. What can be said here from the woman's standpoint? Given a preponderance of women in the sexual ratio, monogamy would require many women to remain unmarried. With polygyny, however, as among some African tribes, a woman acquired a husband who built her a house of her own and saw to the cultivation of her crops. Thus she also gained from the arrangement and, to the best of our knowledge, found it anything but degrading.

THE WESTERN FAMILY: GLIMPSES OF ITS HISTORY

There was a time when marriage was a privilege of the well born. For the Romans it was a legal status reserved for patricians and forbidden to slaves, though not to the lower classes. By the twelfth century the Church succeeded in removing the existing legal restrictions even for slaves, but social-class restrictions remained strong nevertheless. Marriage continued to be a social privilege since not everyone could afford it.

In preindustrial European societies, the typical cultivator had to wait until a house or cottage was accessible on the commons or until he had received permission to build a new one. In order to marry he needed either a plot of land or a place in the trades and crafts. Marriage, in short, presupposed some productive property by means of which to make a living. The early European marriage pattern implied that only propertied or substantial citizens were entitled to the status. In times of economic prosperity or, ironically, when the death toll was especially high, more people could get married, and they did not have to wait so long. But in hard times the waiting period was long, in many cases lasting forever.

In colonial America the situation was quite different. The European constraints on marriage became inoperative where land was practically unlimited and anyone could build himself a cottage or house. In fact, given the need for more working hands, young people were encouraged   213 and even urged to marry. In both colonial America and Europe in that era, marriage resulted in the creation of a nuclear-family household. That fact deserves some emphasis, for it has been widely assumed until recently that the typical kinship unit of the preindustrial era was the extended-family household, and that nuclearization began only with the spread of industrial capitalism. However, recent historical studies have shown that the extended family of the past is largely a myth.

The nuclear family, consisting of parents and children and no other adults, has been the predominant family form from the earliest period about which we have information. Historically, relations between relatives appear to have been pretty much what they are today: Patterns of companionship and mutual help which rarely if ever involved the sharing of room and board. The living together of married siblings or three generations was exceptional then as it is today. Co-residence was a temporary response to economic difficulties and social distress, and it was abandoned as soon as the situation improved.

That the extended family is myth rather than historical fact is something we have learned from the work of several historians. Availing themselves of a wide assortment of quantitative evidence—parish marriage records, local and national censuses, wills, birth and death registrations, and the like—they have successfully reconstructed family arrangements in early European and American communities.

Peter Laslett, of the Cambridge Group for the History of Population and Social Structure, has analyzed studies of 100 English communities between the years 1574 and 1821.[6] The studies disclose that in this period covering almost 250 years, households were predominantly nuclear. The size of the average household was just under five persons; 75 percent of the households included children; and only 10 percent contained any resident relative other than children, such as grandparents and siblings of the married couple. Three generations—children, parents, and one or more grandparent—were found in only 6 percent of the households. Some of the better-off households were larger than others, but they were peopled with servants and lodgers as well as by parents and children. Actually, those households were small manufacturing establishments in which families and nonrelatives lived and worked. It was only with the Industrial Revolution that manufacturing was gradually taken out of the home and the home separated from the factory.

What is therefore relatively recent historically is houses devoted exclusively to family living and not to industrial production as well. Little by little the unrelated working residents surrounding the married couple departed. First the journeymen left to set up their own establishments; then the apprentices followed; finally the servants left since only the best-off families could any longer afford them. Thus in preindustrial England the typical household contained only parents and children. One important reason for the absence of three-generational households was that people did not live long enough. In the sixteenth and seventeenth centuries, life expectancy was low. When the death rate began to decline in the eighteenth century and the population expanded, extended families became possible. The fact that the extended pattern was not adopted shows that it was rejected for other reasons.

In colonial America households were generally larger than they were in Europe. The average household ranged in size from 5.4 persons in New York in 1703 to 7.2 in Massachusetts in 1764.[7] In 1790, at the time of the first United States census, it has been calculated that the average American household had 5.8 persons. Although the American households were larger than the European, they were too small to be accounted for by the presence of extended relatives. Most families in colonial America contained parents and children, and occasionally a lodger.[8]

With the growth of industrialization in the United States, a change in the family did occur. Throughout the entire period of industrialization, the average size of households diminished from 5.6 persons in 1850 to 5.0 in 1890 to 4.1 in 1930. If one compares several nineteenth-century towns varying in the degree of their industrialization, one finds that they do not much differ in the proportions of their households with live-in relatives. In any event, the proportion never rose above 20 percent. In the nineteenth century as earlier, families appear to have taken in a relative as an economic expedient. The extremely low wages of both urban and rural workers might have prompted families to make such residential arrangements as would enable them to live somewhat better.[9] By 1970 the proportion of American family households containing any relatives other than parents and children had declined to 7.5 percent.

However, Americans do tend to expand their households in order to accommodate the needs of certain categories of kin: Young people, both married and unmarried, who are just beginning their careers and who are not yet able to fend for themselves; divorced and widowed persons who suffer emotional as well as economic distress and who therefore need the support of a family; and elderly people who may undergo numerous social and financial difficulties. It appears that co-residence becomes less attractive as economic conditions improve; and that families extend a hand to relatives when the need arises.

HUSBANDS AND WIVES:
IS MARRIAGE GOOD FOR THEM?

It seems to be better for men than for women—and this despite the fact that men continue to rail against marriage as they have for centuries. If one looks at the question objectively and with good humor, one is inclined to agree with Jessie Bernard who writes:

If marriage were actually as bad for men as it has been painted by them, it would long since have lost any future it may ever have had. In the face of all the attacks against it, the vitality of marriage has been quite stupendous. Men have cursed it, aimed barbed witticisms at it, denigrated it, bemoaned it—and never ceased to want and need it or profit from it.[10]

Early in the twentieth century, Emile Durkheim was among the first to document the salutary effects of marriage on men. In his classic study of suicide, he found that the suicide rate was higher for unmarried than for married men. This differential still prevails. In the United States, for example, the rate at which single men take their own lives is almost twice that of married men, while the corresponding differential for single women is only one and a half. Married men also enjoy better mental health. They suffer fewer mental impairments and show fewer serious symptoms of psychological distress. Married men live longer and happier lives. One researcher found that almost twice as many married as never-married men described themselves as very happy.[11] The benefits men derive from a good marriage are evident: love, sex, and companionship, not to mention wifely care. What is more, men know that marriage is good for them. Summarizing the available evidence, Jessie Bernard concludes:

The actions of men with respect to marriage speak far louder than words; they speak, in fact, with a deafening roar. Once men have known marriage, they can hardly live without it. Most divorced and widowed men remarry. At every age, the marriage rate for both divorced and widowed men is higher than the rate for single men. Half of all divorced white men who remarry do so within three years after divorce. Indeed, it might not be far-fetched to conclude that the verbal assaults on marriage indulged in by men are a kind of compensatory reaction to their dependence on it.[12]

The effects of marriage on women, on the other hand, appear to be less salutary. A sizable body of research spanning several decades shows that more wives than husbands report marital frustration, dissatisfaction, and problems, and that more wives than husbands describe their marriages as unhappy. Wives not only have more grievances than husbands;

216

their relative dissatisfaction reflects itself in poorer mental health. A number of studies have indicated that more married women than men feel themselves on the verge of a mental breakdown; they experience more frequent states of depression and distress; and they reveal feelings of inadequacy, often blaming themselves for their marital difficulties.

Doubtless there are several factors contributing to the married woman's sense of malaise. But one factor seems to stand out: the role of housewife. Married women who are exclusively engaged in housework have more mental, emotional, and physical problems than wives who work outside the home. "Working mothers," we are told "are less likely than housewives to complain of pains and ailments in different parts of their body and of not feeling healthy enough to carry out things they would like to do."[13]

In recent years spokeswomen for the feminist movement have alerted us to the existence of a wide range of sexual inequalities, in and out of the home. They have pointed out that sex prejudice and discrimination prevail throughout our society and others, and that one of the main manifestations of sexual inequality is the sex-typed division of labor. It has become deeply rooted in our culture that some occupations are "masculine" and others "feminine." A close examination of that labeling process discloses that it often serves to bolster male supremacy. The sex-typing of occupations is discriminatory in its effects because in virtually all societies and historical epochs, the occupations classed as "feminine" are the low-status, low-paying monotonous tasks which men prefer not to do. Men, on the other hand, reserve for themselves the high-prestige, more interesting jobs with access to positions of power. The traditional division of labor in which men work outside the home and women are restricted to housework is a case in point. That division of tasks has its roots in an earlier historical era. But changing circumstances are rapidly undermining whatever rationale it once may have had.

In colonial America housewives made indispensable contributions to the domestic economy. They produced a wide range of goods and services both for the family and for trade. Although they periodically joined the men at work in the fields, their main responsibilities were in and around the home. They gardened, raised chickens and other barnyard animals, spun, made candles, medicines, and beer—and cared for the children at the same time. Right up to the nineteenth century, a woman's work was for the most part done in the home. However, as manufacturing in the nineteenth century moved out of the household into the factory, the nature of women's work changed, particularly in

217

the cities. Textiles were still produced by women, but by unmarried rather than by married ones. Mothers simply could not care for their children while working in the factories, so they remained at home. Under the new urban-industrial conditions, women for the first time in American history stopped producing food and goods and became entirely absorbed in housework. It was at that time that the changing economic reality expressed itself in the emerging ideology that a married woman should devote herself to the role of wife and mother.

But that state of affairs began to change again in the twentieth century. By 1940, 14.7 percent of American married women had paying jobs outside the home. The percentage of gainfully employed married women rapidly increased both during and after World War II, reaching almost 25 percent in 1950. By 1974 it was up to 42 percent. The high proportion of working wives has raised the question of sexual equality more sharply than ever. Men are almost always hired in preference to women, especially for the most desirable jobs, even when their qualifications are equal. Once hired, women often receive lower salaries than men regardless of their qualifications or performance. And as women do well in their jobs and look forward to advancement, they find that sex discrimination continues to operate. They are bypassed for promotions—especially for positions that would place them over men. Summarizing the evidence, Constantine Safilios-Rothschild writes:

Promoting a woman to a supervisory position over men would, in fact, break down the prevailing sex stratification system and that is why the reluctance to do so is great and the resistance very strong. Several studies of male managers or of bank supervisors showed that not only do they hold stereotypic views of women but they also have less confidence in the ability of a female supervisor than a male supervisor and they would feel uncomfortable with a woman supervisor. These strong resistances to having a woman supervisor explains [sic] why men occupy the top administrative (or all administrative) positions even in occupations dominated by women such as social work, elementary school teaching or librarianship. A recent American study showed that among librarians only in sub-specialties such as school librarianship, in which only six percent are men, women can get promoted to top positions and have power. In other sub-specialties such as academic librarianship in which one-third are men, this minority controls all top positions of power in the field. Similar patterns have been documented for Poland and the U.S.S.R.[14]

In light of the high percentage of working wives, the various sexual inequalities *within* the home also deserve attention. Typically, the wife-mother who works at a full-time outside job returns home in the late afternoon to several more hours of taxing work: housecleaning, preparing dinner, caring for the children, and so on. The husband-father, in contrast, either takes it for granted that the evening hours are his for relaxation or uses them for conventional masculine chores such as making house repairs and cutting the lawn. As a result the most menial household chores are consistently avoided by men and performed by women. Cleaning the toilet and bathroom, for instance, is almost always defined as women's work.

Criticisms of traditional masculine attitudes and conduct have given rise to a reexamination of husband-wife roles in the home. A major aim of the women's-liberation movement has been the elimination of sexual discrimination and of the sex-typing of occupational roles in the world of work. The realization of that aim would require a different masculine attitude toward sex, an attitude in which the sexual act is no longer an expression of a man's general power over a woman. Sexual equality would entail the freeing of the woman from exclusive responsibility for child care and housekeeping chores. Those goals, some representatives of the women's movement argue, are attainable without turning traditional male-female roles upside down. What they propose is not a reversal but a sharing of roles. Men would no longer be the sole providers; women would no longer shoulder the burden of housework and children alone. Instead, both marriage partners would partake equally in all the responsibilities that marriage and the family entail. Children should have the care of both parents; all who benefit from household services should contribute to them; and both partners should share the burden of making a living. Married women and mothers who have yet to work at paying jobs outside the home should be given the opportunity to do so without neglecting their children and homes. Men, on the other hand, who have yet to undertake household management, should also be given the opportunity, without jeopardizing their jobs or reducing their family's income. How might such changes be achieved? Replying to that question, one woman has proposed "a new kind of life-style in which a husband and wife would each work [a] half-day and devote the other half of the day to caring for their children."[15]

The role-sharing lifestyle, it is further argued, would not only be more equitable than the traditional one; it would also be more beneficial to all concerned. Women who heretofore have been unable to work on the outside or to pursue careers would be able to do so under the new arrangement. Children, who under the conventional system of role allocation have been relatively deprived of a father's tender care, would gain that most precious good. Fathers in turn would gain because the new lifestyle promises women greater fulfillment and less frustration in everyday life; and a more fulfilled woman can only be a better marriage partner.

The conventional male-female roles in our culture have been defined as polarized: Women are "feminine" and weak, men are "masculine" and strong. By definition, "feminine" women are passive, yielding, delicate, dainty, emotional, and so on. And as Jessie Bernard has remarked, "Since femininity is the norm to which females are supposed to conform, they do their best. They are punished by rejection if they do not. Women, in brief, are rewarded for being weak, punished for being strong. An egalitarian relationship is hard to achieve under this design." The new lifestyle, however, is designed to create a genuinely egalitarian sexual relationship in which women would no longer be "feminine" in the traditional sense but rather "unintimidatedly female." Such women "can be not only powerful but also fascinating and exciting and even fun. Men who have acquired a taste for them find them delightful and feminine women dull, useful as servants and assistants and underlings but not as companions or even sex partners."[16]

Under the recommended lifestyle, finally, fathers would gain a deeper love from their children. For children would now experience a new side of their fathers and therefore relate to them in a new and more gratifying way. At the present moment, however, it is too early to tell whether men and women will choose to change their relations in the proposed direction.

INTERPRETING FAMILY STATISTICS

According to the 1980 U.S. census, 97.5 percent of all Americans live in "households." The rest, some 5,742,000, reside in "group quarters," including dormitories, barracks, hospitals, nursing homes, and prisons. The term *household,* a basic element in census statistics, is divided into two categories: "family households" and "nonfamily households."[17]

A family household consists of two or more persons living together who are related by birth or marriage. The most common family household is a married couple, with or without children at home. A single parent with one or more children also comes under the heading of a family household. Nonfamily households are comprised of two or more unrelated persons, of the same or different sexes, sharing living quarters. But the census also defines people who live by themselves as "single-person households" of the nonfamily kind.

In the ten-year period from 1970 to 1980, the total number of households rose from 63,401,000 to 79,108,000. This amounts to an increase of 24.8 percent, more than double the general population growth of 11.5 percent. During this period the size of the average household declined from 3.14 persons to 2.75. The drop is accounted for by there being fewer children in each family, more single-parent households, and more people living alone.

During the decade 1970–1980, family households shrank from 81.2 percent of the total to 73.9 percent, while nonfamily households rose from 18.8 percent of the total to 26.1 percent. The increase in the number of married couples (7.7 percent) was one-third less than that of the general population; but families with unmarried heads grew by 52.3 percent.

By 1980, 23.4 percent of all children aged 17 or under were not living with both parents. Such children resided instead either with one parent, another relative or a nonrelative. This was true of 17.3 percent of white children and 57.8 percent of black children.

In 1970, among couples with a husband under 25 years of age, 44.6 percent had no children. By 1980 this category had grown to 52.0 percent. And in 1980, 6,965,000 *more* people were living by themselves than in 1970. The increase resulted from several circumstances, all of them associated with a trend away from family living. Young people were not only marrying late; they were also residing alone while single. In the census category called "never married," the number of men with their own residences grew by 118.3 percent, while the comparable increase for women was 89.3 percent.

Equally noteworthy is the increase in the category of those separated and divorced and no longer residing with their former mates. The number of men in the category rose by 121.8 percent, while the figure for women was 79.4 percent. The discrepancy between the sexes is explained by the fact that when divorced and separated women have children, they generally obtain custody over them. Given that fact, such women fall into the category of "single heads of families," whereas their former husbands are classed as living alone."

Two facts have suggested to observers that the family may now be undergoing radical change. The first is the increasing disruption of marriage by divorce. The second is that men and women are experimenting with new forms of marital and extramarital relationships. Let us examine each of those facts in turn and see how they should be interpreted.

That the divorce rate is rising is indisputable. Twelve percent of the marriages contracted by women born in the years 1900–1904 ended in   221 divorce. In contrast, we can expect that between 30 to 40 percent of the marriages by women born in 1940–1944 will so end.[18] If we look at trends from, say, 1930, we find that the divorce rate rose to a high in 1946, fell until 1958, and then began to rise again. In 1973 the divorce rate surpassed the 1946 peak and continued to rise in subsequent years.

Some observers view that trend with alarm. It is worth reminding ourselves, however, that divorce, though it has its high emotional costs, may have more salutary effects for the parties concerned than sticking together at all costs. Moreover, it is still a minority of marriages that end in divorce, and the vast majority of divorced men and women remarry. On the other hand, only an insignificant proportion of people marry more than twice. The data are therefore inconsistent with the view that today's society has rejected marriage or that it has been replaced by serial, short-term relationships.

Are the problems and conflicts of marriage partners increasing in seriousness? Or has the liberalization of divorce laws allowed for the disruption of marriage by less serious problems? Although we know very little about the reasons for the rising divorce rate, available information indicates that the removal of social and legal obstacles to divorce is related to its probability. However, as Mary Jo Bane notes,

State divorce laws may be a result rather than a cause of changing societal attitudes toward divorce. They do tend to reflect the demographic composition of their states. A California study of divorce rates before and after the 1969 non-adversary (no-fault) divorce law provides an example of one kind of effect which changing divorce laws can have. The study found that divorces did increase the year following the passage of the new law; but the increase occurred because people waited for passage of the law before filing for divorce. The general trend of divorce in California was not affected by the reformed law; divorce did not go down, but it did not go up either.[19]

Most of the factors associated with divorce appear to be unrelated to the liberalization of divorce laws. Teenage marriages, for example, are especially vulnerable. Often they entail premarital pregnancies and financial and other difficulties. Perhaps, as Bane has suggested, "Teenagers make less sensible choices of marital partners."[20] Another factor is low income aggravated by unemployment; still another is a wife's earning power. The higher it is relative to the husband's, the greater is the likelihood of divorce. In such cases the husband's self-esteem as breadwinner may have been undermined, leading to marital tension and instability. Finally, there is the impact of the welfare system in the United States, about which there is considerable disagreement. Summarizing the evidence on the effect of the welfare system, Mary Jo Bane writes:

Logic suggests that a system which provides welfare support to female-headed but not to male-headed poor families should induce men to leave their families. Some studies support this logic. The effect of welfare is always found to be small, however, and other studies find no effect at all. When welfare mothers in New York City were asked why they separated from their husbands, none gave the availability of welfare as a reason, though some cited financial reasons. Much more important, they said, were drugs, alcohol, other women and physical abuse. [21]

What the rising divorce rate tells us, then, is not that marriage as an institution is on the wane, but that people's attitudes toward the permanence of a marital relationship are changing. Unhappy marriages were tolerated in previous generations; today they are terminated.

It is important to note that the remarriage rate has kept pace with the divorce rate. That fact shows that the persons concerned were dissatisfied with a particular marriage partner and not with marriage in general. That marriage as an institution has remained highly popular is undeniable since 90–95 percent of Americans marry at least once, and, if divorced, remarry soon afterwards (within five years). The bulk of adult Americans find themselves in stable, long-term marriages. In sum, the evidence clearly indicates that marriage—in the words of Mary Jo Bane—is "here to stay." Yet there are those who believe otherwise. They contend that extramarital relations as well as the sexual liaisons that are formed between sets of married couples point to a weakening of monogamy.

EXPERIMENTS WITH NEW FORMS

Students of sexual and marital behavior have called attention in recent years to what they perceive as a growing trend. An increasing number of spouses, they claim, permit each other full sexual liberty. They do so on the condition that there be no emotional "involvement" with the extramarital partner, and on the further condition that the extramarital relations remain discreet. The popular media have called such arrangements *swinging*. Some social scientists have adopted that term, but others prefer to speak of *co-marital* relations. Projecting the trend, they see a future society in which couples are still central, but in which they openly engage in a wide range of sexual relations with others without rejecting the primacy of their own relationship.

The same social scientists further argue that new contraceptives, such as the "Pill" and the IUD, have facilitated the trend. Both men and   223 women can now indulge in sex with less apprehension. Indeed, the new contraceptives have enlarged opportunities for women especially to pursue sex without unwanted pregnancies. The result is an increased "incentive for women to seek—as men have always done—sexual variety outside marriage. Among the available ways for both husbands and wives to find such variety, mate swapping is the least threatening and the one most compatible with monogamy."[22]

"Swinging," then, refers to a husband and wife's exchange of sexual partners with other couples or individuals. There are at least two types of swinging, the *recreational* and the *utopian*. The utopian swinger, something of a rebel, personally rejects several of society's conventions and publicly opposes them. This individual tries to replace those conventions with a new style of life, such as communal living, of which swinging is a part. The recreational swinger, on the other hand, accepts society's norms though deviating from them in one respect: the sexual sharing of his or her marital partner. Swingers of the recreational type favor monogamy. They are presumed to engage in such sharing in order to improve their marriages by overcoming the alleged monotony of sex with one partner.

Who are the people who swing? All the studies we have tend to agree that most of them are middle-class, mainly suburban couples living highly conventional lives in all respects but one. They take great pains to keep their one unconventional activity hidden from their very conventional neighbors and from their children. Some attend church regularly. In their general lifestyle they are indistinguishable from their fellow suburbanites; it is only in their sexual activities that they violate prevailing norms. In all the samples studied, the middle-class nature of swinging couples is confirmed. In one sample were found a corporation president, a mathematician, a lawyer, a realtor, a student, an artist, a computer salesman, an office-supply salesman, an engineer, and a restaurant manager. Swingers are young, married four to five years, and most of them have children. The women are mostly housewives and occasionally teachers. The largest concentrations of swingers have been found in the big cities of both the west and east coasts—in San Francisco, Los Angeles, New York, Boston, and Washington—but also in the Chicago area, the Texas-Louisiana-Mississippi area, and in and around Miami, Florida.

The recreational swingers have developed certain "rules of the game." They believe, in the words of one respondent, that "those who swing together stay together." It is an activity agreed to and participated in by the *couple* and not by separate individuals. The most damaging element in extramarital sex, they assert, is lying and cheating. If that element is removed, extramarital sex can strengthen the marital bond. Hence recreational swinging presupposes that the marriage will command paramount loyalty; that the interest in other sexual partners will be strictly physical, not emotional; that all parties concerned will remain discreet; and that the individual members of a married swinging couple will be completely open and honest about their respective sexual activities.[23]

How do couples become involved in swinging? They read about it in the media and they hear about it from acquaintances engaged in it. They are attracted to swinging because they desire sexual excitement and because their own relationships appear to them to be sagging. An important ingredient in becoming a swinger is seeing oneself as such. Some couples are introduced to the activity by friends, but they may also learn of available participants from newspaper advertisements and swinging clubs. In some cities the process has become commercially organized. In Los Angeles, for instance, an organization has issued numbered automobile stickers designating the owners as swingers. One can then contact them by calling a central registry. Several nightclubs cater to swingers.

Some novices become committed to the new arrangements but others do not. The woman may begin to feel like a prostitute; the man becomes impotent. They either discover that they cannot have sex with people to whom they do not relate emotionally, or that emotional involvement inevitably accompanies sex. Quite often they drop out, having decided that swinging is not for them.

A very informative study based on 1,175 dropouts from swinging who were seeing marriage counselors gave the following reasons for dropping out: jealousy; guilt; jeopardizing the marriage; developing emotional attachments with other partners; boredom and loss of interest; disappointment; separation or divorce; wife's inability to "take it"; fear of discovery by the community or by one's children; husband's impotence. Most studies show that it is the husbands who usually propose swinging and the wives who prompt the dropping out. Summarizing the experience of the marriage counselors, Duane Denfeld writes:

The marriage counselors suggested that wives were considerably more troubled by swinging than husbands. Wives expressed such feelings as disgust or repulsion. Husbands were more likely to initiate dropping out not for reasons of distaste but rather [because] they were bothered by their wives' popularity or sexuality. Wives were forced into swinging. . . . Husbands were described as "children with new toys." One counselor reported

225

a common finding among the counselors which is "In all my cases the men initiated the swinging and the women forced the termination."[24]

Apparently for the men swinging was a way of experiencing sexual variety with their wives' permission. But the women often felt exploited by the arrangement and had no enthusiasm for it. They acquiesced temporarily for their husbands' sakes. The data tend to cast doubt on the fashionable theory that swinging is beneficial to marriage and is seen as such by the parties concerned. As Denfeld remarks in his conclusion, "It is clear from the marriage counselors' reports that many couples left swinging psychologically damaged."[25] That finding, he adds, should give pause to those who recommend swinging as a way to help one's marriage.

The present state of our knowledge about such multilateral sexual arrangements strongly suggests that not only do they fail to solve or alleviate marital problems; they aggravate and accentuate them. Moreover, it is altogether doubtful that swinging or any other experiment with group sex noted in the media constitutes a genuinely growing trend, for the dropouts are as numerous as the initiates. None of those experimental patterns appears to have any big future in the United States, either as an alternative to monogamous marriage or even as a legitimate supplement to it.

10

Religion and Society

The roots of an individual's religious experiences may be traced to the unique human capacity for creating symbols, concepts, and abstractions. That capacity makes it possible for humans to posit hypothetical "somethings" that are greater than themselves. Humans hold those "somethings" in awe; they love them, fear them, bless them, and hate them, just as if they were concrete beings.

In the various layers of human consciousness, hypothetical beings are experienced as real, objective presences. What we have here is the perception of something general, a product of the mind and not of any of our special senses. In fact, it is a feeling of reality lying deeper than both the cognitive faculties and the senses. Such is the nature of human mental processes that they can grasp and create unpicturable beings whose existence is more real to those who hold them than any direct sensory experience. When those beings evoke in an individual a solemn attitude, whether glad or sad, one may say that he partakes of what might be called a religious experience.

At the turn of the century, the great American psychologist William James was among the first to describe those characteristics of the human consciousness. He documented the diversity and complexity of religious experiences. Religion, he maintained, involves joy and fear, gladness and sadness, and moods of expansion and contraction. In the "religion of healthy-mindedness," joy and happiness produce in an individual a sort of solemn feeling of gratitude. Most individuals partake at least occasionally of the religion of the healthy-minded type. They see more good than bad in all things and are thankful for it. But there are also individuals with a "radically opposite view, a way of maximizing evil . . . based on the persuasion that the evil aspects of our life are of its very essence, and that the world's meaning most comes home to us when we lay them most to heart."[1] In that attitude, said James, it is the religion of the "sick soul" that emerges. Evil is so radical and general that no rearrangement of the environment or of the inner self can eliminate it. What is required is a *supernatural* remedy.

What is it that gives rise to the belief in the supernatural? In all societies and in all epochs, most human beings have valued life over death, health over sickness, happiness and joy over pain and suffering, and the ready availability of life-enhancing goods over want and privation. But life, health, and happiness are so often bound up with their negations. Indeed, the negations are essential facts of existence. "The fact that we *can* die," wrote William James, "that we *can* be ill at all, is what perplexes us; the fact that we now for a moment live and are well is irrelevant to that perplexity. We need a life not correlated with death, a health not liable to illness, a kind of good that will not perish, a good in fact that flies beyond the goods of nature." This, James wisely observed, is the essence of the religious problem: "Help! Help!"[2] The crying need for help is met by the belief in immortality, and by the belief in benevolent spirits and deities who can overcome the evils besetting human existence. In his awareness of his limited, corporeal, earthly existence, the human individual is unique in the animal kingdom. By means of religious beliefs, he gains a sense of safety and peace. In his religious attitude, he sees the visible world as part of a greater spiritual universe, the real source of his life's meaning, and his hope for salvation.

William James sought to demonstrate that a religious person's sense of being moved by an external power is grounded in experience. For

it is one of the peculiarities of invasions from the subconscious region to take on objective appearances, and to suggest to the Subject an external control. In the religious life the control is felt as "higher"; but since in our hypothesis it is primarily the higher faculties of our own hidden mind which are controlling, the sense of union with the power beyond us is a sense of something not merely apparently, but literally true.[3]

Every introspective person will admit that James's conception of the psychological foundations of religious phenomena is highly convincing. It is no doubt true that the "higher," external, supernatural beings that one experiences are the products of the creative human mind. But the mind in turn derives the form and substance of those higher beings from another source, namely, *society*. That is the view first systematically developed by Emile Durkheim.

THE SOCIAL ORIGINS OF RELIGIOUS PHENOMENA

Durkheim's major study of religion is called *The Elementary Forms of Religious Life*. The main aim of that work was to lay bare the fundamental elements of religion. Such elements, Durkheim thought, could not be easily discovered in the religions of advanced civilizations. Those religions are the product of long, complex historical developments in which the fundamental elements have been obscured. However, if one assumes that small, simple, and economically primitive societies possess correspondingly simple religious forms, then perhaps the basic elements of religious life will be more readily accessible. That was in fact Durkheim's supposition. He attacked the problem by employing data on the most primitive contemporary societies known to anthropologists: the Australian aborigines.

The first task was to define the subject matter. What do we mean by religion, and what does it comprise? The definitions prevalent at the time all stressed a belief in supernatural and spiritual beings. That was the view, for example, of the great British anthropologist E. B. Tylor, who suggested that the best minimal definition of religion is "the belief in Spiritual Beings."[4] That definition is inadequate, Durkheim argued, since it fails to embrace those religions in which the idea of spirits or gods is absent. Buddhism is a case in point. In none of its basic principles does it concern itself with the notion of divinity.

After reviewing several other definitions and finding them deficient, Durkheim offered his own. All known religious beliefs divide the world into two domains: the *sacred* and the *profane*. For Durkheim, it was the totality of beliefs and practices concerned with the *sacred* that constitutes what we call "religion." When members of a society think and act in the same way with respect to the sacred, they share a common religion. They are members of a common "Church"—a moral community formed by all the believers in a single faith. Thus "a religion," wrote Durkheim, "is a unified system of beliefs and practices relative to sacred things, that is to say, things set apart and forbidden—beliefs and practices which unite into one single moral community called a Church, all those who adhere to them."[5]

Armed with that definition, Durkheim confronted the leading theories of his day. They could be divided into two schools. The first put forth

229

the theory of *animism*. Tylor, a leading representative of that school, held that primitive religion is a form of animism, a belief in souls, spirits, and a future state. Where did primitive men and women get the idea of a soul or spirit? They got it, wrote Tylor, by reflecting on two questions: "What is it that makes the difference between a living body and a dead one . . . ? [And] what are those human shapes which appear in dreams and visions?"

> Looking at these two groups of phenomena [Tylor continued], the ancient savage philosophers probably made their first step by the obvious inference that early man has two things belonging to him, namely, a life and a phantom. . . . As both belong to the body, why should they not also belong to one another, and be manifestations of one and the same soul? Let them then be considered as united, and the result is that well-known conception which may be described as an apparition-soul, a ghost-soul.[6]

For Tylor, then, the idea of the soul originated in the experience of dreams and fantasies.

But for Durkheim, Tylor's theory was quite unsatisfactory. For even if one admitted the plausibility of the dream origin of the soul idea, the theory had one crucial defect: It failed to explain why a phantom—"a simple reproduction of the individual"—should have been elevated to the rank of a *sacred* being, as in the ancestor cult, for instance. Animistic theory, Durkheim argued, fails to provide a convincing answer to this all-important question: "If it [the phantom-soul] was only a profane thing, a wandering vital principle, during life, how does it become a sacred thing all at once, and the object of religious sentiments?"[7]

The other school with which Durkheim quarreled may be called *naturism*. Animistic theories had claimed that the divine was derived from internal, mental experiences. Naturistic theories, in contrast, held that the first objects of religious sentiment were external natural phenomena. The things and forces of nature were the first to be deified. Nature presumably presents to primitive peoples numerous awesome spectacles which suffice to inspire religious ideas. Primitives personify and spiritualize those spectacles by means of metaphors and images.

For Durkheim, however, that theory suffered from the same defect as the first. Natural forces are, after all, natural forces, however intense and spectacular they might be. Missing from the theory therefore is an explanation of how they acquired a sacred character. It is doubtful, maintained Durkheim, that the sense of sacredness can be directly derived from natural phenomena. Thus rejecting the conclusions of both schools, Durkheim prepared the way for his own distinctive sociological theory.

Totemism: An Elementary Religion

The small, simple aboriginal societies of Australia afforded the best opportunity for the study of totemic beliefs. Such beliefs, it was widely agreed, formed an elementary religion, the most elementary known to scholars. Totemism was first discovered among the Indians of North America. Evidence for totemic beliefs also existed for ancient Egypt, Arabia, Greece, and the southern Slavs. But none of those societies was technologically as primitive as that of the aborigines, and in none did totemism appear in as pure a form as it did among the Australian groups. That is why Durkheim largely limited his attention to the Australian data.

The typical Australian grouping was a clan, an exogamous unit whose members considered themselves to be descended from a common ancestor. Each clan had its "totem," that is, an *emblem* designating a particular species of animal or plant which in turn represented the clan. The emblems of the kangaroo and crow, for instance, represented two distinct clan groups, and every clan member identified himself by the name of the respective species. The totem, it should be stressed, was not a pictorial representation of a species but rather a sign or "coat-of-arms." Totemic images were placed on the walls of huts, on the sides of canoes, and on the bodies of men. One of the principal initiation rites by which a young man entered into the religious life of the group consisted in painting the totemic sign on his body.

The totem is more than just a name and emblem. It is employed in religious ceremonies and is a part of the liturgy. It has a religious character and, according to Durkheim, "is the very type of sacred thing." That becomes evident from the role of the *churinga,* a ritual instrument which anthropologists call a "bull-roarer." These oblong pieces of wood or polished stone, suspended by a string, were rapidly whirled in the air so as to produce a loud humming sound. The *churinga* was employed in all important rituals. However, ritually profane persons, such as women and boys yet to be initiated into religious life, were prohibited from touching the instrument. The *churinga* was also believed to possess extraordinary properties. By contact it healed wounds and sickness; it gave clan members strength and courage; and it ensured an adequate reproduction of the totemic species.

The *churinga* was distinguished not only by its use in a ritual context but by the totemic mark engraved upon it. Typically such instruments were constructed each time anew, and then, once the rite was over, stripped of the sign, dismantled, and scattered. It was the totemic em-   231 blem which imparted a religious character to the instruments and was sacred.

The next step in Durkheim's inquiry was to examine the clan's attitude toward the totemic species. As an animal or plant its profane use would have been to serve as food, but its sacredness was demonstrated by the fact that all clan members were forbidden to eat it. And yet, surprisingly, while the *churinga* and other objects bearing the clan emblem were never to be touched or seen by ritually profane persons, the totemic animal or plant could be touched or seen. If, therefore, the degree of sacredness of an object may be measured by the pains taken to isolate it from the profane, "we arrive at the remarkable conclusion that *the images of totemic beings are more sacred than the beings themselves.*"[8]

To Durkheim that was a highly significant fact. If the totemic sign is more sacred than the totemic species, that suggests that the sign is so highly sacred not because it represents a species of plant or animal, but because it represents something else. Furthermore, since the totemic species and the clan members are also regarded as sacred, that must mean that the sign, the species, *and* the clan all share some common "principle." It is the common partaking of that principle that makes them all sacred. What is that principle? *It is an anonymous impersonal force.* It is independent of all subjects in whom it incarnates itself; it precedes them and survives them. That force is the *divine.* In a sense, said Durkheim, "it is the god adored by each totemic cult. Yet it is an impersonal god, without name or history, immanent in the world and diffused in an innumerable multitude of things."[9]

The divine principle is a "force" in both the physical and the moral sense. An individual failing to take proper ritual precautions receives a shock comparable to the effect of an electric charge. On the other hand, an individual observes his rites not merely out of fear of such physical effects. Rather, he observes them because his ancestors have always done so and because he feels a strong moral obligation to behave likewise. Thus the totemic cult, though it may appear to be addressed to plants, animals, or other objects, is actually directed to the "power" that permeates them. If a species of plant or animal, or even the sun, moon, or stars are adored, it is not due to their intrinsic nature, but to the fact that they partake of the sacred power. The believers themselves have only a vague notion of that power. But an awareness of its existence is evident in more advanced polytheistic cultures. The Greeks, for example, called it *Moira* or Fate, and even the most powerful gods were powerless before it. Yet the gods partake of that force when they produce rain or wind or crops. Zeus, Poseidon, Hades, and

232

the other Greek gods all retain marks of their original impersonality.[10]

Moreover, the impersonal power lying behind the later personified gods is the chief cause of all the movements occurring in the universe. Hence what we find in Australian religion is the first form of the idea of "force" as it was later conceived of in Western philosophy and science. Students of ancient Greek culture have shown that Greek philosophical ideas such as Necessity, Cause, Substance, Nature, and Matter are all rooted in the much more ancient religious conceptions of a sacred, all-powerful, impersonal, cosmic force.

But the most important question remains: What is the ultimate origin of the idea of a divine cosmic force or god? Durkheim's reply to that question constitutes his most important contribution to the sociology of religion.

Are Society and God One?

We have said that the totemic emblem was so highly sacred because it was, above all, a symbol of something else. If we can discover what that "something else" actually is, reasoned Durkheim, then we will have found the real basis for the idea of the divine. If we have followed Durkheim's analysis thus far, we can see that the totem symbolized two things: (1) the impersonal divine force, or "god," and (2) a specific society called the clan. The totem is the clan's "flag," the sign by which one clan distinguishes itself from another. "So," wrote Durkheim,

if it is at once the symbol of the god and of the society, is that not because the god and the society are only one? How could the emblem of the group have been able to become the figure of this quasi-divinity, if the group and the divinity were two distinct realities? The god of the clan, the totemic principle, can therefore be nothing else than the clan itself, personified and represented to the imagination under the visible form of the animal or vegetable which serves as totem.[11]

And upon reflection, argued Durkheim, it seems quite evident that society has all the attributes necessary to inspire a sense of the divine. Society is, after all, experienced as a superior force on which everyone depends. Members submit to society's authority even when it is felt to be repressive. They yield to its rules not only because it is strong enough to overcome them, but also because it is an object of respect. The social pressure brought to bear on individuals by society is largely of a spiritual kind. Ultimately it is the *social* reality that gives individuals the idea that there exists a superhuman principle, all-powerful and moral, on which everyone depends. It is the experience of society, therefore, that   233 gives rise to what Durkheim called a "collective representation"—a collective intuiting of the divine.

But it is not the profane, everyday social experiences that achieve that effect. It is rather those special, sacred ritual occasions in which individuals find themselves dominated and carried away by an external power. Often lasting days on end, such ceremonial occasions transport the participants from the gray world of everyday life into the extraordinary and effervescent world of the sacred. It "is out of this effervescence itself," wrote Durkheim, "that the religious idea seems to be born. The theory that this is really its origin is confirmed by the fact that in Australia the really religious activity is almost entirely confined to the moments when these assemblies are held." And Durkheim continued: "Since religious force is nothing other than the collective and anonymous force of the clan, and since this can be represented in the mind only in the form of the totem, the totemic emblem is like the visible body of the god."[12]

Hence when individuals believe in a moral power on which they depend, that is no illusion. For that power exists; it is society. Sacred assemblies serve the apparent function of strengthening men's bonds with the divine; but at the same time they serve the real function of strengthening the bonds of an "individual to the society of which he is a member, since the god is only a figurative expression of the society." "Religious force," concluded Durkheim, "is only the sentiment inspired by the group in its members, but projected outside of the consciousnesses that experience them, and objectified."[13]

From that perspective the nature and origin of the *soul* is also illuminated. It follows from Durkheim's analysis that the "soul" is no mere phantom, dream image, or mental reproduction of the individual. Rather, it is the experience each clan member has of the totemic principle incarnate in him. Or, in sociological terms, the soul is what society implants in every individual. The individual soul is a particle of the great collective soul of the group.

Similarly, the idea of immortality also originates in the reality of the group or society. Though individuals die, the clan survives. There must thus exist some principle or force that enables the clan group to possess eternal life. Finally, the idea of spirits and deities, far from being directly aroused by natural spectacles, is awakened in us by the sociomoral processes of the social world.

In sum, the ultimate source of the religious experience, for Durkheim, was society. Religion accordingly reflects both the good and bad sides of society, its just ideals and practices as well as its moral ugliness.

There are gods of theft and trickery, of lust and war, of sickness and death. Christianity itself, howsoever high the idea which it has made of the divinity may be, has been obliged to give the spirit of evil a place in its mythology. Satan is an essential piece of the Christian system; even if he is an impure being, he is not a profane one. . . . Thus religion, far from ignoring the real society and making abstraction of it, is in its image; it reflects all its aspects, even the most vulgar and the most repulsive.[14]

Criticisms of Durkheim's Theory

For Durkheim and his followers, therefore, society and God were one. Society, unconsciously divinized, is the stuff all religions are made of. Since Durkheim rested his theory on evidence from primitive cultures, it will be instructive to hear the critical comments of Bronislaw Malinowski, one of the outstanding anthropologists of all time.

It is evident, Malinowski observed, that Durkheim rested his entire case on the behavior of the collectivity, not the individual. Yet anyone who has experienced religion profoundly will agree that some of the strongest and most meaningful religious moments come in solitude, not in the group or crowd. That is no less true of primitive than of modern man.

Among primitives the novice is often secluded at initiation, and he undergoes a personal ordeal including a communion with spirits and deities. It is hard to see the social basis of those sacred powers in such lonely spots. It is equally difficult to see how the belief in immortality can be explained without considering the state of mind of the *individual* facing his inevitable death in fear and sorrow. Though Durkheim's theory virtually ignores the role of the individual, evidence is plentiful that in primitive religion prophets, seers, interpreters, and other practitioners play a key role. Those facts strongly indicated to Malinowski that the stuff of religion cannot be regarded as purely social.

Also questionable is Durkheim's central argument that the idea of the divine is somehow derived from primitive ceremonies and festivities. The religious idea, he maintained, "is born out of their effervescence." Durkheim thus tended to place the entire weight of his argument on the emotional excitement one feels while participating in such gatherings. To that argument Malinowski replied that just

a little reflection is sufficient to show that even in primitive societies the heightening of emotions and the lifting of the individual outside of himself are by no means restricted to gatherings and to crowd phenomena. The lover near his sweetheart, the daring adventurer conquering his fears in the face of real danger, the hunter at grips with a wild animal, the craftsman achiev-

235

ing a masterpiece, whether he be savage or civilized, will under such conditions feel altered, uplifted, endowed with higher forces. And there can be no doubt that from many of these solitary experiences where man feels the forebodings of death, the pangs of anxiety, the exaltation of bliss, there flows a great deal of religious inspiration. Though most ceremonies are carried out in public, *much of religious revelation takes place in solitude.*[15]

As for the presumed connection between the religious idea and collective *effervescent* festivities, that too seems dubious. There are numerous exciting and effervescent occasions of a collective sort in primitive societies which are nevertheless lacking in the faintest religious coloring. Malinowski cited the collective work in the gardens of Melanesia, "when men become carried away with emulation and zest for work, singing rhythmic songs, uttering shouts of joy and slogans of competitive challenge, [and which] is full of this 'collective effervescence.' But it is entirely profane, and society which 'reveals itself' in this as in any other public performance assumes no divine grandeur or godlike appearance."[16] Collective effervescence is also evident in battle, in sailing expeditions, in tribal gatherings for trading purposes, and in numerous other occasions, all of which generate no religious experience. It seems clear, then, that religious inspiration must take account of the solitary experiences of an individual; and that social effervescence may have no religious meaning at all.

There is still another objection raised by Malinowski. How can society be the prototype of the divine when so large a portion of what we inherit socially—traditions, knowledge, customs, norms, and skills—is profane, not sacred? Society as keeper of both the sacred and profane traditions cannot be the basis of divinity, for it is in the sacred domain only.

Therefore, though there is an element of truth in Durkheim's view, it is going too far to say, as he did, that society is the author of religious truths and that at bottom, the concepts of society and divinity are different aspects of the same notion. The main shortcoming of Durkheim's theory is its one-sidedness. In expanding the role of the group to an extreme, Durkheim effectively eliminated the role of the individual. As Malinowski rightly insisted, "Without the analysis of the individual mind, we cannot take one step in the understanding of religion."[17] Or to return to William James, for a moment, he likewise argued that personal religion is

more fundamental than either theology or ecclesiasticism. Churches, when once established, live at second-hand upon tradition; but the *founders* of every church owed their power originally to the fact of their direct personal

236

communion with the divine. Not only the superhuman founders, the Christ, the Buddha, Mahomet, but all the originators of Christian sects have been in this case;—so personal religion should still seem the primordial thing, even to those who continue to esteem it incomplete.[18]

A final criticism of Durkheim's theory is that it tends to dissolve the sacred and divine in the "social." For Durkheim, god and society were one because what people call *god* is actually the symbolic manifestation of the powers of society. Each individual is right, said Durkheim, in believing that there exists a power greater than himself, for it is the "moral power upon which he depends and from which he receives all that is best in himself: this power exists, it is society." The divine is thus *reduced* to the social. Implying that there is no essential difference between one religion and another, and between religions and national assemblies, Durkheim asked: "What essential difference is there between an assembly of Christians celebrating the principal dates of the life of Christ, or of Jews remembering the exodus from Egypt or the promulgation of the decalogue, and a reunion of citizens commemorating the promulgation of a new moral or legal system or some great event in the national life?"[19] The answer, of course, is that there exists a world of difference in the minds of the participants.

Hence the inevitable result of Durkheim's approach is that it entirely ignores the *subjective meaning* which religious beliefs and acts have for the actors concerned. But one cannot grasp the authentic meaning of such acts without recognizing that they are directed to the divine, not society. Grasped authentically, religious acts must be understood in their own right, and not reduced to the social. That is a point to which we shall return in our discussion of Max Weber, for whom the meaning and motives of actions were of paramount importance.

MAGIC, SCIENCE, AND RELIGION

Although Durkheim distinguished between the profane, workaday activities of everyday life and those of holidays, he had only little to say about the social settings of the respective spheres. That is an interesting question which Malinowski addressed with respect to primitive society. In his famous volume of essays bearing the title *Magic, Science and Religion,* Malinowski demonstrated that primitive societies possessed "scientific" knowledge as well as religious beliefs. He thereby laid to rest a prevalent but misguided notion that "primitive man" is "prelogical" and steeped in the supernatural.

237

By *science* Malinowski meant the large store of knowledge primitives have of their natural surroundings—knowledge based on experience and reason. The Melanesians, for example, whom Malinowski knew best owing to his many years of field work among them, had an extensive knowledge of the natural conditions on which their agriculture depended. They distinguished different qualities of soil and various types of cultivated plants as well as the interdependence of the two. They understood that there is a time "for clearing and burning the scrub, for planting and weeding, for training the vines of the yam plants. In all this they are guided by a clear knowledge of weather and seasons, plants and pests, soil and tubers, and by a conviction that this knowledge is true and reliable, that it can be counted upon and must be scrupulously obeyed."[20]

At the same time, however, their rational activities were accompanied by magic. Even gardening was often conducted under the leadership of a magician. Does that mean that the rational and supernatural actions were so mixed up together that the natives could not distinguish one type from the other? Certainly not, replied Malinowski, for the natives knew very well that there were natural conditions and causes which were controllable with the appropriate knowledge and efforts. They knew that for such knowledge and efforts magic was no substitute. "If you were to suggest to a native that he should make his garden mainly by magic and scamp his work, he would simply smile on your simplicity." Then why the magic? The answer is that

his experience has taught him . . . that in spite of all his forethought and beyond all his efforts there are agencies and forces which one year bestow unwonted and unearned benefits of fertility, making everything run smooth and well, rain and sun appear at the right moment, noxious insects remain in abeyance, the harvest yields a super-abundant crop; and another year again the same agencies bring ill luck and bad chance, pursue him from beginning till end and thwart all his most strenuous efforts and his best-founded knowledge. *To control these influences and these only he employs magic.* (Italics added.)[21]

Thus magic entered where "science" and work seemed inadequate.

That the rational and supernatural ran side by side without ever mixing was also evident in numerous other activities, notably in deep-sea fishing. The natives understood quite well that the outrigger canoe had to be constructed according to definite specifications if it was to be serviceable. They could explain to a Westerner why the span of the outrigger had to be just the right width in relation to the length of the dugout. They possessed a full system of sailing principles. They knew that in the face of a sudden gale the outrigger must always be on the weather side. They had also learned that certain forms of social cooperation were most effective in such undertakings, and they had a good knowledge of the range of weather conditions at sea. In those terms they were altogether rational, basing their conduct on a body of scientific knowledge.

Yet they had also learned from experience that their scientific knowledge was not enough to ensure their success and their safe return home. Although they carefully took into account and controlled all the factors they could, unanticipated disasters befell them. Occasionally they embarked on an expedition when the sea was as calm as an inland lake on a summer day. Then, at sea, a bolt from the blue, an unexpected storm and disaster. Hence, noted Malinowski, it was the task of the canoe magician to cope with those forces and to forestall them. Magic was resorted to only under such circumstances and for such purposes.

Malinowski tested his hypothesis by investigating lagoon fishing among the same people. That activity involved a method of poisoning the fish which entailed no special hazards or uncertainties. And, indeed, Malinowski found that in the case where men had complete control and where rational knowledge was sufficient, magic was *not* employed.

From a Western rational standpoint, canoe magic may appear to be superstitious nonsense. Yet even modern seamen, sailing huge power-driven, steel-built ships provided with all sorts of safety devices, have their nonrational beliefs and practices. Both the native and the modern seaman resort to the supernatural in the hope of mastering the elements of chance and luck. And for both there is a favorable effect: the allaying of anxiety in the face of potential dangers.

Some scholars draw a hard and fast line between magic and religion; others make a more fluid distinction. Magic and religion are especially evident among primitives in times of mental and emotional stress: life crises, sickness, hazardous situations, unfulfilled pursuits, and impending death. Both magic and religion appear to offer the individual a supernatural way out of crises and impasses for which he has found no rational solutions. Yet Malinowski discerned a significant difference between magic and religion.

Magic is a *practical* art consisting of acts directed toward definite aims. It is a matter of circumscribed technique: The practitioner employs a special spell or rite to achieve a definite effect. But religion is broader and much more complex. It creates for primitive man an entire supernatural reality, a second world consisting of a pantheon of spirits and demons, the beneficent powers of the totem, and the hope for future life. Religious ceremonies revitalize and reintegrate society, as   239

Durkheim maintained. But religious faith also strengthens the spirit of the individual, raising his morale and giving him courage in his struggle with difficulties and troubles. Thus Malinowski delineated the respective spheres of magic, science, and religion and their effects on both the individual and society.

TOTEM = FATHER = GOD

Primitive religion, and particularly totemism, impressed still another outstanding student of human behavior, Sigmund Freud. Freud based his analysis on the same ethnographic sources as Durkheim, but scrutinized totemism from a psychoanalytic viewpoint. He did so mainly in two well-known studies, *Totem and Taboo* and *Moses and Monotheism*. He reflected on religion in general in a third study called *The Future of an Illusion*.

In reviewing the vast literature on totemism, Freud noted one particular fact with special interest: The totemic animal is sometimes called the father of the men of the clan. The totem is an ancestor, tutelary spirit, and protector from whom the male clan members believe they are descended. That belief, Freud assumed, had some connection with the two most basic taboos of totemism: to avoid killing the totem animal and to abstain from sexual intercourse with females of the same clan.

Yet there were sacred occasions in which the men engaged in a sacramental killing of the totem animal and consumed it in common. That was an essential and recurring trait of totemic religion. No fully initiated male was allowed to exclude himself from either the killing or the sacrificial feast. By consuming the totem animal in common, they apparently strengthened their identification with it and with one another. Afterward the slain animal was bewailed and mourned in a rite intended to exculpate them from responsibility for the slaying. But the mourning period was soon followed by a loud and gay festival in which the sexual and other taboos were violated and every impulse gratified. What does the mourning over the dead totem animal have to do with the holiday spirit? If they are happy over the slaying of the animal, why do they mourn it? Reflecting on those matters in light of his psychoanalytic experience, Freud wrote:

Psychoanalysis has revealed to us that the totem animal is really a substitute for the father, and this really explains the contradiction that it is usually forbidden to kill the totem animal, that the killing of it results in a holiday and that the animal is killed and yet mourned. The ambivalent emotional attitude which today still marks the father complex in our children and so often continues into adult life, is also extended to the father substitute of the totem animal.[22]

How did the totem animal become a "father substitute"? To shed light on that question, Freud postulated a pretotemic social state which, following a suggestion by Darwin, he called a *primitive horde*. That is a hypothetical construct for which there is no direct historical or anthropological evidence. However, faint echoes and hints of such a condition may be found in the ancient mythologies of many peoples.

For Freud, the first human group was formed and maintained by the enforced rule of one highly powerful male. He was the *father*—that is, the man who possessed the most desirable women and who with them sired sons and daughters. The father effectively monopolized those women for himself. The fate "of the sons was a hard one; if they excited the father's jealousy, they were killed or castrated or driven out. They were forced to live in small groups and to . . . provide themselves with wives by stealing them from others." Thus excluded from the supreme pleasure, the sons were also compelled to shoulder the burden of whatever labor had to be done. The social order of the horde survived the death of the primal father: "One or the other son might succeed in attaining a situation similar to that of the father in the original horde. One favored position came about in a natural way: it was that of the youngest son, who protected by his mother's love, could profit by his father's advancing years and replace him after his death."[23]

The primal patriarchal despotism proved effective for a while. In due course, however, the hatred and resentment it generated were too strong for it to withstand, and it was destroyed. The exiled brothers joined forces, killed and ate the father, and put an end to his domination once and for all. The brothers who had so envied and feared the violent primal father identified with him by devouring him and acquiring a portion of his power. Though they hated the father who had thwarted their sexual desires, they also loved and admired him. With his removal and their hate satisfied, the tender impulses began to assert themselves. Remorse and guilt set in. The prohibitions which the father had physically enforced were now psychically internalized. As memories of the original patricide were painful, it was commemorated with a symbolic father substitute in the totemic feast. Thus the brothers "created two fundamental taboos of totemism out of the *sense of guilt of the son,* and for this very reason these had to correspond with the two repressed wishes of the Oedipus Complex."[24]

Totemism repressed everything hateful in the father and brought out   241 everything the sons had wished for: protection, care, and forebearance. In return the sons pledged to honor his life,

that is to say, not to repeat the act against the totem through which the real father had perished. Totemism also contained an attempt at justification: "If the father had treated us like the totem, we should never have been tempted to kill him." Thus totemism helped to gloss over the real state of affairs and to make one forget the event to which it owed its origin.[25]

The first totemic feasts were therefore the beginning of fraternal social organization, moral restrictions, and religion. The new brotherly solidarity of the clan meant that no one was to treat the other as the primal father had treated them—nor as they, in the end, had treated him. The father totem was transformed into a wholly benevolent being and his perfection of power was placed out of reach. It is out of that process that the idea of "god" emerged. Justifying his inference on the basis of his clinical experience, Freud wrote:

Psychoanalytic investigation of the individual teaches with special emphasis that god is in every case modelled after the father and that our personal relation to god is dependent upon our relation to our physical father, fluctuating and changing with him, and that god at bottom is nothing but an exalted father. Here also, as in the case of totemism, psychoanalysis advises us to believe the faithful, who call god father just as they called the totem their ancestor.[26]

For Freud, mythologies, properly analyzed, often revealed the repressed wish fantasies of a people. Myths stand in the same relation to a people as dreams do to an individual. With the introduction of agriculture in antiquity, Freud noted, there are clear expressions of the father-son tension. In numerous myths the son's incestuous desires find symbolic expression in laboring over mother earth. There appear youthful male divinities like Attis, Adonis, and Tammuz who enjoy the favors of the maternal deities and who commit incest with them in defiance of the father. However, those youths did not go unpunished. They suffered either short life or castration at the hands of the father-god appearing in animal form. Adonis was killed by a boar, the sacred animal of Aphrodite; and Attis, the lover of Cybele, died of castration. In that way the terrible sense of guilt following upon incest and murder is allayed.

Echoes of the primordial tension may also be heard in the myths of Christianity. But there the son allayed the sense of guilt by sacrificing his own life, thereby redeeming all his brothers from the primal sin. In the Christian myth man's original sin was an offense against God the Father. And Freud assumed that if the redemption of mankind required the sacrifice of the son, then the sin must have been murder. It is deeply rooted in human feeling, said Freud, that a murder can only be atoned for with another life. The self-sacrifice of Jesus indicates a blood guilt. Psychoanalysis reveals, Freud added, that "the suicidal impulses of our neurotics regularly prove to be self-punishments for death wishes directed against others."[27]

Yet in the very self-sacrifice of the son, which was to achieve reconciliation with the father,

the son also attains the goal of his wishes against the father. He becomes a god himself beside or rather in place of his father. The religion of the son succeeds the religion of the father. As a sign of this substitution the old totem feast is revived . . . in the form of communion in which the band of brothers now eats the flesh and blood of the son and no longer that of the father, the sons thereby identifying themselves with him and becoming holy themselves. . . . At bottom, however, the Christian communion is a new setting aside of the father, a repetition of the crime that must be expiated.[28]

From his general analysis Freud concluded that the beginnings of religion and the idea of the divine may be traced to the Oedipus complex. Freud acknowledged that in the absence of direct evidence, the primordial act of patricide cannot be proven. However, even if such an act never had occurred, "the mere impulses of hostility towards the father and the existence of the wish fantasy to kill and devour him may have sufficed to bring about the moral reaction which has created totemism and taboo."[29]

RELIGION AND REASON

Freud found the tenacity with which religious beliefs are held to be psychologically remarkable. Though they are not amenable to rational or empirical proof, people cling to them anyway. How shall we account for the fact? Freud's answer was that people cling so tenaciously to their religious beliefs because they are "fulfillments of the oldest, strongest and most urgent wishes of mankind. The secret of their strength lies in the strength of those wishes."[30] Religion is an "illusion" not in the sense that it can definitely be proved to be false; it is an illusion because it is the product of wish fulfillment and the need for consolation.

In those terms Freud, with his rational-scientific outlook, was an heir of the eighteenth-century Enlightenment. The philosophes—Holbach, Diderot, La Mettrie, Voltaire—were among the first to look upon re- 243 ligion as an illusion. The advance of science and reason, they believed, would eventually put an end to religion, which they regarded as just so much nonsense. In the nineteenth century God was pronounced dead several times by rationalistic philosophers. By the early twentieth century, however, serious thinkers had begun to have their doubts. Like William James, others also recognized that though many religious creeds may be absurd, yet religion on the whole is one of mankind's most vital activities.

Freud agreed that historically religion has met the basic human need for consolation. But the time has arrived, he thought, "to leave God out altogether and honestly admit the purely human origin of all the regulations and precepts of civilization. Along with their pretended sanctity, these commandments and laws would lose their rigidity and unchangeableness as well. People could understand that they are made, not so much to rule them as, on the contrary, to serve their interests."[31] But as we reflect on the history of the twentieth century, we see clearly that Freud's prognosis has not been borne out. Contrary to his expectations, the rigidity, unchangeableness, and oppressiveness of rules and social relationships have not diminished in militantly atheistic societies.

Freud was not alone in making that error. Indeed, Durkheim shared Freud's rationalistic assumptions. Durkheim also agreed that the gods are "growing old or already dead."[32] He believed that his demonstration of the social basis of the divine had pulled aside the veil with which the religious imagination had covered up the underlying social reality. He hoped that thanks to his analysis, society would take the place of the old gods, and that science would replace religion. For though science and religion pursue the same ends, "scientific thought is . . . a more perfect form of religious thought. Thus it seems natural that the second should progressively retire before the first, as this becomes better fitted to perform the task."[33] In those terms both Durkheim and Freud and, indeed, all those who shared their assumptions, may have failed to appreciate the deep-rootedness of the religious experience.

RELIGION AS ALIENATION

As we have said, the thinkers of the Enlightenment claimed that religion was an "illusion." "God," they argued, was simply a symbolic expression of human yearnings for perfection. As Voltaire once cleverly remarked, "God created man in his own image, and then man returned the compliment."

In the nineteenth century several thinkers elaborated the ideas of the Enlightenment. Among the most influential of those thinkers was Karl Marx, who as a young man shaped his own distinctive conception of religion in response to several of his older contemporaries. One of them was the German materialistic philosopher Ludwig Feuerbach. In a book called *The Essence of Christianity,* Feuerbach presented a view of religion quite similar to that of the French Enlightenment. God, Feuerbach maintained, is simply a creation of the human imagination. The divine is a symbolic expression of humanity's unfulfilled promise and aspirations. Humans unconsciously project their ideals onto hypothetical beings which they then treat as sacred and divine. They thus come to worship the product of their own minds.

When Marx was a young university student, such atheistic theories were quite fashionable in radical circles. He found them convincing as far as they went, but in his judgment they did not go far enough. They remained on the psychological level and ignored what he regarded as the key sociological question: *Why* do people project the best part of themselves onto the cosmos? What are the *social* conditions that prompt people to externalize their own powers and values and to attribute them to hypothetical, superhuman beings? Marx's reply to that question was that religion is the product of *social alienation.* That means that historically humanity has been divided against itself by the class cleavages of society. It is the domination, oppression, and exploitation of man by man that has given rise to religion. Religious ideas are an expression of human suffering and a protest against it as well. In the words of the young Marx:

Religion is the sigh of the oppressed creature, the sentiment of a heartless world, and the soul of soulless conditions. It is the opium of the people.

The abolition of religion as the *illusory* happiness of men, is a demand for their *real happiness.* The call to abandon their illusions about their condition is a call to abandon a condition which requires illusions.[34]

Reflecting on the degraded condition of the industrial proletariat and other oppressed classes of his time, Marx concluded that by itself a demonstration of the illusory character of religion is not likely to have liberating effects. So long as domination and sharp inequalities prevail, people will continue to create comforting illusions. The main task, therefore, is to change the social order and to eliminate the circumstances which require illusions. For Marx it was industrial capitalism in particular which imposed great suffering on the majority of the people, diminishing their humanity and distorting their self-understanding. Religion   245 is an "opium" because it so often persuades people to seek happiness not in the human world but in the divine hereafter. In that light we can more readily grasp Marx's meaning when he wrote that religious criticism

has plucked the imaginary flowers from the chain, not in order that man shall bear the chain without . . . consolation but so that he shall cast off the chain and pluck the living flower. The criticism of religion disillusions man so that he will think, act and fashion his reality as a man who has lost his illusions and regained his *reason*.[35]

It is only through the rational reconstruction of society—so that it should meet the human needs of each and every member—that alienation can be overcome.

Thus Marx, like Freud, wished to rid humankind of its great illusion. But unlike Freud, Marx traced that illusion directly to society's ills. And like Durkheim, Marx provided a sociological view of the origin of religion. Both Marx and Durkheim discerned the roots of the divine idea in the structure and processes of society. Indeed, Durkheim occasionally grasped an aspect of alienation. He remarked in passing that the "power of souls is increased by all that men attribute to them, and in the end men find themselves the prisoners of this imaginary world of which they are, however, the authors and the models. They fall into dependence upon these spiritual forces which they have created with their own hands and in their own image."[36]

But Durkheim did not go on to link that phenomenon with the shortcomings of society. On the whole, society was a positive entity for Durkheim—which enabled him to equate society with God. For Marx, in contrast, society in its prevailing form was an oppressive and alienating entity. It frustrates more than it fulfills human needs.

Marx thus stressed the ideological function of religious institutions and ideas. Religion served the interests of the privileged and legitimized their rule. Ignoring the other side of religion, he never expressly acknowledged that religiomoral visions and teachings have frequently inspired movements for freedom. Nor did he acknowledge that such visions were the true wellspring of his own secular-humanist ideals. As a distinguished theologian has recently observed:

Marx never made a careful study of religion. He never asked himself the question whether religion had always and everywhere exercised the same legitimating function. Two generations later, another German sociologist, Max Weber, made a detailed study of the social role of religion and discovered that while in most ages the successful religions offered an ideological defense of the existing order, there were also periods when religious trends were a source of social criticism and, by offering a new vision of human life, actually affected the transformation of culture and society.[37]

It is Weber's contribution to the sociology of religion to which we now turn.

RELIGION AND SOCIOECONOMIC CHANGE

In his now famous essay entitled *The Protestant Ethic and the Spirit of Capitalism,* Max Weber addressed this question: Did religion play any significant role in the emergence of capitalism in Europe? In part, Weber's essay was written in response to the dominant school of Marxism at the time, which had subscribed to a form of economic determinism.

When the capitalist economy first made its appearance in sixteenth- and seventeenth-century Europe, it entailed a sharp break with the past. It involved a new code of economic conduct and new social relations at odds with the accepted conventions and laws of church and state. How did the pioneers of the capitalist system overcome the resistance of the old order and elbow their way to success?

The answer to that question typically given by the Marxists was roughly this: The successful emergence of the new system was made possible by changes in the economic world. The influx of precious metals from America, capital accumulated in commerce, expanding markets, the growth of population, and new technology resulting from the advance of natural science—such were the major factors.

Weber did not deny the importance of those conditions, but he believed that the Marxian answer was one-sided and incomplete. For there were countries in which all of the conditions enumerated above were present but which failed to give birth to capitalist industry. France of Louis XIV, for example, commanded immense resources by the standards of the time, but dissipated them in luxury and war. Hence there is something missing in the economic explanation, Weber insisted, and one must look outside of economics to find it. If the first entrepreneurs engaged in their economic pursuits with a special vigor and dedication, that fact may be traced to the *Protestant ethic*—to the new moral values that emerged with the religious changes of the sixteenth century, the Reformation.

247

Protestantism and Capitalism

Weber began by drawing attention to certain significant cultural differences between Protestants and Catholics. In their education Protestants were more inclined to study technical subjects; they were also more prominent as proprietors of industrial enterprises. Catholics, on the other hand, seemed to prefer more traditional humanistic studies and nonindustrial occupations such as crafts. Such differences between Protestants and Catholics remained evident when one controlled for the social-class background of the two religious categories. Protestants, whether from upper or lower strata, "have shown a special tendency to develop *economic rationalism* which cannot be observed to the same extent among Catholics" (italics added).[38] What is the source of that more pronounced economic rationalism among Protestants?

At first glance it might appear that they have been more worldly and hedonistic than Catholics. But closer examination shows that that has not been the case. The "English, Dutch and American Protestants," wrote Weber, "were characterized by the exact opposite of the joy of living."[39] Indeed, they adhered to a strict religious and moral code of self-denial. They were, in a word, *ascetic*. And it was, ironically, their ascetic Protestant ethos which made them especially receptive to the rational spirit of capitalism.

To document that thesis, Weber employed the figure of Benjamin Franklin. From *Necessary Hints to Those That Would Be Rich* and *Advice to a Young Tradesman,* Weber selected some typical sayings which illustrate Franklin's commitment to industry, frugality, hard work, and punctuality. Franklin was important to Weber because, though he said "Time is money," his attitude toward wealth was different from that of the rich men of earlier eras. Franklin's motives for making money, Weber argued, were devoid of hedonism. They were rooted in his strict Calvinist upbringing. Why should men make money, and why should "money be made out of men"? To that question Ben Franklin replied by quoting the Bible: "Seest Thou a man diligent in his business? He shall stand before kings." Apparently, Franklin's business interests were thus religiously motivated and justified.

But how did Calvin's doctrine of predestination lead to worldly activity such as business? Actually, Weber explained, it was not the teachings of Calvin himself but rather of his followers that yielded that result. Calvin, though certain of his own election, rejected the principle that one could learn whether one was chosen or damned as an attempt to force God's secrets. But Calvin's doctrine proved to be too heavy a

248

psychological burden for ordinary individuals who required a "sign." Thus Calvin's followers increasingly gave way to the expressed need for "infallible criteria by which membership in the *electi* could be known."[40] The original doctrine was therefore modified. It now stressed the

duty to consider oneself chosen, and to combat all doubts as temptations of the devil, since lack of self-confidence is the result of insufficient faith, hence of imperfect grace. The exhortation of the apostle to make fast one's own call is here interpreted as a duty to attain certainty of one's own election and justification in the daily struggle of life. In the place of the humble sinners to whom Luther promises grace if they trust themselves to God in penitent faith are bred these self-confident saints whom we can rediscover in the hard Puritan merchants of the heroic age of capitalism and in isolated instances down to the present. On the other hand, in order to attain that self-confidence, intense worldly activity is recommended as the most suitable means. It and it alone disperses religious doubts and gives the certainty of grace.[41]

Worldly activity, then, though useless for the attainment of salvation, became a possible *sign* of election. Such activity served to allay the fear of damnation. Hard work in the morally dutiful pursuit of a worldly calling and absolute avoidance of anything which detracts from an ascetic way of life—that was the Protestant ethic. It was embodied, in varying degrees, in Puritanism, Pietism, Methodism, and the Anabaptist sects, and it had the "greatest significance for the development of the spirit of capitalism."[42]

Ascetic Protestantism thus produced the highest appreciation of the sober, middle-class, self-made man. The new religious principles brought with them "the ethos of the rational organization of capital and labor" and "turned with all its force against one thing: the spontaneous enjoyment of life and all it had to offer." Restless, continuous, systematic work in a worldly calling, as proof of genuine faith, "must have been the most powerful conceivable lever for the expansion of that attitude toward life which we have here called the spirit of capitalism."[43] Protestant asceticism thus provided a religious sanction for the emerging economic system: It eased the employer's conscience and at the same time gave the worker religious motives for treating his labor as a calling.

In that way Weber documented his thesis that the earliest capitalist entrepreneurs were motivated by Protestant values. He answered his original question in the affirmative: Religion did play a significant role in the rise of capitalism in Europe. However, because there has been   249 considerable confusion concerning Weber's thesis, we should be perfectly clear about what he was and was not saying.

Weber was not saying that Protestantism had created capitalism. Nor was it his view that capitalism would not have arisen in the absence of Protestant religious values. Weber's thesis was rather more modest, namely, that the markedly energetic character of early capitalism may be attributed to the affinity of two normative patterns: the Protestant ethic and the rational capitalistic spirit. Highly congruent, those patterns mutually reinforced each other to produce a methodical devotion to work and business. A vigorous capitalist beginning was the result.

Since the publication of Weber's thesis in 1904–1905, numerous objections have been raised against it. Only a few can be briefly mentioned here. The first is that several elements of modern economic rationality which Weber attributed to Protestantism were already present during the Renaissance. A second objection relates to Weber's assignment of antitraditional norms primarily to Protestantism. Some scholars have maintained that those norms were much more general than Weber's essay would have us believe, and that they were found among Catholics as well. Finally, critics have objected that in tracing the effects of religion on economics, Weber neglected the reverse influences. Weber had in fact intended to study the impact on religion of economics and politics in the same historical era, but he never published an essay on the subject. That has sometimes led to the impression that his Protestant-ethic thesis is one-sided. But if we remember that the whole point of his study was to correct the one-sidedness of the economic determinists, we can see that he accomplished his purpose. Despite the many criticisms which have been raised, there is wide agreement today that the Weber thesis contains a large element of truth.

Religions East and West

Weber's interest in religion and economics soon led him to a larger question: Why had capitalism originated in Western civilization and not in the East? That question prompted him to undertake his voluminous studies of the religions of China, India, and ancient Israel.

The rise of capitalism in the West meant, among other things, that economic changes were transforming the entire social fabric of Europe. In their attempt to explain why the new economic system had first emerged in Europe and not elsewhere, scholars had stressed two factors: the large influx of precious metals and a substantial growth in population. Weber observed, however, that in China the same factors

250

had been at work, with quite different effects. Instead of weakening and shattering the traditional order to make way for a new one, those factors strengthened traditionalism. They stimulated no capitalist development and left the Chinese economy intact. To explain that fact Weber compared the many facets of Chinese society and religion with those in the West.

In the West, Weber reminds us, the ancient and medieval cities were commercial and financial centers.[44] It was in such cities that the earliest forms of rational exchange economy took hold and expanded. In China, in contrast, there were no real cities in the European sense. Unlike the *polis* of antiquity and the commune of the Middle Ages, the Chinese "city" had neither political privileges nor military power of its own in the form of a self-equipped militia. The Occidental city eventually became sufficiently strong to repel an assault of knights; it had merchant associations and craft guilds. But those institutions were nonexistent in the Chinese "city," which always remained an appendage of the centralized bureaucracy. In short, there never emerged in imperial China a relatively independent bourgeoisie centered in autonomous cities. So although there existed something of a money and commercial economy, it never led to the Western type of industrial-capitalist enterprise. For Weber, a basic obstacle to China's development of capitalism was the nature of China's religious culture.

Chinese culture, Weber maintained, had not undergone "rationalization" as had the culture of the West. In the West rationalization had begun with the struggle against magic by the ancient Hebrew prophets, and it was carried through most consistently by ascetic Protestantism. That does not mean that the Puritans had rid themselves of all superstitious beliefs; that they had not is evident from their witch trials. It means, rather, that they came to regard "all magic as devilish." To Weber, then, one criterion for the rationalization of religion was the degree to which it has freed itself of magic.

But another criterion was a religion's attitude toward the world. Thus while Puritanism resulted in a tension with the world and hence its transformation, Confucianism required its adherents *to adjust* to the cosmic order. It is true that the Confucian exercised self-control and repressed all irrational passions that might disturb the cosmic harmony. But he did not oppose the magicians. On the contrary, he took the powers of magic for granted, while the masses were altogether steeped in it.[45] Thus despite its wealth and the advanced state of its scientific knowledge, Chinese society remained an "enchanted garden." That, for Weber, was the chief cultural obstacle to capitalism. We can best under- 251 stand what Weber intended by the term *enchanted* if we now catch a glimpse of the religions of India.

Indian religions, including Buddhism, all led to an extreme devaluation of the world. None of them enjoined their followers to act upon the world, for the highest good was a contemplative flight from it. Indian asceticism never translated itself into a methodical, rational way of life that tended to undermine traditionalism. India, like China, remained an "enchanted garden" with all sorts of animistic and magical beliefs and practices: spirits in rivers, ponds and mountains; highly developed word formulas and spells; finger-pointing magic; and so forth. Like the Confucian Mandarins the Indian Brahmins recognized the powers of magic and made numerous religious concessions to the magicians.

Weber therefore concluded that Far Eastern religions never provided their adherents with the motive and justification for the domination of nature by rational scientific means. Instead, Asiatic religions became "the means of mystical and magical domination over the self and the world . . . by an intensive training of body and spirit, either through asceticism or, and as a rule, through strict, methodologically-ruled meditation."[46] In contrast to the soul-saving doctrines of Christianity, no emphasis was placed on "this life." Asiatic religion led to a relative otherworldliness, an outlook that had a profound influence on economic conduct. Magic was employed to achieve all sorts of earthly aims. There were spells against sexual and economic rivals, spells to win legal suits, spells to compel one to pay his debts, spells for the securing of wealth and for innumerable everyday undertakings. It was the depth and scope of the magical mentality that precluded the rise of the economic system which Weber called modern rational capitalism. Missing from Asiatic religious culture was a process which could break the hold of animism and magic. Missing was a "rational this-worldly ethic." In the history of Western culture, Weber perceived the beginnings of that process in ancient Israel.

The Religion of Ancient Israel

According to the Israelite world view, God (Yahweh) not only created the world but intervened periodically in history. The world in its existing state reflected God's response to the way in which humans conducted themselves. God was a personal, transcendental, all-powerful being that ruled the universe. Above all he was an ethical deity. The present condition of trouble and suffering would ultimately give way to a God-ordained good life and happiness. The attainment of that future depended entirely on the moral conduct of the Israelites and their faithful devotion to God's commandments. Ancient Judaism was therefore "a highly rational religious ethic of social conduct; it was free of magic and all forms of irrational quest for salvation; it was inwardly worlds apart from the paths of salvation offered by Asiatic religions."[47]

Class divisions in Israel became especially marked under the monarchy, under Solomon and afterwards. Tensions and conflicts were now evident between indebted peasants on the one hand and large landlords and urban creditors on the other. It was against this social background that the prophets of social justice emerged. To them it appeared that the kings were turning Israel into an oppressive *corvée* state not unlike the Egyptian "house of bondage" from which Yahweh had liberated the people. The prophets spoke out against the kings, voicing sharp criticism against both their public policies and their private sins. Amos, Isaiah, and others thus expressed the sentiments and defended the interests of the peasants and other oppressed groups which were increasingly subjected to debt bondage, forced labor, and heavy taxes.

However, that does not mean, Weber emphasized, that the prophets were simply ideological spokesmen for the poor and the oppressed. Doubtless they had sympathy for the oppressed, but their primary motives were essentially religious. Yahweh and his commandments were being forsaken and his covenant violated. It was this consideration that prompted their denunciations of the kings and the patricians.

Hence the growing rationalization of Israelite religious culture, Weber maintained, resulted from the increasing struggles of the prophets and other devout elements against any and all violations of Yahweh's will— whether such violations be magic, idolatry, or social injustices. What the prophets demanded was an unswerving devotion to Yahweh.

Such devotion was based on the unique relationship of Israel to its God, a relationship expressed and guaranteed in a momentous historical experience—the exodus from Egypt and the conclusion of a covenant with Yahweh. The prophets and the people always hearkened back to that miraculous historical event in which God liberated the Israelites from Egyptian bondage. That was proof not only of God's power but of the absolute dependability of his promises. Israel, then, owed a lasting debt of gratitude to serve Yahweh and to have no other gods beside him. The rational covenant, unknown elsewhere, created an unconditional ethical obligation on the part of the people.

More, the rational character of Israelite religion lay in the *worldly* character of God's promises to Israel. Not some supernatural paradise or nirvana was held out but rather that "the people would have numerous descendants . . . and that they would triumph over all enemies, enjoy rain, rich harvests, and secure possessions." And to Moses was held out the hope of leading his people out of bondage and into a promised land in *this* world. "The god," wrote Weber, "offered salvation from Egyptian bondage, not from a senseless world out of joint. He promised not transcendent values but dominion over Canaan . . . and a good life."[48]

The worldliness and rationality of Israelite religion also emanated from the covenant relation. When

Yahweh was angry and failed to help the nation or the individual, a violation of the *berith* [covenant] with him had to be responsible for this. Hence it was necessary for the authorities as well as for the individual . . . to ask which commandment had been violated. Irrational divination means could not answer this question, only knowledge of the very commandments and soul searching. Thus the idea of the *berith* . . . pushed all scrutiny of the divine will toward . . . a relatively rational mode of raising and answering the question. Hence, the priestly exhortation . . . turned with great sharpness against soothsayers, augurs, day-choosers, interpreters of signs, conjurors of the dead, defining their ways as characteristically pagan.[49]

A rational knowledge of Yahweh's will thus became the paramount concern of Israelite religion. In contrast to the Far Eastern religions, the Israelite faith was from the very first worldly and "disenchanted." It was the beginning, for Weber, of the "disenchantment of the world" which had been carried out more thoroughly in the West than elsewhere.

Subsequently the entire Western culture underwent rationalization, so that by the time of the modern era there were in principle no mysterious, unknowable, or inscrutable powers, and humanity could master all things through rational methods. Virtually all facets of Western culture—economics, politics, large organizations, law, and even music—have been molded by the rational mentality. Ultimately it was that mentality that enabled capitalism to emerge first in the West. Thus Weber sought to document the important role of religion in shaping modern civilization.

THE SECULARIZATION THESIS

The classical theories reviewed here together with the results of some contemporary studies have given rise to the *secularization thesis*. The notion of "secularization" was first systematically set forth by the eighteenth-century Enlightenment thinkers. The application of the scientific-rational spirit to ever more social spheres, they believed, would result in the decline and ultimate disappearance of religious myths, rites, and beliefs. As a consequence modern culture was destined to become more and more pragmatic-minded and worldly—or in a word, secular.

Perhaps the best way to begin an examination of the secularization thesis is to ask whether the classical theorists considered in this chapter would have agreed with it. The answer seems to be that Marx and Freud were proponents of secularization while James and Durkheim were not. As for Weber, his attitude was complex and ambivalent.

For Marx, as we have seen, religion was the product of human self-alienation. Individuals create and worship fantastic beings because they are divided against themselves. The suffering they experience is the result of oppression and sharp social inequalities. Under those circumstances individuals are determined, manipulated, and unfree. They have no rational control over their lives. So long as human beings continue to live in societies based on class domination, they will continue to need the consolation of illusory beings. However, if humans could reconstruct their social relations so as to create a new kind of society founded on freedom and equality, alienation would disappear and with it the otherworldly fantasies that humans have engaged in from the beginning of history. In Marx's view, then, postcapitalist society would become increasingly secular.

Freud too viewed religion as illusion. Whatever truth is contained in religious doctrines is highly distorted and disguised. What is done in religion is "similar to what happens when we tell a child that newborn babies are brought by the stork." Freud vehemently disagreed with those who described as "religious" anyone with a feeling of impotence and insignificance in the face of the universe. No, said Freud, "the essence of the religious attitude is not this feeling but only the next step after it, the reaction to it which seeks a remedy for it." And a *real* remedy, Freud believed, can only be found by scientific means. For "science is no illusion. But an illusion it would be to suppose that what science cannot give us we can get elsewhere."[50] Thus Freud also saw the advance of science as a secularizing agency which would (and should) dislodge and supplant religion.

William James, in contrast, viewed religion as endemic to the human species. Every human feels on occasion that he is passing through the valley of the shadow and longs to come out in sunlight again. Every man experiences an uneasiness, "a sense of something wrong about us," 255 and longs to be "saved from the wrongness by making proper connection with the higher powers."[51]

While James never expressly discussed the secular theories of Marx and Freud, it seems clear that he would have rejected their conclusions. For there are forms of "wrongness"—sorrow, pain, and illness—for which social reconstruction and science can provide only a partial remedy; and another wrongness—death—for which they can find no remedy at all. In these terms James, like some existentialist philosophers, would have seen definite limits to secularization. Some forms of religious experience will never disappear because the human need for comfort and solace cannot be met by alternative agencies.

Although Durkheim anticipated a decline in the traditional religions, he would have agreed with James that the total disappearance of religion was unlikely. But while James traced the reasons for the perpetuation of religion to the nature of the individual, Durkheim attributed them to society. There "is something *eternal in religion*," wrote Durkheim,

which is destined to survive all the particular symbols in which religious thought has successively enveloped itself. There can be no society which does not feel the need of upholding and reaffirming at regular intervals the collective sentiments and . . . ideas which make its unity and personality. Now this moral remaking cannot be achieved except by the means of reunions, assemblies and meetings where the individuals, being closely united to one another, reaffirm in common their common sentiments; hence come ceremonies which do not differ from regular religious ceremonies, either in their object, the results which they produce, or the processes employed to attain these results.[52]

What Durkheim seems to be saying here is that however rational and scientific societies are bound to become, they will never be able to dispense with religious ceremony. It is in the very nature of society that it reconstitutes and revitalizes itself by religious means—rites, assemblies, doctrines, etc. Traditional religions may decline and even disappear, but they will inevitably be replaced by new national or civil religions—a point to which we shall presently return.

Finally, there is Weber's attitude toward secularization. Weber defies easy classification in that regard. It is true that he saw all major spheres of modern social life succumbing to a kind of formal, technical rationality. From that angle, he would seem to accept elements of the secularization thesis. On the other hand, however, he believed that rationality has its limits and that fundamental human values cannot be derived from science. For Weber, in fact, there was an unbridgeable gulf between reason and values. With respect to moral conscience, Weber once wrote: "Here we reach the frontiers of human reason, and we enter a totally new world, where quite a different part of our mind pronounces judgments about things, and everyone knows that its judgments, though not based on reason, are as certain and clear as any logical conclusion at which reason may arrive."[53] In the same context, he wrote that "the idea of the Good and the contemplation of the Beautiful rest on laws which are basic for human nature." Those statements suggest that for Weber the realm of ultimate human values possessed a validity that one could never test by scientific-rational means. Human beings derived their sense of life's meaning and purpose from that realm. Secularization, therefore, can never totally eliminate fundamental religiomoral experiences.

The Contemporary Debate on Secularization

What is the state of the secularization thesis among sociologists today? Some scholars maintain that the advance of science and industrialization inevitably leads to a decline in religion, and others hold that that claim has not at all been demonstrated. There is no firmly established relation, the latter insist, between religion and modern industrial institutions.

Peter L. Berger, a well-known American proponent of the secularization thesis, has defined secularization as the "process by which sectors of society and culture are removed from the domination of religious institutions and symbols." In the history of the West, that process has manifested

itself in the evacuation by the Christian churches of areas previously under their control or influence—as in the separation of church and state, or in the expropriation of church lands, or in the emancipation of education from ecclesiastical authority. . . . As there is a secularization of society and culture, so is there a secularization of consciousness. Put simply, this means that the modern West has produced an increasing number of individuals who look upon the world and their own lives without the benefit of religious interpretations.[54]

Brian Wilson, a British scholar, is another leading proponent of the secularization thesis.[55] His argument rests on statistical evidence showing that the power and influence of the Church of England has steadily diminished over the past 100 years. One indication of diminishing church influence is the declining number of baptisms, confirmations, parish enrollments, Sunday-school classes and teachers, and Easter

257

communicants. Another indication is the growing proportion of civil over church marriages. Both Wilson and Berger agree that secularization is closely associated with the industrialization of society: As people become more involved in industrial production and its offshoots, they become less religious.

There is considerable truth in the secularization thesis as applied to Western history. But it is not a completely valid generalization since the United States, the mostly highly industrialized society, stands out as an exception. Studies of church membership in the United States reveal a quite opposite trend. In 1960, 63 percent of the American population were church members as compared with only 20 percent in 1880.[56]

Wilson and Berger accept the reliability of those figures, as do most sociologists. But both of them have tried to dismiss the importance of the American case. Berger writes, for instance, that "the situation is different in America, where the churches still occupy a more central symbolic position, but it may be argued that they have succeeded in keeping this position only by becoming highly secularized themselves, so that the European and American cases represent two variations on the same underlying theme of global secularization." And Wilson, in a similar vein, tries to make light of the popularity of American religion by describing it as superficial. However, as Gregory Baum has observed, the arguments of Berger and Wilson are unacceptable on methodological grounds. For if "the theory of secularization intends to express a law regulating religion and secular society, then it is inadmissible to introduce qualifications that exclude some forms of religion from being admitted as evidence."[57]

The American case must be viewed as a real exception and ought to be explained. Andrew Greeley, in his valuable book *The Denominational Society,* has offered an explanation. He begins by examining the concepts of "church" and "sect," concepts generally employed by sociologists for an analysis of religion. Those concepts, derived from European experience, are misleading, Greeley notes, when applied to the United States, which never had an established church. American religion may best be understood as organized in *denominations.* It is in and through denominations that adherents gain a sense of belonging and meaning. Comparing Europe and the United States, Greeley observes that

in the disorganization, personal and social, that occurs as a part of the pilgrimage from the peasant communalism to the industrial city, man attempted and still attempts to compensate for the deprivation he endured and for the absence of social support and the intimacy of the village by evolving quasi-

Gemeinschaft institutions—the nationality groups, the lower-class religious sects, the radical political party and, in the United States and Canada, the denomination.[58]

In contrast to a "church," a denomination recognizes that it is only one among many diverse religious groups; and in contrast to the closed "sect," a denomination is typically open. But there is no good reason for looking upon American denominational religion as either "superficial" or "secular."

The secularization thesis is problematic for another reason. It takes for granted that the decline in religious affiliation in Europe, and wherever else it might occur, necessarily implies a secularization of individual consciousness. But that assumption remains unproved. There is no reason to suppose that a necessary relation exists between institutional and personal religion. To return once again to William James, any adequate understanding of religion must include *"the feelings, acts, and experiences of individual men in their solitude, so far as they apprehend themselves to stand in relation to whatever they may consider the divine."*[59]

Finally, the proponents of secularization have apparently overlooked Durkheim's conclusion that "there is something eternal in religion." Applying that dictum to the United States, a contemporary sociologist, Robert N. Bellah, has coined the term *civil religion*. He notes that despite the secular appearance of American society, its public life is characterized by a religious dimension. He cites the frequency with which references to God are made during the inauguration of presidents and on other solemn occasions. Bellah recognizes that some observers might be inclined to dismiss these references to God as having only a superficial ceremonial significance. A cynic may go even further and say that the semblance of piety is an unwritten qualification for office, "a bit more traditional than but not essentially different from the present-day requirement of a pleasing television personality."[60]

However, Bellah contends that when a president or any other high American official makes reference to "God," he does so because that general term is most consistent with (1) the denominational character of American religious life, and (2) the constitutional separation of church from state. And Bellah asks, "Considering the separation of church and state, how is a President justified in using the word 'God' at all?" His reply is

that the separation of church and state has not denied the political realm a religious dimension. Although matters of personal religious belief, worship, and association are considered to be strictly private affairs, there are at the   259 same time certain common elements of religious orientation that the great majority of Americans share. These have played a crucial role in the development of American institutions and still provide a religious dimension for the whole fabric of American life, including the political sphere. This public religious dimension is expressed in a set of beliefs, symbols, and rituals that I am calling the American civil religion.[61]

The inclusion of references to God indicates that the people and their representatives consider themselves responsible to a higher principle. Properly understood, that is not a deification of American society. It is not a form of national self-worship but rather "the subordination of the national to ethical principles that transcend it and in terms of which it should be judged."[62]

Of course, the appeal to higher principles can be abused and employed for nationalistic, oppressive, and other self-interested purposes. In such cases the religious symbolism becomes ideological. On the other hand, the appeal to higher ethical principles and ideals can become the ground for overcoming injustices and widening the boundaries of freedom. It seems fairly clear that however secular societies may become, human beings will continue to sustain themselves individually and socially by believing in higher principles and appealing to them.

11

Social Movements
and Collective Behavior

Throughout history there is evidence of concerted social action on the part of large numbers of people. Religious, political, and social movements appear again and again. Often a movement begins with a single leader of striking qualities and a small band of followers. At first such a band is scarcely distinguishable from numerous similar ones in the vicinity. In time, however, it does distinguish itself from the others. Its utterances and practices begin to appeal to ever-larger numbers of people, and the originally minute circle finds its doctrine spreading far and wide. It has somehow fostered a mass movement sweeping across borders and capturing the imagination of broad masses of converts. That is the remarkably dynamic process which social scientists call a *social movement*.

Not all mass actions are social movements. Mobs, crowds, and wildcat strikes are relatively spontaneous and short-lived. They may be symptoms of social discontent and unrest; they may be facets of an incipient movement or its tactical devices. But in and of themselves, they do not qualify as social movements. For they are forms of collective behavior that are lacking in durability and in the *distinctive structure* of genuine social movements.

We may begin to grasp the "distinctive structure" of a social movement by noting its essential elements. Speaking in general terms, every social movement owes its growth to four sociological factors: (1) *ideology*, (2) *organization*, (3) *charismatic leadership*, and (4) *the prevailing social context*.[1] The last factor is fundamental, for it is the   261 social context that constitutes the fertile ground from which the movement rises and from which it continues to draw its nourishment. However, it is all four elements taken together that account for the comparative durability and success of a social movement. The analytical usefulness of the four factors may best be demonstrated by applying them to definite historical examples: (1) the Nazi movement, which led to the Second World War; (2) the Chinese Communist movement, which resulted in the consolidation of Communist rule over the Chinese mainland; and (3) the American civil rights movement, which has affected the consciousness of all Americans.

THE RISE OF NAZISM IN GERMANY

With Nazism as with other social movements, it is the prevailing social context that enables us to understand why and how the movement arose.[2] Nazism emerged in Germany in the aftermath of World War I, following the country's defeat and surrender. In the months prior to the defeat, the German army had suffered several decisive setbacks which came as a shock to the majority of the people, for the monarchy had misled them into believing that victory was ensured. The people at home, war weary and enduring hunger and privation, now suffered a severe blow to their morale. Food riots and mob violence were daily occurrences, and political street meetings and mass demonstrations became ever more frequent. The kaiser's regime increasingly lost prestige in the eyes of the people, while authority broke down both at home and at the front, where soldiers deserted and sailors mutinied. The majority of the people burned with anger and resentment toward the government which had brought so intolerable a situation upon them. The result was that when defeat finally came in November 1918, the kaiser was forced to abdicate. The old regime fell, and a republic was proclaimed in its place. That was the famous Weimar Republic, so named because the new democratic constitution was drawn up in the city of Weimar.

Chaos continued to reign in Germany despite the proclamation of the Republic. One major reason for the continuing unrest was that the main pillars of the old regime—the army, police, civil bureaucracy, etc. —looked upon the Weimar government with contempt. The Republic, from their standpoint, was made up of Socialists and other plebian upstarts who deserved no respect and even less loyalty. They yearned for the restoration of the old order and sought to undermine the Republic by force from the moment of its birth.

The other major reason for the prevailing violence and anarchy was the conflict between the political parties of the center and left. The Social Democrats favored the Republic and a peaceful, parliamentary evolution toward socialism. In the elections of 1919 to the National Assembly, they received 13,800,000 votes out of 30,000,000 cast, and they won 185 out of 421 seats. Hence they were considerably short of a majority. Yet they agreed to govern alone because no other party, whether working-class or middle-class, would share the burden. Although the two middle-class parties, the Catholic Center and the Democratic party, professed support for a moderate republic, there was considerable sentiment among them for the restoration of the monarchy.

At the same time there was a deep and growing split in the ranks of the left-wing parties. The Social Democrats were vehemently opposed by the German Communists or Spartakists, as they were called. Led by Karl Liebknecht and Rosa Luxemburg and aided by Soviet Russia, the Spartakists called for an armed insurrection. With the recent Russian Revolution in mind, they turned to the numerous soldiers' and workmen's councils that had sprung up in Germany and called upon them to rise up and to form a dictatorship of the proletariat. Civil war broke out and bloody battles ensued in Berlin, Halle, and several other cities. Soon afterward the Spartakists were decisively defeated by both regular army units and volunteers who rallied to the support of the republican government.

Moreover, the councils themselves rejected the revolutionary path. At their Berlin congress of December 16–20, 1918, the delegates voted overwhelmingly in favor of the Republic and the National Assembly. That outcome reflected the mood of the German people at the time. They were bitter toward the old order, but they had no wish to transform the entire structure of society. Nor did they want violent insurrection and a prolonged civil war. Rather they longed for social stability—to get back to work and to live peaceful, normal lives. That was not only true of the large and powerful middle classes but of the majority of workers as well.

However, peace was not forthcoming. Instead, the society remained in a state of turmoil, with the opponents and supporters of the Republic constantly at odds. The old guard, the military caste, and all others who had a vested interest in the old regime did everything they could to discredit and sabotage the new government. The Republic, 263 on the other hand, did little to defend itself against its assailants. In fact, the new government continued to employ most of the officials of imperial Germany—the very elements that were most hostile to it. Using those elements to combat the extreme left, the Weimar government found itself largely defenseless against its violent enemies on the right. The latter went so far as to attempt a coup d'etat in 1920 (the so-called Kapp *Putsch*), resulting in the assassination of several government leaders. Right wing forces also created all sorts of illegal military organizations with which they harassed and intimidated government and trade-union officials. In sum, the new Republic proved itself incapable of putting an end to the violent conflict and disorganization which the majority of the people found intolerable.

The situation was soon further aggravated by an event that produced a general and profoundly hostile reaction against the Republic.

The Versailles Treaty

The Kaiser's imperial government had provoked the First World War; and it was the officials of that government who finally surrendered to the Allies. Yet it was the Weimar Republic that was blamed for the disastrous consequences of the Kaiser's policy.

In May 1919 the terms of the peace treaty, which the Allies had decided to impose on Germany without negotiation, were published in Berlin. The news came as a staggering blow. Many Germans had come to believe that they were entitled to easy peace terms now that they had rid themselves of the royal family, squashed the Communist revolution, and instituted a republican form of government. But instead of easy terms, they were being subjected to conditions which they regarded as "harsh," "unjust," and "intolerable."

However, if we take a close look at what was to become the Versailles treaty and compare it with others—the Treaty of Brest-Litovsk, for example, which the Germans had imposed on the Russians only one year earlier—then the terms of the Versailles treaty do not appear especially harsh at all. It returned Alsace-Lorraine to France and Schleswig to Denmark, both of which territories Bismarck had forcibly annexed in the late nineteenth century after defeating those countries in war. It returned to the Poles the area taken by Germany in the partition of Poland in the eighteenth century. The treaty also provided that some 800 war criminals were to be turned over to the Allies and that the first payment for war reparations, $5 billion in gold marks or in equivalent goods, be paid between 1919 and 1921. Finally, it disarmed Germany, restricting the army to 100,000 volunteers and prohibiting it from having planes or tanks. The navy was also reduced to a token force.

It is instructive to compare those terms with the conditions imposed on a defeated Russia by the German High Command on March 3, 1918. The Treaty of Brest-Litovsk, writes William L. Shirer,

deprived Russia of a territory nearly as large as Austria-Hungary and Turkey combined, with 56,000,000 inhabitants, or 32 percent of her whole population; a third of her railway mileage, 73 percent of her total iron ore, 89 percent of her total coal production; and more than 5,000 factories and industrial plants. Moreover, Russia was obliged to pay Germany an indemnity of six billion marks.[3]

It is clear, then, that if the Versailles treaty was "harsh" and "extreme," it was certainly less so than the treaty which the Germans themselves had earlier foisted on their defeated enemy.

Nevertheless, the terms of the Versailles treaty were perceived by virtually all segments of German society as a humiliating assault upon the German nation. The treaty, writes Theodore Abel, "cut to the quick the prevalent, highly developed sentiment of nationalism. Inasmuch as it curtailed the power of the nation, deprived it of its prestige, attacked its traditions, and impaired its integrity, it was regarded as a fatal thrust against social values held and shared by the vast majority of Germans."[4] Indeed, opposition to the treaty was nearly unanimous and included not only the pan-German reactionaries but the Social Democrats and Communists as well.

President Ebert and his Social Democratic colleagues consulted with the army. Could the army resist an Allied attack from the west if the treaty were rejected? The answer of Field Marshall von Hindenberg and the High Command was in the negative. They recognized that a resumption of the war would end in the destruction of the German officer corps and, indeed, of Germany itself. Meanwhile the Allies, growing impatient, delivered an ultimatum: Either the treaty will be accepted or the armistice will be terminated and the Allied Powers will "take such steps as they think necessary to enforce their terms." Faced with the ultimatum, Ebert once again urgently consulted with the High Command. He promised to do his utmost to secure the rejection of the treaty by the National Assembly if the High Command saw the slightest chance of successful military resistance. The reply he received was the same as before: "Armed resistance is impossible." In light of that reply the National Assembly approved the signing of the treaty by a large majority and communicated its approval to the French government 265 minutes before the Allied ultimatum expired. Four days later, on June 28, 1919, the peace treaty was signed in the Hall of Mirrors in the Palace of Versailles. Ultimately it was the army that had made the final decision to sign. But that fact never became widely known, and the blame was placed squarely upon the Republic.

General dissatisfaction with the Republic became evident in the elections of June 6, 1920. The parties that had voted for the treaty—the Social Democrats and the Center party—lost a total of 11 million votes, most of which were transferred to the Nationalists and others who had voted against it. The tide had turned against the Republic, and it gathered new momentum from the events of the early 1920s. The mark dropped precipitously from 75 to the dollar in 1921 to 7,000 by the beginning of 1923. Already in 1922 the German government had requested a moratorium on reparations payments and received a blunt refusal from the French. When Germany defaulted in the delivery of timber, French forces occupied the Ruhr, thus cutting German industry off from 80 percent of its coal and steel production. The workers of the Ruhr, encouraged and supported by the German government, responded with a general strike, thus further aggravating the state of the German economy. By August 1, 1923, it required one million marks to buy a dollar; by November it took four billion, and thereafter trillions. German currency had become worthless; the purchasing power of both the middle and working classes was destroyed, and their life savings were wiped out. Who was to blame? In the minds of the vast majority of the German people, the answer was clear: The responsibility for the disastrous state of affairs lay squarely with the Republic. For it had surrendered to the enemy and had accepted the harsh burden of reparations.

Those were the general social conditions which led to the deep and widespread discontent of the German people. And it was that discontent, in turn, which provided the highly fertile ground for Adolf Hitler and the Nazi movement.

The Origins of the Nazi Party

In the general period under discussion, 1919–1923, numerous political circles had sprung up in Germany. In most cases they consisted of acquaintances who met to express their dissatisfaction with existing conditions. One such circle, made up of a handful of workers, met regularly in a Munich beer hall. Led by a machinist named Anton

Drechsler, that circle was the nucleus of what eventually became the National Socialist (i.e., Nazi) party.

Drechsler found no ready-made niche for his political views. In 1918 he had written articles for a Munich newspaper urging the workers to support the imperial government and to prosecute the war until victory was achieved. As a worker he favored the reforms advocated by the Socialists; as an ardent nationalist, however, he opposed the Social Democrats for their internationalism. The aversion for proletarian internationalism was shared by the several other workers who joined Drechsler. The group soon attracted a few intellectuals, also dedicated to German nationalism. Among them was Dietrick Eckart, a poet, journalist, and rabid anti-Semite; Alfred Rosenberg, a disciple and enthusiastic supporter of the racist theories of H. Stuart Chamberlain; Count von Bothmer, an advocate of a form of socialism in which the individual is strictly subordinated to the nation; and other men with similar ideas. Drechsler's circle of nationalistic socialists soon came to the attention of the military authorities, who eagerly subsidized their nationalistic activities. Adolf Hitler, whose regiment was stationed in Munich at the time, was instructed to contact the circle, which called itself "the German Workers' party."

Hitler's recollections of his first meeting with the group are given in his book *Mein Kampf* [My Struggle]. They met in an old dilapidated room in a cheap beer hall. Their discussions struck him as quite boring. There was nothing distinctive about the group; it was just like the many others that appeared everywhere. Hitler participated in the discussion, directing his remarks against a member who advocated the separation of Bavaria from Prussia. At the meeting's end, as Hitler got ready to leave, Drechsler approached him and pressed a pamphlet into his hand, urgently requesting his opinion of it. Hitler accepted it on the chance that he might learn more about the group without having to attend further meetings.

The following morning Hitler read the slim brochure and, much to his astonishment, found it quite interesting. Its author recounted how he had gone through a welter of Marxian ideas only to return to the sound principles of nationalism. Hence the title *My Political Awakening*. What so impressed Hitler about Drechsler's story was that it reflected his own experience 12 years earlier. Hitler goes on to relate how he wrestled for days with the question of whether or not to become a member of the group. Like Drechsler, he felt there was no place for him in the major parties; Drechsler's circle, on the other

267

hand, was unimpressive. Yet upon reflection Hitler sensed that it might offer one distinct advantage: "It had not yet frozen into an organization." And he continues:

The more I thought about it the more convinced I became that it was just such a tiny movement as this that might be made a harbinger of the future welfare of our people. The same end could never be achieved through the standing parliamentary parties that either placed too much weight on old concepts or deliberately profited through the new lords. For the thing that was needed was not a new campaign slogan, but a new world philosophy.[5]

Hitler then gives us some insight into his real motives. He was not only poor but also lacking a higher formal education. He had no credentials with which to impress the leaders of the established conservative and nationalist parties. But the thought was unbearable that he should remain one of the countless numbers whose lives and deaths went unnoticed. After days of painful indecision, he finally made up his mind and joined Drechsler's German Workers' party.

Hitler's entry was a turning point both for him and the group. For there can be little doubt that it was his energy, ideas, ambition, and personality that soon distinguished his group from countless others. As a condition for joining the group, he demanded and received full control of its propaganda activities. The right kind of propaganda, he sensed, would lift this insignificant circle out of obscurity. By exploiting the huge wave of discontent in Bavaria and particularly in Munich, the party would grow and prosper while he became the leader of a movement to be reckoned with.

Munich had just experienced the bloodiest social upheaval in its history. It began with the establishment of a Communist government and ended after a period of terrorism, assassinations, executions, and general disorder. As a consequence the people turned against the Republic. The reaction against communism, together with the treatment accorded Germany by the Allies, led to a powerful revival of nationalist sentiment. Since Jews were conspicuous among the leaders of both the Communists and the Republic, anti-Semitism become more widespread. That was the general mood which Hitler ably exploited. While other agitators attracted only small street audiences, Hitler succeeded by 1921 in recruiting thousands of members to the German Workers' party and thousands more to his weekly meetings.

How did he accomplish those results? By employing propaganda devices which were novel in Germany at the time. As he acknowledges in *Mein Kampf,* he imitated the tactics which the Allies had used during the war. He announced his meetings with large glaring placards; he employed trucks to distribute propaganda material throughout Munich; he made his party highly visible by means of distinctive uniforms and badges; and he organized dramatic parades and loud street demonstrations. He compelled the opposition to take notice of him and his organization by staging its meetings in Social Democratic strongholds and by disrupting their gatherings by means of heckling and violence.

Soon he was able to purchase a daily newspaper and to give it his peculiar stamp. Where did the funds come from to pay for Hitler's expanding organizational and promotional activities? Partly from public collections and a few wealthy sympathizers. But the bulk of his funds came from the military authorities, who were interested in the development of a strong nationalist party.

No less important than his propaganda techniques were Hitler's aggressive tactics. These especially appealed to many young men who had no better prospects under the circumstances. University students and unemployed young men from both middle- and working-class backgrounds became part of his faithful following. They were formed into well-disciplined, quasi-military fighting units; and shows of force and actual violence figured more and more prominently in the party's public activities. Those fighting units were the nuclei of what later became Hitler's infamous storm troopers and special guards, the SA and the SS.

It was soon evident to all concerned who the real leader of the party was. Hitler became a favorite orator. Huge crowds eagerly awaited his appearance and gave him long and enthusiastic ovations. It was not long before his followers bestowed upon him the title *der Fuehrer* [the leader]. A veritable Hitler cult developed. He became a charismatic leader par excellence. That is the man, his followers ardently believed, who would save Germany.

By 1923, as noted earlier, German banks were paying four billion marks for a dollar, and the French army was occupying the Ruhr. The German government now aroused greater animosity than ever for doing nothing in the face of a foreign invasion and a ruinous inflation. Early in the fall of 1923, the several nationalist organizations of Bavaria formed a united front and planned a *Putsch* or coup d'etat. Up to that time Hitler had steadfastly avoided cooperation with other groups. All of his energy was devoted to building his own party and his own personal image. But now he succumbed to pressure and joined the coalition, placing his storm troopers at the disposal of the military. He was appointed leader of the united front and hailed as the German Mussolini. The actions of the front were supposed to precipitate a   269 massive nationalist uprising with the support of the military. But the uprising never materialized. On the eighth and ninth of November, Hitler led his troops in a march against the government leaders of Munich and attempted to arrest them. However, a volley of bullets by troops loyal to the government brought a quick end to the *Putsch.* Hitler fled but was soon afterward apprehended, tried, and sentenced to imprisonment.

Although those events were a setback for the movement, Hitler's name now became nationally known. Before the attempted *Putsch,* the German press had practically ignored him and his organization. Afterward, however, the press was forced to give space to the event and to Hitler's speeches during the trial. He became a national figure overnight. Hitler had also learned important lessons from the experience of the abortive *Putsch.* As Theodore Abel has observed,

Hitler became convinced that his aim could not be realized unless he had the majority of the nation behind him. Consequently he intensified his propaganda, and directed his organization in conformity with legal procedure. Hence the participation of the Party in the Reichstag elections, and his own candidacy for the presidency. The *Putsch* furthermore convinced Hitler that he could not count on the support of other groups. The policy of no compromise was adopted as a standing principle, and afterwards became the germ of the totalitarian state.[6]

After the Putsch

There was another unanticipated consequence of Hitler's setback at Munich. His imprisonment afforded him the opportunity to pull his ideas together into an integrated ideology. It was during his prison term that he wrote or, more correctly, dictated, his autobiography *Mein Kampf* to Rudolf Hess. In that book Hitler formulated his movement's ideology, strategy, and tactics.

When Hitler was freed from prison in November 1924, he faced a discouraging situation. His followers were scattered among other groups; the remains of his party had been badly defeated at the polls; and the nationalist tide had temporarily subsided. Nevertheless, he was determined to rebuild the movement. Now that he was committed to legal means of struggle, he persuaded the Bavarian authorities of that fact and they withdrew the ban on the National Socialist party. On January 27, 1925, he called a meeting at the Munich beer hall which had been the scene of his past triumphs. Over 4,000 people showed up to welcome him, most of them his former followers and lieutenants who, in Hitler's absence, had fought among themselves for his crown.

Hitler now won them over once again and exacted a pledge of exclusive allegiance to him. Thereafter he was the supreme and unquestioned leader. The meeting clearly demonstrated that he had lost none of his appeal despite the failure of the *Putsch.*

Yet in the new social conditions of the period between 1925 and 1929, his movement made very little headway. Germany had come upon better times and now enjoyed some material prosperity thanks to enormous American investments. Between 1924 and 1930 the German government borrowed some $7 billion, which was used to pay its reparation debts and to develop a vast system of social services which evoked admiration throughout the world. State and municipal governments borrowed to finance airfields, theaters, sports stadiums, and public swimming pools. German industry borrowed billions to modernize its productive plant. In 1923 its output had dropped to 55 percent of that in 1913; by 1927 it rose to 122 percent. By 1928 unemployment, for the first time since the war, fell below 650,000. Retail sales expanded and real wages rose. Not only the workers but the lower middle classes—millions of small shopkeepers and small-salaried employees—shared in the general prosperity. The Republic now gained prestige in their eyes, and they had little patience for Hitler, or any other anti-republican agitator for that matter.

William L. Shirer was an American correspondent stationed in Germany in those years. Describing the mood of the people, he writes:

A wonderful ferment was working in Germany. Life seemed more free, more modern, more exciting than in any place I had ever seen. Nowhere else did the arts or the intellectual life seem so lively. In contemporary writing, painting, architecture, in music and drama, there were new currents and fine talents. And everywhere there was an accent on youth. One sat up with the young people all night in the sidewalk cafes, the plush bars, the summer camps, on a Rhineland steamer or in a smoke-filled artist's studio and talked endlessly about life. They were a healthy, carefree, sun-worshipping lot, and they were filled with an enormous zest for living to the full and in complete freedom. The old oppressive Prussian spirit seemed to be dead and buried. Most Germans one met—politicians, writers, editors, artists, professors, students, businessmen, labor-leaders—struck you as being democratic, liberal, even pacifist.[7]

It is not surprising, therefore, that those were very lean years for Adolf Hitler and the Nazi movement.

However, the springtime soon came to an abrupt end. In 1929 the Great Crash occurred. That worldwide economic crisis of unprecedented proportions hit Germany with catastrophic force. Unemployment rose rapidly and soon attained the staggering total of five million.

Thousands upon thousands of workers were driven to join the relief rolls; and large numbers of peasants lost their land for defaulting on their mortgage payments. With the declining purchasing power of the masses, many small businesses fell by the wayside while the general standard of living in the middle classes sank dramatically. Bitterness and discontent mounted steadily. The government found itself powerless in the face of those circumstances and lost the respect it had earlier commanded.[8]

Not surprisingly, the crisis brought new life to the radical parties of the right and left, and thousands of new members daily swelled their ranks. The National Socialist and Communist parties gained most in that period; but the latter was soon dramatically outstripped. By 1930 a rising volume of "Heil Hitler" greeted the Nazi parades. But there was widespread opposition to the Nazis as well, with the result that violent clashes were common throughout the turbulent period from 1930 to 1932. Street wars were frequently waged between the Nazis on the one hand, and the prorepublican, trade-union, and Communist forces on the other. In time it became clear that the odds were overwhelmingly in favor of the Nazis. They had become a formidable political-military force capable of fomenting a prolonged and bloody civil war.

From January 1930 to December 1931, the Nazi movement grew from about 400,000 members to well over 800,000. The increased volume of public support for the movement became evident in the results of the many local elections of that period. By 1932 Hitler felt strong enough to run for the presidency against Hindenburg, the widely respected, legendary war hero, supported by both the conservative anti-Nazi right and the democratic parties. At the close of the election on March 13, 1932, the results were:

Hindenburg	18,651,497	49.6%
Hitler	11,339,446	30.1%
Thaelmann (Communist)	4,983,341	13.2%
Duesterberg (Nationalists)	2,557,729	6.8%

The old President Hindenburg had defeated the Nazi leader by over seven million votes; but he fell .4 percent short of the required absolute majority. A second election was therefore necessary. Hitler had increased the Nazi vote by almost five million since the 1930 elections, but he nevertheless fell far behind Hindenburg. In the second election the Nationalists withdrew Duesterberg from the race, appealing to their followers to vote for Hitler. The results of the second election, on April 10, 1932, were:

Hindenburg	19,359,983	53%
Hitler	13,418,547	36.8%
Thaelmann	3,706,759	10.2%

Despite Hitler's electoral defeat, the Nazis held several trump cards: their formidable organization and massive support; their determination to fight to the finish, by violent means if necessary; and the fact that they were the strongest single party in the Reichstag. Hindenburg and his advisers understood all that. After many long conferences and agonizing indecision, they decided that Hitler had to be given something in the hope of staving off a coup d'etat and a long and bloody civil war. Thus on January 30, 1933, Hitler was appointed chancellor, that is, chief minister of the Reich.

From that strategic vantage point Hitler was able to aggrandize even more power for himself and his party and to begin the nazification of Germany. On the morning of August 2, 1934, Hindenburg died in his eighty-seventh year. Some three hours later, it was announced that in accordance with a law enacted on the *preceding* day, the offices of the chancellor and president had been combined. Adolf Hitler had become the head of state and the commander-in-chief of the armed forces. The title of president was abolished, and Hitler became known as fuehrer and Reich chancellor. Hitler had thus established himself as dictator of Germany. With those acts he had, of course, violated the Constitution, which called for the election of Hindenburg's successor. But the army refrained from interfering and even swore an oath of allegiance to Hitler personally. To lend his usurpation of power an aura of legitimacy, Hitler scheduled a plebiscite for August 15th. He had somehow managed to prevail upon the deceased president's son, Colonel Oskar von Hindenburg, to broadcast the following message on the eve of the voting: "My father had himself seen in Adolf Hitler his own direct successor as head of the German State, and I am acting according to my father's intention when I call on all German men and women to vote for the handing over of my father's office to the Fuehrer and Reich Chancellor."[9] However, historians are generally agreed that the son's statement was false, and that Hindenburg's last wish was for the restoration of the monarchy. It is almost certain that the portion of Hindenburg's last will and testament conveying that wish was suppressed by the Nazis.

On August 19, 90 percent of the German voters registered approval of Hitler's usurpation of total power. Now, with total control of the means of violence and coercion in their hands, the Nazis could deal mercilessly with all their opponents, who were thrown into concentration camps, murdered, or exiled. Then, by launching a mammoth rearmaments program, Hitler largely wiped out unemployment by 1936. He thus sustained the support of the privileged classes as well as the middle and working classes. The industrialists and financiers enthusiastically supported the profitable business of arms and munitions manufacturing; and the masses were happy to have jobs again, even at the cost of their personal freedom and an austere diet ("Guns before Butter").

The rest of the story is well known. The Third Reich, which Hitler promised would endure for 1,000 years as the master of all the "lower races," lasted a mere 13. But in that comparatively short span of time, the Nazi "beast from the abyss" inflicted untold suffering and death on humanity.

The Key Factors Accounting
for Hitler's Rise to Power

We can now return to the four main sociological factors which help us understand why and how the Nazis became a mass movement while other groups either disappeared or were absorbed. The objective conditions existing in postwar Germany bred massive privation and bitterness. The factor of prevailing social context may therefore be summed up under the catch word *discontent*. Discontent was a necessary but not sufficient condition. It explains why many radical groups sprang up, but not why the Nazis distinguished themselves by their phenomenal growth. Only when we add three more factors do we see clearly why the Hitler movement in particular prospered.

The factor of *ideology* was crucial, as we have seen. The conservative parties of the right, representing the privileged classes of Germany, were highly nationalistic. But their brand of nationalism failed to appeal to the masses of the people because it ignored their plight. The Socialist and Communist parties, on the other hand, espoused an internationalism which refused to recognize the powerful nationalist sentiments of the people. It seems evident, then, that the major reason for the ideological appeal of the Nazis was their combination of nationalism and socialism in their party's name and program. The concept of National Socialism was not new: It had first appeared in the late nineteenth century. But as Theodore Abel has noted, "The fact that a mass movement was created in its name . . . [was] an original accomplishment. . . . The timeliness of the idea, in view of prevalent conditions, favored [Hitler's] effort and in part accounts for his success."[10]

The third factor was *organization,* which includes the Nazi party's structure, propaganda techniques, strategy and tactics. At the core of the Nazi movement was a highly disciplined, aggressive militaristic organization. By employing modern advertising methods, the Nazis achieved the mass psychological effects they desired. At the same time they intimidated their opponents with shows of force and readily engaged in violence against them.

The fourth factor was *charismatic leadership.* Max Weber, who was among the first to employ that analytical concept, defined "charisma" as "a certain quality of an individual personality by virtue of which he is considered extraordinary and treated as endowed with supernatural, superhuman, or at least specifically exceptional powers or qualities."[11] There can be little doubt that Hitler was a charismatic leader. The members of his movement were "followers" in the true sense of the word. They submitted to him obediently and unquestioningly as to a demigod. The charisma of its leader was therefore one more decisive advantage which the Nazi movement had over all others in Germany at the time. Taken together, then, the concepts of social context, ideology, organization, and charisma appear to be invaluable for the analysis of social movements. Let us now see how those concepts can be applied in an analysis of other movements.

THE CHINESE COMMUNIST REVOLUTIONARY MOVEMENT

Massive discontent, we have seen, is often a basic factor in the emergence of a major social movement. What were the social conditions underlying such discontent in imperial China?

The earliest beginnings of a Western type of industrialization did not occur in imperial China until the second decade of the twentieth century. Indeed, China has only recently begun to lay the foundations of a modern industrial economy. Why there was no Industrial Revolution in China as in the West is an interesting question in its own right. In his monumental study *Science and Civilization in China,* Joseph Needham has shown that between the third and thirteenth centuries B.C., China maintained a level of scientific knowledge which surpassed 275 that of the West. However, in subsequent centuries China gradually began to fall behind in science and technology. By the nineteenth century, when several Western European countries and the United States had undergone significant industrialization, China suddenly found itself in a state of scientific-industrial backwardness. It was that economic backwardness that enabled the Western powers from the mid-nineteenth century on to invade China and to maintain it in a state of subjection. In the Opium War, for example, Chinese musketeers, mounted archers, and warjunks proved themselves no match for Western gunboats.

Scholars have offered several explanations for China's relative technological backwardness in modern times. We have seen in the preceding chapter how Max Weber sought to account for the absence of indigenous industrial capitalism in China and India. The nature of Far Eastern religious culture, he believed, was a major inhibiting factor. A related thesis is that there was a pronounced strain of conservatism in Chinese thinking. Daily practical activities were entirely governed by ancestral norms and traditions, and thus Chinese peasants and artisans continued to employ in the nineteenth century the same type of tools and techniques which had been in use several centuries earlier. Furthermore, Chinese Confucian intellectuals were entirely absorbed in literary and humanistic studies, paying no attention whatsoever to science. In contrast, the highly educated man of post-Renaissance Europe experienced a great surge of interest in scientific questions.

Other important reasons for the absence of an Industrial Revolution in China are of an economic nature. That point may be best brought out by comparing imperial China's situation with that of England, the first industrial country. England's industrial capitalism was made possible by the accumulation of capital from the seventeenth century on through lucrative foreign trade, maritime shipping, and the exploitation of colonies. The ever-increasing foreign market for goods presented England with fabulous profit-making opportunities if only it could produce those goods in the required volume. Every effort was therefore made to raise the productivity of labor. The application of science to production led to the invention of steam-powered machinery and other innovations that soon became the foundation of a new type of industrial economy. China, in contrast, had neither substantial foreign trade nor profitable colonial possessions. The domestic demand for manufactured goods was adequately met by handicrafts.

When we take these several factors—religious, cultural, and economic—into consideration, it is not difficult to see the causes of imperial China's relative backwardness.

The Agrarian Economy and the "Population Problem"

Throughout most of Chinese history, the traditional economy sufficed to feed and clothe the population. There were of course the periodic crises characteristic of all agrarian economies—famine caused by drought, flood, other natural calamities, and war. But there was no marked population pressure on resources. In the mid-eighteenth century that state of affairs began to change. The Chinese population increased from approximately 150 million in 1700 to 313 million in 1794, more than doubling in one century.[12]

The enormous population increase has been attributed, first, to the introduction in China of drought-resistant crops such as peanuts, corn, and sweet potatoes; and, second, to the prolonged period of peace that began in the latter half of the seventeenth century. The dramatic growth of the Chinese population created new economic problems for the society. In the absence of scientific-technological advances in agriculture, the population increase could only result in a progressive lowering of the standard of living.

However, rapid population growth was only one facet of the problem. Equally important was the class structure of the countryside. Throughout the Manchu or Ch'ing dynasty—as it is known among Sinologists—there existed in China a class of wealthy landlords as well as a stratum of powerful mandarin officials and scholars.[13] The two groups were not identical though there was considerable overlap between them.

Landlords and officials were often connected by family ties. Fortunes acquired in the imperial service through tax farming were invested in land. On the other hand, large landlords with aristocratic pretensions had to have a degree holder in the family. As a rule, wealth was a prerequisite for entry into the Confucian schools, though occasionally a bright but poor young man was sponsored by a wealthy family. Hence there were links between family, land, and office. The Chinese countryside was dominated by a landlord-official-scholar class, which we may call a *gentry*.

With the huge growth of population that began in the eighteenth century, formerly well-to-do peasants became poor while poor peasants became impoverished. The vast majority of China's inhabitants earned their living by working for rich landlords in tenancy arrangements not unlike those found until recently in the southeastern United States. They were "sharecroppers" and hired laborers. The lord played no apparent role in agriculture but was rather a man of leisure. This socio- 277 economic arrangement was enforced by the imperial government. Indeed, the emperor was himself a super landlord, collecting grain from his subjects.

The Chinese landlord had a definite material interest in the existence of a large mass of land-hungry peasants. Their competition for access to the land raised his rents. However, it was not population pressure alone that created the land hunger of the peasants. The absence of primogeniture, that is, the exclusive right of inheritance belonging to the eldest son, led to an extreme parcellization of the land of the poor. Equal division of the inheritance among several sons meant that a family could become impoverished in a few generations. The well-to-do were able to prevent such a misfortune by sending a son to the Confucian schools to become an imperial official. The son's income from tax farming enabled the family to purchase land and to replenish its holdings. But that avenue was closed to the poor peasant.

Thus we see the conditions that led to the impoverishment of the Chinese peasant masses in the nineteenth century. Not surprisingly, they became increasingly bitter toward the landlords whom they looked upon as parasites—particularly as they made no productive contribution to the village economy. The lords thrived in that system so long as the state could guarantee their property rights and ensure the collection of rents. But with the rising tensions between lord and peasant, the state found it more and more difficult to keep the social peace. Banditry and local insurrections became endemic, and the lords lived in constant fear of revolutionary upheavals.

There were no social mechanisms in the imperial system to prevent the lords from squeezing the peasants to the utmost; yet the more they did so, the more they inevitably generated peasant resentment. Nor were there any mechanisms by which to halt the deterioration of the government's administrative control over the country. A modern governmental bureaucracy with modern tax-collection methods has considerable control over its officials. They are salaried and therefore directly dependent on the government for their livelihood. However, the far-flung agrarian bureaucracy of imperial China was much less efficient both in collecting taxes and controlling its officials. Corruption was rampant among the tax-farming officials who pocketed anywhere from four to nineteen times their "salary." The weakening of the imperial apparatus coincided with, and was accelerated by, the entrance of the Western powers.

The Western Encroachment on China

In the 1830s British merchants peddled large quantities of opium in Canton and other Chinese cities. After several futile protests in 1838, the Chinese government stepped in, confiscated the drug, and destroyed it. The British government then decided to retaliate by force. The result was the "Opium Wars," which ended in a British victory and in an "open-door" policy. One Western power after another forcibly imposed "unequal treaties" upon China, depriving it of all its dependencies and snatching large portions of Chinese territory—an area in excess of four million square kilometers. In addition, China was compelled to grant concessions and privileges such as the following:

1. The opening of over 60 "treaty ports" for foreigners and foreign trade.
2. The creation of "concessions" within certain Chinese cities such as Shanghai, Hankow, and Tientsin.
3. The designation of some Chinese places such as Kowlow, Tsingtao, Kwangchowwan as "leased territories."
4. The granting of the "most favored nation" treatment to foreign countries, i.e., whenever one country obtained a new privilege from its treaty with China, all the other countries would enjoy the same benefits.
5. The deprivation of China's rights to adjust her tariff rates, to administer her customs, and to appropriate her customs revenues.
6. The right of coastal trade and the right of inland navigations.
7. The right to send missionaries anywhere in China.
8. The right to station foreign troops in the legation areas of the Chinese capital and along the railroad from Peking via Tientsin to Shanhaikwan, as well as in a number of other places.
9. The employment of foreign postal employees and the establishment of foreign postal offices.
10. The right to establish factories in the treaty ports and the right to secure reduced taxes or complete tax immunity for the articles manufactured by such factories.
11. The right to clear the channels of inland waterways and the privileges of employing foreign pilots and erecting buoys, navigating marks, lighthouses and watchtowers.
12. The right to build railroads, to exploit mineral resources, and to issue currency.
13. The denial to the Chinese of the right to fortify certain strategic seaports.
14. The payment of large sums of indemnity.
15. The establishment of foreign consular jurisdiction (extraterritoriality) in China.[14]

China thus came under the colonial domination of both the Western powers and Japan. The numerous "concessions," "treaty ports," and 279 "leased territories" were powerful outposts from which the foreigners conducted military operations against China and kept it in a state of subjection. The unequal treaties and the denial to China of an independent tariff policy meant that the country was soon flooded with foreign merchandise. As a result China's traditional domestic handicraft industry was undermined. Since handicraft manufacturing was a vital part of the peasant economy, its rapid decline markedly reduced the income of the peasants, further aggravating their plight. Consequently, from the mid-nineteenth century on they turned again and again to insurrection and rebellion. The following are the most notable: the Taiping Rebellion, 1850–1864; the Nein, 1848–1868; the Miaotze, 1855–1872; the Mohammedan Rebellion in Yunnan, 1855–1873; and the Mohammedan Rebellion in the northwestern provinces, 1861–1878.[15]

The Revolution of 1911 and the Rise of the Republic

Following these insurrections and the defeat of China in the first Sino-Japanese War (1894–1895), the Manchu court made some half-hearted and unsuccessful attempts at reform.[16] The helplessness of the regime in the face of the country's impoverishment and subjection by foreign powers was plain for all to see. Chinese intellectuals now also became revolutionary. They organized themselves in 1905 under the leadership of Dr. Sun Yat-sen, and in the ensuing years they drew to themselves a large number of urban, middle-class followers. On October 10, 1911, they staged an uprising in Wuchang which promptly inspired anti-Manchu revolts in several other cities.

When the premier, Yuan Shih-kai, was called out from his enforced retirement to suppress the rebels, he betrayed the court. Instead of crushing the revolution and saving the dynasty, he temporarily allowed the rebels to hold their positions. That enabled him to persuade members of the court that abdication was the only course open to them. At the same time he made clear to the rebels the futility of further struggle with his vastly superior forces. The rebels came to terms with Yuan, and the price they paid for his assistance was his election to the presidency of the Republic in February 1912.

However, the price soon proved to be even dearer. For it soon became evident that the so-called Republic was a "phantom." Yuan maintained control of the former imperial army, which remained intact; the bulk of his administration continued to consist of the corrupt bureaucracy of the fallen dynasty; and he totally ignored the aims of the revolutionary party which now called itself the Kuomintang (KMT). With a majority in the legislature, the KMT made several efforts to restrict Yuan's power. The constant friction between the executive and the legislature finally led to Yuan's illegal dissolution of both the parliament and the KMT in November 1913.

Soon afterward Yuan sought to restore the monarchy with himself as emperor. That action precipitated a series of revolts in south and central China protesting the proposed change. The movement ended with the death of Yuan in June 1916, an event that was a mixed blessing. Yuan's regime had been the sole stabilizing force in the country, and his death plunged the society into internecine warfare and chaos. Power fell into the hands of local satraps constantly at war with one another. The "warlord" period had begun.

The Kuomintang and the Chinese Communist Party

Meanwhile the Kuomintang (KMT) continued to build a coalition which eventually included the newly formed Chinese Communist party (CCP). To understand how and why the Kuomintang turned to the Communist party, we need to say a word about China's situation in the aftermath of the First World War.

In light of Woodrow Wilson's "Fourteen Points" and the Allies' fine-sounding declarations on "war aims," Chinese intellectuals looked forward to a new era following the war. Since China had been one of the Allied powers, the country naturally expected that the Paris peace conference would restore territory and interests lost to Germany and Japan. It therefore was a profound shock to learn that the peace conference had decided to legitimize the Japanese seizure of Shantung province. Demonstrations protesting that decision broke out throughout the country in May 1919. The May Fourth Movement, as it came to be called, resulted in the Chinese delegation's refusal to sign the treaty.

Those events had considerable impact on China's educated youth, notably Ch'en Tu-hsiu, Mao Tse-tung, and Chou En-lai, who were highly impressed with the efficacy of the May Fourth Movement. Disillusioned with the policies of the Western powers, they took an active interest in the recent Bolshevik Revolution in Russia. Indeed it was Soviet Russia's friendly gestures toward China which persuaded Ch'en Tu-hsiu and his colleagues to accept Marxism and to form the CCP. In contrast to the behavior of the Allies, the Soviet government announced in July 1919 that it would return the Chinese Eastern Railroad and all other concessions seized from China by the czarist regime. That

281

policy generated considerable pro-Soviet sentiment among Chinese intellectuals. A society for the study of Marxian theory was instituted at the National Peking University in 1920, and the Communist Manifesto was translated into the Chinese language for the first time. The fascination with Marxism soon affected the Kuomintang leaders, who began to use Marxian theories in their interpretation of Chinese history.

It was at that propitious moment that the Communist International (Comintern) dispatched Gregor Voitinsky, the secretary of its Far Eastern Bureau, to China. He made contact with Ch'en Tu-hsiu and his colleagues, who in turn introduced Voitinsky to a group of Socialists, anarchists, and Marxists. In May 1920 the group formed the Chinese Communist party. A vigorous organizational campaign was then launched in several major Chinese cities.

The First United Front with the KMT

At its party congress in May 1922, the CCP joined the Communist International. The official ideological position of the Comintern at the time was that China was not yet ready for a proletarian Socialist revolution. As an economically backward, semicolonial country, it had yet to develop the industrial capitalist system which was a precondition for the transition to socialism. Therefore in the first stage of the Chinese revolution, it was the duty of the CCP to give "active support" to a national movement against both foreign powers and domestic exploiters. Such support entailed the cooperation of the CCP with the bourgeois-republican elements. At the same time, however, the Communists were to develop their own proletarian movement in preparation for the second stage of the revolution, the struggle *against* the bourgeoisie and *for* socialism.

In accordance with Comintern directives, the CCP in 1922 called for a democratic united front of the proletariat, poor peasantry, and petty bourgeoisie against imperialism and warlordism. The KMT was invited to a joint conference with the CCP for the purpose of building the united front. The CCP at first insisted on creating a two-party alliance with the KMT, with the two groups remaining distinct and independent. However, when Dr. Sun Yat-sen rejected those terms but agreed to have CCP members join the KMT as *individuals,* his offer was accepted. Sun's requests for aid had been recently rebuffed by the West, and he hoped that working with the Communists might secure Soviet Russia's assistance for the KMT in its struggle against the warlords. The common ground between the CCP and the KMT at that time was the

282

recognition that the defeat of the warlords was a precondition for the economic development of China.

Sun's expectations were partially fulfilled. A leading military leader of the KMT, a young general named Chiang Kai-shek, was invited to Moscow in the summer of 1923, where he studied the organization of the Red Army. When he returned he formed the Whampoa Military Academy, which became the nucleus of the Kuomintang army. In the years 1923–1926, Communist influence in the KMT grew steadily. CCP members came to occupy key positions in the organization, a fact which caused great concern among the more conservative KMT leaders. Yet open intraparty conflict was avoided so long as Sun Yat-sen was alive. With his death, however, in March 1925, the KMT leaders launched an anti-Communist campaign with the aim of expelling all Communist elements from the Kuomintang. By that time Chiang Kai-shek had emerged as one of the two most powerful leaders of the KMT. He had been suspicious of the CCP from the very beginning of the collaboration agreement. But the time was not yet ripe, he believed, for a full split with the CCP. As commander of the KMT army he knew that Soviet assistance was necessary for the projected military campaign against the northern warlords.

The turning point came in the course of the Northern Expedition, which began in July 1926. In the territories taken over from the warlords, the Communists established their leadership by forming peasant associations under their control. They armed workers and peasants and trained them for insurrectionist activity. Matters came to a head soon after Chiang entered Shanghai with his troops. Securing the support of bankers, merchants, and anti-Communist Kuomintang leaders, he decided to strike at the CCP and its supporters. On April 12, 1927, he smashed the Communist stronghold and executed a large number of Communist and left-wing workers and students. In the following months similar executions were carried out in other cities.

The Policy of Urban Uprisings

Despite that disaster for the Communists, Stalin and his supporters in the Comintern instructed the CCP to initiate a series of urban uprisings. The Comintern policy plunged the CCP into a number of costly adventures in the cities, each of which met with a resounding defeat. Instead of responding en masse as Stalin had predicted, the urban workers and peasants simply refused to cooperate.

One of those unsuccessful uprisings was led by Mao Tse-tung in   283

Hunan province, where he and his 5,000 followers attempted to seize the provincial capital of Changsha. Poorly armed and rejected by the local populace, the rebels were quickly overwhelmed and forced into disorderly retreat. For his failure Mao was repudiated by the Central Committee of the Party and dismissed from the Politburo. Having suffered heavy losses in Canton, Nanchang, Hunan, and Swatow, the CCP faced extinction in the urban areas. But Stalin and the Comintern took no responsibility for the debacle. On the contrary, they placed the blame for it squarely on the shoulders of the CCP leadership. Moreover even in the face of the setback the Comintern instructed the CCP to continue to base its revolutionary activity in the *cities*.

Mao and the Chinese Revolution

But Mao had learned an important lesson from the failure of the urban uprisings. The experience had taught him that in an agrarian society like China, it was not the city but the countryside that held out revolutionary promise. Following his defeat in Hunan in 1927, Mao retreated into the mountainous region between Hunan and Kiangsi provinces. Soon after his arrival in Chingkangshan, he launched a policy designed to create a rural military base. He did so by exploiting the discontent of the poorest sections of the peasantry and by organizing them into peasant militias. Relying on the poorest peasants and on the landless hired hands, he confiscated the land of the rich landlords and distributed it among the poor.

In that way Mao gradually developed a highly disciplined Red Army indoctrinated with his own version of Marxism. The army relied exclusively on principles of guerrilla warfare, mobilizing the support of the poor peasants wherever it moved. Mao taught his followers to avoid pitched battles with superior enemy forces and, on the other hand, to concentrate superior forces so as to encircle and annihilate enemy units whenever the opportunity presented itself. Thus Mao was primarily responsible for the formation of the chief units of the Red Army and for the creation of the largest and most powerful Communist district in southern Kiangsi. It was there that he maintained his provisional Central Soviet government from 1931 to 1934.

At first the Central Committee of the CCP paid little attention to Mao's new strategy. They looked upon it as incidental to the main business at hand: the organization of the urban workers. But after repeated setbacks and a disastrous reduction of Communist influence in the cities, they reluctantly lent more legitimacy to Mao's activities. His startling successes led them to define the peasant bases as the "driving force," the "determining factor," the "wellspring" of the revolution.

Mao's military and organizational successes in that period were greatly aided by two objective circumstances. The first was the fact that Chiang Kai-shek was preoccupied with dissident Kuomintang generals and former warlords. The second was the Japanese invasion of Manchuria in 1931. Although Chiang had launched several campaigns against the Soviet districts in the years 1931–1933, his offensives were called off each time owing to those other preoccupations. Then in 1933 after the battle of the Great Wall, which ended in a temporary truce between the Chinese and Japanese forces, Chiang gave the Communists his un-divided attention.

Following the counsel of his German advisers, Chiang developed a new strategy. Instead of leading his troops into Communist strongholds, he threw a tight blockade around the entire Central Soviet District. He thus deprived the Communists and the local communities of vital sup-plies, including salt, without which foods could not be preserved. The blockade consisted not only of the KMT army but of a network of stone fortresses and pillboxes. In his attempt to break through the blockade, Mao was compelled to abandon his favorite guerrilla tactics for positional warfare. As the Communists suffered mounting casual-ties, their position in southern Kiangsi and the neighboring provinces became increasingly precarious. They recognized that remaining in Chiang's circle would lead to their total annihilation. At the start of Chiang's offensive, Mao's Red Army numbered well over 300,000 men. When he finally made the agonizing decision to abandon the bases in Kiangsi, his forces had been reduced to about 120,000 men.

In October 1934 the Communists broke out of the KMT encircle-ment and began their "Long March," an extraordinary human feat. In forced retreat they traveled on foot a distance of more than 6,000 miles. They crossed 12 provinces, marching through forests, jungles, and marshland, and passing over 18 mountains, 5 of them perma-nently snow-capped. In many of those places there was nothing to eat but wild herbs and no shelter but bushes. Added to those difficulties were the skirmishes and pitched battles they fought on the way.

Throughout their forced march Mao's followers spread the Red Army doctrine. When they finally arrived at their northern destination, Yenan, they numbered a mere 40,000. Yet Mao was confident that the entire experience was only a temporary setback and, indeed, a "vic-tory." For the KMT failed to annihilate the Red Army as Chiang had intended, and the Long March, which Mao likened to a "seeding   285 machine," was a success. On December 27, 1935, reflecting on the results of the experience, Mao wrote that the seeding machine "has sown many [revolutionary] seeds in eleven provinces, which will sprout, grow leaves, blossom into flowers, bear fruit and yield a crop in future. To sum up, the Long March ended with our victory and the enemy's defeat."[17]

The initial successes of Mao's movement from 1931–1934 may be accounted for by his realistic grasp of the Chinese situation following the failure of the urban insurrections of 1927. Mao recognized what the Comintern and other CCP leaders had failed to see: The chief revolutionary force in China was the peasantry. He based his movement on the discontent of the impoverished masses who found that their lot had improved not at all under KMT rule. Thus Mao temporarily succeeded in building a formidable political-military movement.

Yet at that particular moment, following the Long March, the CCP fortunes were at their lowest. As resentful and rebellious as the peasant masses were, their organization under the Communists had thus far proved inadequate for changing the social order. It was Mao's genius to recognize that the war with Japan offered a new opportunity to the CCP.

The War Against Japan and the
Second United Front with the Kuomintang

The Communists noted the growing indignation over Japan's increasing encroachments on China. Chiang Kai-shek had been doing little to oppose the foreign enemy so long as he was occupied with his domestic opponents. In those circumstances Mao perceived an opportunity to bring the civil war to an end and to achieve reconciliation with the KMT. Directing their propaganda at all segments of the Chinese population, the Communists soon struck a responsive chord with such slogans as "Fight Japan and not the Communists," "Chinese must not fight Chinese," "The Kuomintang and the CCP Must Cooperate to Save the Nation." Almost two years elapsed before the KMT responded to the Communist overture. It issued a resolution reiterating its determination to eradicate Chinese communism but added that a reconciliation was possible if the CCP accepted these conditions: the abolition of the Red Army and the Soviet government; an end to the call for class struggle; and the cessation of all Communist propaganda. Eager to remove all obstacles to an early war of resistance against Japan, the CCP accepted the conditions.

On July 7, 1937, Japanese forces launched an attack at the Marco Polo bridge near Peking. When the Chinese resisted, full-scale war between the two countries ensued. The CCP's call for a united opposition to Japan was prompted by both patriotic and Party considerations. Mao recognized that his proposed policy would win respect and sympathy for the CCP. Yet he also understood that the truce with the KMT could never become a real peace. Chiang Kai-shek, as an implacable enemy of the CCP, eventually was bound to resume his wars of annihilation against it. For Mao, then, the patriotic war presented an opportunity for the CCP to recover and expand and to prepare for the inevitable confrontation with his domestic enemy. Mao had no intention of abandoning his revolutionary aims once and for all. On the contrary, he viewed the anti-Japanese war as a means of advancing the revolution. According to the well-known American correspondent Edgar Snow who was in northern Shensi in 1936, the Communist leader Chou En-lai had told him that "the first day of the anti-Japanese war will mean the beginning of the end for Chiang Kai-shek."[18]

Nevertheless, the Communists showed a spirit of accommodation during the first few months of the war. They coordinated their operations with Chiang's troops and fought pitched battles against the Japanese invaders. But after the Japanese conquest of Shanghai and other victories in late 1937, the accommodation quickly evaporated. With the KMT forces seriously depleted as a consequence of those battles, Mao resumed independent Communist activities. The Communists expanded their power behind the Japanese lines through organization and recruitment. A full-scale struggle for territory now developed between the KMT and the Communist forces.

Employing the principles of guerrilla warfare at which they had gained considerable experience, the Communists most often succeeded in overpowering the KMT although the sole source of arms for the expanding Communist armies was the captured weapons and ammunition of the enemy. The victorious Communists soon acquired exclusive control of the disputed areas and established regional governments there. They also instituted a more moderate land policy which enabled them to win over large sections of the rural population. Under the new policy only landlords who had collaborated with the Japanese were fully expropriated. All others were permitted to retain their land on the condition that the rent did not exceed 37.5 percent of the crop. In their sustained efforts to gain maximal popular support, the CCP in March 1940 also instituted a more moderate political system than it originally envisioned. Only one third of the elective posts in any local community were filled by Communists, while the remaining two thirds were filled by non-Party leftists and middle-of-the roaders, respectively. At the same time the Party made it widely known that it had no intention of confiscating all private property for distribution among the masses. Only the property of big banks, big industry, big business, and big landlords would be expropriated. That policy, conspicuously put into effect in the areas under their sway, induced most of the minor parties and many businessmen to cooperate with the Communists.

The rising fortunes of the CCP throughout the 1940s also enhanced Mao's personal reputation. He established himself as chief interpreter of Marxism-Leninism in China. Through his resounding successes, he acquired more and more charisma until he came to be regarded by his followers as the source and arbiter of the Chinese Communist movement. The war with Japan had in fact provided the CCP with the favorable circumstances which Mao had anticipated. For the KMT, however, the same war had disastrous effects. Having borne the brunt of the Japanese assault in the first phase of the war, in which it lost more than a million men, it became militarily ineffectual thereafter. Moreover, owing to its economic and political policies, the KMT soon found itself alienated from most segments of the urban population. Intellectuals and manual workers increasingly opposed the Nationalist government, which did nothing to curb the self-enriching practices of high KMT officials. Small businessmen were also alienated by the KMT's government monopolies and government-operated enterprises.

Thus by the time the Japanese surrendered to the United States and its allies in August 1945, the prestige of the Nationalist government had burnt itself out. At the same time the Communists found themselves in control of 19 provinces with a total population of 100 million people. They had a regular army of a million men and a people's militia of 2.2 million men.[19] Mao was now ready for a final showdown with the debilitated Nationalist forces.

The Conquest of Power

The final clash between the two contending parties began as a struggle for the territory, arms, and supplies which the Japanese army was to surrender. Because of American support for Chiang Kai-shek, all but a small fraction of the Japanese forces in China surrendered to the KMT army. In Manchuria, however, a quarter of a million Communist troops acquired huge stockpiles of weapons and ammunition which the

Russian army had captured following the Soviet Union's declaration of war on Japan on August 8, 1945. A ferocious civil war now ensued.

For the first year or so it seemed that the KMT forces had the advantage. They dislodged the Communists from numerous cities and towns, but those victories proved to be illusory. They were very costly to the Nationalists in men and equipment, and the Communists fled without incurring heavy losses. Furthermore, while the KMT was intent upon seizing and holding as many cities as possible, Mao cared little about the cities at that stage of the war. He now applied his strategy of "encircling the cities." By concentrating his superior forces and overwhelming the enemy in surprise attacks, Mao's forces extended their sway throughout rural China and encircled the KMT in a few key urban centers. By 1948–1949 the Communists had laid siege to several key cities, including Peking, while the Nationalist government took refuge in the southern city of Canton. With the fall of Canton on October 14, 1949, the government retreated to Chungking and then to Chengtu. By the end of December, the government was compelled to flee the mainland for Taiwan. The Communists had for all practical purposes completed the conquest of the mainland. The People's Republic of China was proclaimed.

Discontent, Ideology, Organization, and Charisma

The reasons for the success of the CCP may be summed up with the same four catchwords that were employed in our analysis of National Socialism.

The profound discontent that was characteristic of the Chinese masses under the imperial government continued to prevail under the KMT, which made no significant reforms and left the traditional property and power arrangements intact. The landed upper classes in particular continued to engage in exploitation as before. That should come as no surprise since the main components of the KMT coalition were in fact landlords, warlords, and, to a lesser extent, merchants.

Massive discontent opened the door to a radical alternative to the KMT. But the early organizational activities of the CCP were centered in the cities. The Party had treated as dogma the Marxian proposition that the urban proletariat was the chief agency of revolution. Even after the great massacre of 1927, in which thousands of Communists and their sympathizers were executed, the Comintern and the Party leadership persisted in their urban-centered policy.

289

That is where the ideology, organization, and strategy of Mao made the difference. He was among the first to maintain that in a society such as China it was not the urban proletariat but the impoverished peasant masses that constituted the major potential revolutionary force. By appealing to the land-hungry peasant masses, he built a revolutionary agrarian movement. He organized peasant communities and employed them as rural bases from which to wage guerrilla-type warfare.

Yet the defeat that Mao's forces suffered in 1934, the setback of the "Long March," suggests that his strategy was not invincible. Mao's comeback was aided by other objective circumstances, namely, the Japanese invasion. It is a measure of Mao's political and organizational sophistication that he knew how to turn the invasion to his advantage; and it was his political sophistication and success that gave rise to his charisma.

THE AMERICAN CIVIL RIGHTS MOVEMENT

As late as World War II, black American soldiers were fighting in segregated units. Only after the war and an increasingly energetic campaign by black leaders was segregation abolished in the armed forces. That was the second world war which white and black Americans had fought in the name of democracy, yet following the war blacks found that their status in public life had changed scarcely at all.[20] Jim Crow continued to prevail, and leaders of the black community decided that the time had arrived for an open assault on segregation.

The National Association for the Advancement of Colored People (NAACP) played a key role in that campaign. An opportunity presented itself in professional-school cases in Texas and Oklahoma. In the Texas case (Sweatt v. Painter), the state at first assumed that the NAACP was seeking a "separate but equal" law school. The legislature met and changed the name of the Negro state college to "Prairie View University" but allocated no funds for a law-school building. When it became apparent after preliminary hearings that the NAACP was mounting an attack on segregation itself, the legislature reconvened and appropriated $2,600,000 for a new black law school. Thurgood Marshall, the chief counsel for the NAACP, then presented expert testimony from anthropologists and psychologists that "race" was an unreasonable classification of students. On June 5, 1950, the Supreme Court handed down three decisions which seriously weakened the legal foundation for segregation. In the Sweatt case the Court ruled that equality involved more than physical facilities. In the G. W. McLaurin case, the Court held that a Negro student, once admitted, may not be segregated. Finally, in the Elmer W. Henderson case, the Court prohibited dining-car segregation.

On the basis of those rulings, NAACP attorneys filed suits against elementary- and high-school segregation. When the cases reached the Supreme Court, Marshall contended that segregation is itself discrimination. He argued that state laws requiring "separate but equal" facilities for black and white school children were unconstitutional and therefore should be abolished. On May 17, 1954, the Supreme Court handed down its epoch-making decision. "We cannot turn the clock back to 1868," the Court said, "when the [Fourteenth] amendment was adopted, or even to 1896, when Plessy versus Ferguson was written. . . . We conclude that in the field of public education the doctrine of 'separate but equal' has no place." A year later the Court ordered public school desegregation "with all deliberate speed."

Although Baltimore, Louisville, and Washington, D.C., took some first steps toward integration, the Deep South breathed defiance. White Citizen Councils spread across the South, declaring that they would resist school integration "by every lawful means." The leader of the resistance was Tom Brady, a circuit judge in Mississippi. In his book, *Black Monday,* he described the Supreme Court decision as a deep, dark plot which would lead to the tragedy of miscegenation. "Oh, High Priests of Washington," he wrote, the

decision which you handed down on Black Monday has arrested and retarded the economic and political and, yes, the social status of the Negro in the South for at least one hundred years. . . . When a law transgresses the moral and ethical sanctions and standards of the mores, invariably strife, bloodshed and revolution follow in the wake of its attempted enforcement. The loveliest and purest of God's creatures, the nearest thing to an angelic being that treads this terrestrial ball is a well-bred, cultured Southern white woman or her blue-eyed, golden-haired little girl.

Brady then went on to propose that the Court ruling be disobeyed:

We say to the Supreme Court and to the Northern world, "You shall not make us drink from this cup." . . . We have, through our forefathers, died before for our sacred principles. We can, if necessary, die again.[21]

Brady's words reflected the Southern white mood at the time. Concerted opposition to school integration was organized. National Guard units had to be called out to protect little black children as they passed through the lines of angry crowds. A mob prevented blacks from en-   291 rolling at Mansfield High School in Mansfield, Texas. The Tennessee National Guard was sent to Clinton to quell violent demonstrations against integration. In Nashville the new Hattie Cotton Elementary School was destroyed by a dynamite blast, apparently because it had admitted one black pupil. Meanwhile, Southern lawyers were flooding the federal courts with suits and appeals.

Enter Martin Luther King, Jr.[22] He was the man who carried the struggle into the Negro church, the cultural foundation of the black community since slavery. Fusing the teachings of Jesus and Gandhi and adding spirituals and symbols of deep significance to the black community, King created an awesome passive-resistance movement. His ideology and strategy struck a responsive chord among laborers, businessmen, and professionals alike. It fired the imagination of Northern urban blacks as well as Mississippi fieldhands.

Martin Luther King, Jr., was a third-generation Baptist preacher. His father was an Atlanta minister, militant in his defense of Negro rights. His grandfather, A. D. Williams, helped lead a boycott against an Atlanta newspaper which spoke disparagingly of Negro voters. Later he organized a group which successfully pressed for the establishment of the first Negro high school in Atlanta. Young King attended that school, named after Booker T. Washington, and went on to study at Atlanta's Morehouse College.

In 1947 King was ordained in his father's church in Atlanta. Later, at Crozier Seminary near Philadelphia, he discovered Walter Rauschenbush's *The Social Principles of Jesus,* a book that burned the moral teachings of early Christianity into his soul. Equally influential in King's spiritual development was Gandhi's method of passive resistance, which was brought to his attention at the time. King continued his studies in Boston where he met and married Coretta Scott. After obtaining his Ph.D. in systematic theology from Boston University, he and Coretta moved to Montgomery, Alabama, where he began preaching the social gospel to the members of the Dexter Avenue Baptist Church.

But King had not yet translated his ideas into political practice. The opportunity to do so was furnished by a tired black seamstress named Rosa Parks, riding home from work on a bus. On December 5, 1955, Mrs. Parks decided not to obey the old custom which required blacks to yield seats to whites. She was arrested, and the black community staged a one-day boycott in protest. The one-day boycott grew into a movement which swept across the South.

292 King, elected president of the Montgomery Improvement Associa-

tion, now undertook the task of mapping strategy. A huge community-wide car pool was organized, in which professionals and domestic servants alike participated. Periodic rallies were held with spirituals and hymns. For 13 months Montgomery blacks stayed off the buses. They achieved victory on December 21, 1956, when, following a federal court order, the buses were integrated. King now became a national and even international figure.

After the victory in Montgomery, King moved to Atlanta where he set up the general headquarters for the Southern Christian Leadership Conference (SCLC), a nonviolent, direct-action group whose influence was felt in every corner of the South. As chief spokesman, organizer, and strategist of the passive-resistance movement, King traveled several thousand miles a week explaining his philosophy to all segments of the black community. He advocated four methods of struggle: direct, non-violent action; legal redress; ballots; and economic boycotts. The legal basis of segregation has been destroyed, he argued. The point now was to implement the Court's decision.

In time King and his movement created a new mood and a new consciousness in the black community. The movement's influence was felt in Little Rock, Arkansas, where Daisy Bates, president of the Arkansas NAACP, outmaneuvered state and city officials and kept nine black children in school. It was felt in Greensboro, North Carolina, where four Agricultural and Technical College students inaugurated the "sit-in" age. The influence was also felt in the "jail-in" and "freedom-ride" movements. The net result of the new militance was a kind of second Reconstruction. It began on September 25, 1957, when in accordance with President Eisenhower's orders, soldiers of the 101st Airborne Division escorted nine children into Little Rock's Central High School. That was the first time in 81 years that a president dispatched American troops to the South to defend the constitutional rights of black Americans.

That the militant mood had become general was demonstrated on February 1, 1960, when the four North Carolina freshmen sat down at a lunch counter in a Woolworth store in Greensboro.[23] Unwittingly they started an unprecedented student-protest movement which spread throughout the South. Not long afterward thousands of black students were seating themselves at "white" lunch counters and requesting service. If arrested, they submitted without resistance. If attacked, they refused to fight. The movement quickly hit chain stores, department stores, libraries, supermarkets, and movie houses. Demonstrations, picket lines, and boycotts occurred daily. By September 1961, more   293 than 100 cities in 20 states had been affected. At least 70,000 black and white students had participated, and a sizable number of establishments in Southern and border states had been desegregated.

But no less important than the immediate practical outcome of those direct actions was their impact on the minds and mood of black youths. Black consciousness underwent a profound change. Organizations like the NAACP, hitherto regarded as radical, were prompted by events to reevaluate their positions and to step up their pace of activities. New organizations came into being, such as the Student Nonviolent Coordinating Committee (SNCC, pronounced "snick"). With SNCC, the sit-in movement increased its effectiveness. That organization was heavily involved in the jail-in movement which began in Rock Hill, South Carolina, on February 6, 1961, when students refused to pay fines and went to jail instead.

SNCC also played a key role in the freedom-ride campaign, which was initiated by 13 members of the Congress for Racial Equality (CORE). They set out from Washington, D.C., on May 4, 1961, for an integrated bus ride through the South. Testing compliance with integration orders of the Interstate Commerce Commission and federal courts, the freedom riders crisscrossed the South. They were beaten and manhandled at several stops. After they were attacked at Montgomery, Alabama, on May 20, 1961, Attorney General Robert Kennedy dispatched several hundred United States marshals to protect them.

In the North the movement was being felt as well. In the Northern cities, however, it was de facto segregation that was challenged. By July 1961 suits had been filed in more than a dozen communities, including Chicago, Philadelphia, Newark, New Jersey, and Kansas City, Kansas. And sit-ins had occurred in Chicago (schools and beaches), Cairo, Illinois (restaurants and recreational facilities), and Englewood, New Jersey (schools).

At the same time the North witnessed different forms of protest in the black communities. There were black nationalist groups such as Elijah Muhammed's "Black Muslims," who bore similarities to Marcus Garvey's "Back to Africa" movement of the 1920s. Like Garvey, Muhammed glorified blackness and denigrated whiteness. He also organized a chain of cooperative business ventures as had Garvey. Where Muhammed differed was in his championing of a black state in America. The Black Muslims have shown their revulsion for the legacy of slavery by abandoning their "slave" names (e.g., John Jones, Tom

294

Washington) and adopting the letter X (e.g., John X, Tom X, and the famous Malcolm X).

In subsequent years, as we shall presently see, the protests in the North assumed a quite different form. But in the South Martin Luther King's passive-resistance patterns remained dominant. On April 3, 1963, King announced that he would lead racial demonstrations in Birmingham, Alabama, until "Pharaoh lets God's people go." King believed that a breakthrough in Birmingham was essential because that city had long been regarded as an impregnable fortress of bigotry. In response to King's call there was a massive turnout on the part of Birmingham's black youths. They assembled in the 16th Street Baptist Church with the objective of marching downtown and desegregating all the establishments. But access to the downtown area had been closed off by firemen with high-powered water hoses and by police with K-9 dogs and nightsticks. In command was one Eugene ("Bull") Connor, a tough segregationist who had become a national symbol of Southern intransigence.

All at once it happened: Hundreds of young people burst through the door of the church, singing, shouting, clapping, and dancing. They made their way toward the barricades, stopping short in the face of the hoses, dogs, and nightsticks. The next day the scene was repeated, but this time "Bull" Connor gave the order. The teenagers were blasted and bowled over by the hoses while police moved in with clubs and dogs.

On the following day the demonstrations were reinforced, with more than 2,000 youths participating. Most of them succeeded in surging through the police lines and reaching the downtown area where they entered the "white" stores. With tension mounting between blacks and whites, the sheriff called an emergency meeting of the city's leading businessmen and professionals. He told them that unless the demonstrations were peacefully ended, it would be necessary to impose martial law. The city's notables immediately agreed to meet with black leaders for negotiations. Three days later the mayor's office announced that a tentative agreement had been reached on the following points: the desegregation of lunch counters, restrooms, fitting rooms, and drinking fountains in planned stages during the next 90 days; the hiring and upgrading of black workers on a nondiscriminatory basis; the release of demonstrators on nominal bond (some 2,400 had been arrested); and the establishment of a biracial committee.

In the following two to three years, King's nonviolent resistance   295 movement enjoyed similar successes elsewhere in the South. Born in the South, it was a movement tailored to Southern conditions. It had a limited aim: free and equal access to all establishments and facilities to which white citizens had access. The segregation of those facilities had been a matter of policy. Therefore, all that was required to desegregate them was to change the policy—something which King's strategy could achieve. But could such a strategy work in the North?

In retrospect it is clear that it could not. The reason is that the de facto segregation of schools and other institutions in the large Northern cities is not a result of this or that discriminatory policy. It is rather a direct corollary of the ghettoization of blacks and their underclass status. King's strategy was suited to the desegregation of buses and lunch counters, but how could nonviolent resistance change the condition of the underclass and melt away the gigantic ghettos of America's largest cities? Little wonder, then, that as the movement shifted to the North, it changed its ideology, organization, strategy, and leadership. The Southern movement had been dominated by a middle-class group of well-trained nonviolent rebels. The Northern movement, in contrast, reflected some of the bitterest sentiments of the most depressed strata of the black community.

Civil Disorders

On August 11, 1965, the nation was astonished to learn that large-scale rioting had erupted in Watts, a black ghetto in Los Angeles, California. The rioting continued until August 17th, terrifying the city with the most destructive civil disturbances in the nation's peacetime history. Perhaps as many as 10,000 individuals participated in the disorder, attacking white motorists, overturning and burning automobiles, exchanging gunshots with police, and looting and setting fire to stores while stoning and otherwise obstructing firemen. Accompanying the destruction was the frequently heard cry: "Burn baby, burn!" When order was restored, it was found that 34 persons had been killed (25 of them black) and 1,032 wounded or hurt. 3,952 had been arrested, and 977 buildings had been looted, damaged, or destroyed.[24]

An investigating body, the McCone Commission, was appointed soon afterward to inquire into the immediate causes of the disorder. It seemed that a white California highway patrolman was riding his motorcycle somewhat south of the Los Angeles city boundary when a passing black motorist told him that he had just seen a car being driven

296

recklessly. The patrolman gave chase and pulled the car over in a predominantly black neighborhood, but not in Watts. The driver, Marquette Frye, was a 21-year old black, and his older brother, Ronald, was a passenger. The patrolman asked the younger Frye to get out of the car and to take the standard Highway Patrol sobriety test. When he failed the test, the patrolman informed him that he was under arrest. He radioed for a car to take Frye to jail and for a towtruck. Ronald Frye, having been informed that he could not take the car, went to get his mother, the car's owner, so that she could claim it.

Those events transpired two blocks from Frye's home. It was a hot night, and many of the area's residents were out-of-doors. Soon a crowd gathered. Marquette Frye then quickly moved into the crowd, cursing at the officers and shouting that they would have to kill him to take him to jail. A patrolman pursued Frye, who, aided by his brother and mother, resisted. The crowd became hostile, and the patrolman radioed for help. Reinforcements arrived, all three Fryes were placed under arrest, and the patrol car with the prisoners left the scene. As the officers left, someone spat at them. They stopped long enough to arrest a young man and woman who appeared to be inciting the crowd to violence. False and exaggerated accounts of those events spread fast—that the young woman was pregnant, that the Fryes were maltreated, and more. Soon the crowd began to stone automobiles and attack white motorists. Sporadic acts of vandalism continued until the following morning, when it appeared that the situation was simmering, and by which time 29 persons had been arrested. By evening it began to boil again, exploding in a massive riot. That was only the first of such upheavals.

In 1966 there were many more ghetto riots than in any previous year. It is quite likely that civil-rights agitation tended to embolden blacks in those years. Yet, as Benjamin Muse has observed, "The ghetto rioting had been, and remained on the whole, a separate phenomenon from the organized civil-rights movement."[25] However, attempts were made to relate the two in 1966.

In that year SNCC and CORE broke away from the mainstream of the civil-rights movement. They adopted a more militant rhetoric and program designed to appeal to the now turbulent black underclass. The character of those two organizations now changed significantly as "moderates" were ousted from their posts and replaced by militants. SNCC, for example, replacing James Forman, elected Stokely Carmichael as chairman. Carmichael, a highly articulate young man of 24, 297 was a native of Trinidad who had settled in New York with his parents at the age of 11. He received his education at the Bronx School of Science and at Howard University. Politically his name was first linked to the term *Black Panther,* a symbol of the all-black political movement which he had been vigorously promoting in Lowndes County, Alabama. Several white civil-rights workers had been murdered in the area, and black churches had been bombed and burned. Carmichael, himself the butt of daily abuse and harassment, was outraged that the perpetrators of those crimes had gone unpunished. He began to challenge the traditional strategy and ideology of the movement and to propose what he regarded as a more effective one.

The split within the movement first became evident with the "Meredith March." James H. Meredith had become an internationally known figure in 1962. He was the first black student to gain admission to the University of Mississippi, an admission secured only after U.S. marshals and soldiers had fought an all-night battle against a segregationist mob. Now in June 1966, Meredith decided to set out on a march through the Deep South. The point was to overcome "that all-pervasive and overriding fear that dominates the day-to-day life of the Negro in the United States, especially in the South and particularly in Mississippi."[26] He hoped that his example would inspire thousands of Mississippi blacks with the courage to exercise their right to vote. Meredith had walked only ten miles when he was ambushed by a white man who stepped out of the roadside underbrush, firing several rounds from a shotgun. Meredith went sprawling in the dust. A press release first reported him "shot dead," but it later turned out that he had been superficially wounded in the head, neck, back, and legs.

Upon his release from the hospital, Meredith announced his intention to resume the march. By now he had become even more of a heroic figure. The number of marchers swelled. Fifteen hundred joined him, while thousands of others gathered around them along the way. As the march was getting under way, a "manifesto" was issued, stating that the march would be a "massive public indictment and protest of the failure of American society, the government of the United States, and the state of Mississippi" to fulfill the basic civil rights of black Americans. Martin Luther King reluctantly signed the statement out of concern for movement unity, but he soon found himself critical of some of Carmichael's utterances. While King held steadfast to his doctrine of nonviolence, cooperation with white sympathizers, and racial integration, Carmichael called for "black power."[27]

Black Power

In his co-authored book Carmichael began with a call for redefinition on the part of black Americans. Historically, he argued, it is white people who have imposed their definitions on black people. Owing to certain cultural developments in Western society which originally had nothing to do with racism, the term *black* has acquired a negative connotation. One speaks of "black forces of evil," "black market," and the like. On the other hand, *white* is associated with goodness and beauty. With slavery and racism, *black* became a mark of servility, inferiority, laziness, stupidity, shiftlessness, etc. Black Americans, Carmichael contended, must once and for all reject such definitions. Carmichael thus reflected the attitudes and practices of many young blacks of the mid-1960s who began to reject the term *Negro* and to replace it with *black*. "Black is beautiful," they insisted, in an ever-growing chorus. "Black is intelligent, energetic, determined." They thus achieved a basic semantic transformation by investing the term *black* with a meaning fundamentally opposite to the traditional one.

The political corollary of self-definition, for Carmichael, was self-determination. Defending himself against the charge that black power is "racism in reverse" or "black supremacy," Carmichael emphatically denied any parallel between them. Racism, he argued, is a form of subjugation of one group by another; the goal of black power, in contrast, is simply the full participation by blacks in the decision-making processes affecting their lives.

Carmichael was among those black leaders who held that given the mammoth size of urban black ghettos in America, residential integration appears unrealizable in the foreseeable future. Under the circumstances, Carmichael contended, self-determination is the more realistic goal. Self-determination or black power meant that the people of each black community should strive for control of all the key community institutions, whether they be the schools or the police force. Organizations should be formed to compel slumlords, by means of rent strikes, to provide decent facilities and adequate services. Similarly, all businesses in the ghetto should be pressed to make sizable contributions to the community. Such contributions could take several forms: "providing additional jobs for black people, donating scholarship funds for students, supporting certain types of community organizations."[28] Finally, black communities must develop their own political organizations and parties.

299

The black-power enthusiasts scored some limited successes, but mainly in the sphere of self-definition. It soon became clear that the term *Negro* was being generally discarded, just as the term *colored* had been rejected much earlier. Black youths increasingly scorned the old practice of imitating whites in personal dress and appearance, and the "Afro" hairdo became more and more prominent. But the black-power political program never struck roots among the mass of ghetto dwellers.

Moreover, for the general public the label *black power* served to identify the conspicuous militants who espoused an antiwhite, anti-nonviolent, anti-integration line. Worse still, it came to be applied to all forms of black violence. "Black power" had a warlike ring, in spite of Carmichael's numerous clarifications. In the eyes of the white public, the riots of the summer of 1967 appeared as the beginning of a violent black revolution. Up to July 12th the riots were not much different from the racial eruptions of the three previous summers; and none was of the magnitude and fury of the Watts upheaval. Then all records were broken by Newark and Detroit. It has been estimated by the Senate Permanent Committee on Investigations that from 1965 to 1967, there were 101 major riots which resulted in 130 persons killed, 3,623 injured, and 28,932 arrested. The property damage and other costs reached $714.8 million.[29]

In the following year, 1968, the *Report of the National Advisory Commission on Civil Disorders* (the Kerner Commission) was published. It firmly rejected the theory that the riots were the result of a conspiracy on the part of black-power advocates or other militants. The report noted, however, that "militant organizations, local and national, and individual agitators, who repeatedly forecast and called for violence, . . . helped to create an atmosphere that contributed to the outbreak of disorder." The commission placed the blame for the situation on white racism, which "is essentially responsible for the explosive mixture which has been accumulating in our cities since the end of World War II. . . . Pervasive discrimination and segregation in employment, education, and housing have resulted in the continuing exclusion of great numbers of Negroes from the benefits of economic progress."[30]

The main cause of the outbursts, the report went on to say, was the deep and widespread discontent in the ghetto. The deepest level of grievances revolved about unemployment, inadequate housing, inadequate education, poor recreational facilities and programs, and the ineffectiveness of the political structure in dealing with those grievances.

It therefore seems clear that the civil-rights movement and the summer rioting were fairly independent of each other, despite the fact that many perceived them as connected. The riots were in the main spontaneous outbursts—though individuals and groups helped create an atmosphere conducive to disorder. The summer disorders were typically precipitated by an "incident" and further provoked by rumor. Violence usually occurred immediately following the precipitating incident and then escalated rapidly. The violence generally subsided during the day and flared up quickly again at night. The disorders most often began with rock and bottle throwing and window smashing. Once store windows were broken, looting usually ensued. Hence the sociological concepts required for a comprehension of rioting and other forms of crowd behavior are quite different from the concepts we have applied to social movements.

Factors Accounting for the Success of the Civil-Rights Movement

In the civil-rights movement, as in the Nazi and Chinese Communist movements, it was massive discontent that gave rise to large-scale collective action. Given such discontent, the winning of adherents depended upon the appeal of an ideology. The effectiveness and perpetuation of the collective action rested on the organization of the movement and on its strategy and tactics. Finally, charismatic leadership played a significant role in gaining the devotion of the movement's followers and winning their confidence.

12

Crime and Other Forms of Deviant Behavior

The pioneering sociological analysis of deviant behavior is found in the writings of Emile Durkheim. It is he who first set forth the proposition that deviance is no less firmly rooted in social conditions than conformity. Deviance, he maintained, is neither morbid nor pathological but rather normal. In any society there exists an inevitable diversity of human conduct as well as a variation in moral values. No social act is universally regarded as immoral or harmful to the social body. How, then, does one distinguish "normal" from "deviant" behavior?

For Durkheim, a scientific reply to that question rested on a statistical criterion. The "normal," he wrote, refers to "those social conditions that are most generally distributed." The social norm consists of the most frequent forms of behavior, and other, less frequent forms which depart from the norm are deviant. Yet, paradoxically, although crime, for example, is a form of deviance, it is nonetheless normal because it is present "in all societies of all types. There is no society that is not confronted with the problem of criminality. Its form changes; the acts thus characterized are not the same everywhere; but, everywhere and always there have been men who have behaved in such a way as to draw upon themselves penal repression."[1] If a society utterly devoid of crime is unknown, then crime must be a normal and integral facet of every social order.

Durkheim's conception of things should not be misunderstood. When he asserted that crime is necessary, he did not mean that *specific types* of crime are inevitable or that crime *rates* cannot be decreased by appropriate social measures. He meant instead that wherever human beings congregate, they display diverse forms of behavior; and, second, that some of those forms will be seen as departing from established norms and punished accordingly. "Crime" thus ranges all the way from minor infractions of decorum at one end of the scale to major felonies on the other.

Imagine a society of saints, a perfect cloister of exemplary individuals. Crimes, properly so called, will there be unknown; but faults which appear venial to the layman will create there the same scandal that the ordinary offense does in ordinary consciousness. If, then, this society has the power to judge and punish, it will define these acts as criminal and will treat them as such. For the same reason, the perfect and upright man judges his smallest failings with a severity that the majority reserves for acts more truly in the nature of an offense.[2]

Individuals in every society differ in respect to the social and cultural milieux in which they find themselves. For everyone to be socially alike is impossible because no two individuals occupy the same sociocultural space. No society, therefore, is capable of achieving perfect moral uniformity. The diversification of behavior is a social process which results in both extraordinary and ordinary deviants. It produces individuals who may be geniuses and "criminals" at one and the same time. Thus Durkheim observed that

according to Athenian law, Socrates was a criminal, and his condemnation was no more than just. However, his crime, namely, the independence of his thought, rendered a service not only to humanity but to his country. It served to prepare a new morality and faith which the Athenians needed, since the traditions by which they had lived until then were no longer in harmony with the current conditions of life. Nor is the case of Socrates unique; it is reproduced periodically in history. It would never have been possible to establish the freedom of thought we now enjoy if the regulations prohibiting it had not been violated before being solemnly abrogated. At that time, however, the violation was a crime, since it was an offense against sentiments still very keen in the average conscience.[3]

It is the process of social diversification that yields the higher order of deviant such as Socrates; and it is the same process that brings forth ordinary deviants and common criminals. Among all the divergent actions that one finds in any given society, some will inevitably acquire a criminal character. It is not the intrinsic quality of the actions themselves that confers a criminal character upon them. It is rather the definition that is placed on those acts by the dominant consensus.

For Durkheim, then, crime was fundamentally bound up with the conditions of social life. Crime, far from being an unmitigated evil or   303 pathology, was a normal phenomenon, indispensable for the development of morality and law.

CRIME AND PUNISHMENT

We have said that the criminal character of an act does not reside in the act itself. What, then, makes an act criminal? Durkheim believed that the small, primitive society presents the clearest and most direct reply to that question. In such a society, he noted, there are numerous acts which are considered crimes, e.g., touching a tabooed object, failing to make a traditional sacrifice, departing from a precise ritual formula, and so forth. Upon reflection it is clear that those diverse acts have only one thing in common: They are universally disapproved of by the members of the society in question. Those acts are crimes because they shock the collective conscience. In order for an act to qualify as a crime, it must offend strong and intense sentiments and break precise rules. A crime, therefore, is an act which antagonizes the powerful and well-defined sentiments of a collectivity. In Durkheim's words, "We must not say that an action shocks the common conscience because it is criminal, but rather that it is criminal because it shocks the common conscience. We do not reprove it because it is a crime, but it is a crime because we reprove it."[4] An act is a crime because it offends the transcendent authority of society.

If such a conception of crime is sound, Durkheim reasoned, it ought to account for the nature of punishment. Punishment is first and foremost a passionate social reaction against the offender. That is especially evident in a primitive setting. Punishment is a form of vengeance which may appear socially useless and unnecessarily cruel. But actually it enables the community to do something vital for itself. By means of punishment society heals the wounds inflicted upon it by the offender; it restores its moral integrity and reaffirms its most fundamental values.

In modern society the essence of punishment remains much the same. It is still "at least in part, a work of vengeance."[5] We may attempt to rationalize our treatment of the offender in terms of rehabilitation and the like, but we find it just that he should expiate his offense through suffering. For us as for our ancestors, Durkheim convincingly argued, punishment remains a passionate reaction by means of which we reaffirm the validity of our rules and laws. That the reaction is passionate is evident from the conduct of both the prosecutor and the defense attorney in the modern courtroom. The former strives to awaken in the jury the sentiments that have been violated by the defendant, while the latter tries to rouse sympathy for him. Today punishment is carried out by the institutions of the state rather than by the collectivity as a whole, but the essence of punishment continues to be a more or less zealous reaction against those who have violated our basic rules of conduct.

Crime therefore wounds the common conscience while punishment heals and restores it. Crime furnishes the community with an opportunity to revitalize itself by reacting intensely against the criminal offender. "Crime," wrote Durkheim,

brings together upright consciences and concentrates them. We have only to notice what happens, particularly in a small town, when some moral scandal has just been committed. They stop each other on the street, they visit each other, they seem to come together to talk of the event and to wax indignant in common. From all the similar impressions which are exchanged, from all the temper that gets itself expressed, there emerges a unique temper . . . which is everybody's without being anybody's in particular.[6]

The offended sentiments derive their peculiar force from the fact that they are common to everybody. They are unanimous, uncontested, and commonly respected. An act is a crime precisely because it damages that unanimity. To do nothing in the face of a crime, to let it go unpunished, would therefore result in the enfeeblement of the collective sentiments. For it is only by acting in common against the offender that the community can reinforce itself and its basic values.

Hence Durkheim believed that the main object of punishment is certainly not to chasten or correct the offender nor even to deter others from following in his path. Its true object is to maintain the vitality of the community's fundamental values and to safeguard its social cohesion. Punishment enables society to repair the "evil" which the crime has inflicted upon it. There is thus a continuity between crime and punishment. The criminal violates the cherished standards of the community and the upright retaliate "to heal the wounds made upon collective sentiments."[7] Thus Durkheim laid the foundation for a sociological understanding of crime and punishment.

However, since Durkheim's intellectual interests were not primarily criminological, he gave no systematic attention to a central question in criminology: Why and how does an individual become a criminal? Yet there is an implicit rejection throughout Durkheim's work of the criminological theories that were rampant in his day. Crime, it was widely believed, is the result of original sin or of innate depravity. It is caused by certain instinctual or racial predispositions; it is rooted in one's personality or physiological structure. Clearly, Durkheim's conception of the normality of crime not only has nothing in common with such theories but is a repudiation of them. One example will suffice to show just how far apart he was from some of his contemporaries.

One of the most famous of nineteenth-century criminologists was a physician and psychiatrist named Césare Lombroso (1836–1909). While serving as an army doctor, Lombroso thought he noticed that recalcitrant offenders differed from the disciplined troops by the greater prevalence and indecency of their tattoos. Later he employed experimental methods in studying insane patients. Comparing the insane patients with convicted criminals and those two groups in turn with "normal" persons, he measured their skulls and their sensitivities to touch. Once while performing a postmortem examination of a notorious bandit, Lombroso found a distinct depression at the rear of the skull, in the opening in which the spine and the skull are connected. Earlier he had found a similar depression in animals. From that Lombroso concluded that a criminal is an atavistic being possessing the ferocious instincts of primitive humans. The physical stigmata of atavism, Lombroso believed, were a low forehead, a receding chin, ears standing out from the head, too many fingers, unusual wrinkling of the skin, atypical head size or shape, and eye peculiarities.

In response to criticism Lombroso eventually revised his "atavistic" theory. In his last book, published in 1911, he conceded that there were environmental factors at work and listed a host of them from climate to religion. Retaining his original view that the "born criminal" and "insane criminal" are major types, he added a third category, the "criminaloid," who engages in vicious criminal behavior though he is born with neither physical stigmata nor mental aberrations. The "born criminal," who comprised about a third of all criminals, he explained as a reversion to an earlier evolutionary stage. As for the "insane criminal," that was a mixed category of offenders suffering from paralysis, dementia, pellagra, alcoholism, epilepsy, idiocy, and hysteria—all of which Lombroso regarded as causes of crime.[8]

Today there is scarcely a criminologist who continues to subscribe to Lombrosian views. Indeed, few criminologists take any theory seriously which attempts to explain crime in terms of the alleged "organic inferiority" or "degeneracy" of criminals. Even the most sophisticated studies purporting to demonstrate the physiological basis of criminality are methodologically defective and lacking in scientific validity.[9] It is now increasingly recognized that early criminologists were deceived.

They took the unattractive appearance of prisoners as a sign of their mental deficiency. "Abnormality" was reflected in their shaved heads, ungainly uniforms, and bitter facial expressions in reaction to harsh discipline.

Durkheim, however, was not fooled by appearances. He recognized the normality of criminals. He rested his theory solidly upon *social* and not upon physiological or psychological foundations. But it remained for others to extend and elaborate his theory. Contemporary sociologists have shed new light on the key question: How do individuals come to engage in criminal behavior?

A SOCIOLOGICAL THEORY OF CRIMINAL BEHAVIOR

A systematic sociological reply to that question was first put forth by Edwin H. Sutherland. Sutherland believed that criminal behavior is fully explicable in social terms, just as is any other form of human conduct. Accordingly Sutherland's opening proposition states that:

"(1) *Criminal behavior is learned.*"[10] Today that proposition may strike the reader as a statement of the obvious. But when Sutherland first published his theory in the 1930s, it was not obvious at all. Indeed, the view still prevails in certain academic disciplines that crime is the result of some inner biological drive. Hence Sutherland's first proposition is as important for what it denies as for what it affirms. It denies that criminal behavior is biologically inherited; one must learn how to become a criminal just as one must learn any other form of social conduct.

"(2) *Criminal behavior is learned in interaction with other persons in a process of communication.*" The main form of that communication is verbal. Crime is not learned through imitation but through symbolic interaction.

"(3) *The principal part of the learning of criminal behavior occurs within intimate personal groups.*" The third proposition is aimed against those who claim that crime is learned from impersonal agencies of communication such as movies, TV programs, comic books, and the like. Such agencies may reinforce the patterns learned in small groups, but they play a comparatively minor role in the genesis of criminal conduct.

"(4) *When criminal behavior is learned, the learning includes (a) techniques of committing the crime, which are sometimes very compli-* 307 *cated, sometimes very simple; (b) the specific direction of motives, drives, rationalizations, and attitudes.*

"(5) *The specific direction of motives and drives is learned from definitions of legal codes as favorable and unfavorable.*"

We can better grasp the import of these propositions if we imagine the concrete situation they attempt to describe. If we think, for instance, of a poor neighborhood, we can picture teenage boys in the street. Every youth has his small circle of friends and acquaintances. Actually he moves in more than one circle. Some of his friends are honest and law-abiding. They want the good things in life, but they hope to obtain them by earning a living in some working-class occupation. They look upon the legal code as rules to be observed. However, a boy may also have other friends who look down on the "honest Johns." Those youths view the legal code as rules that may be broken. Typically, a teenager moves in such mixed company and finds himself torn between the two attitudes toward the legal code.

The crucial next step in a youth's development is described in Sutherland's sixth proposition:

"(6) *A person becomes a delinquent because of an excess of definitions favorable to violation of law over definitions unfavorable to violations of law.*"

Our young man, as we have seen, with both law-abiding and law-breaking friends, has experienced a cultural tension with regard to the legal code. But now he resolves that tension in favor of violating the law. Having had both criminal and anticriminal associations, he throws in his lot with the criminal group and accepts its norms and attitudes. Of course, his decision may not be permanent and irretrievable. Indeed, most such decisions are temporary, for most juvenile delinquents eventually turn away from crime. But it is the process which Sutherland calls *differential association*—i.e., associating with different groups that hold different attitudes toward the legal code—which ultimately accounts for the fact that one individual becomes a criminal while another does not. In that light the remaining three propositions are self-explanatory.

"(7) *Differential associations may vary in frequency, duration, priority, and intensity.*

"(8) *The process of learning criminal behavior by association with criminal and anti-criminal patterns involves all of the mechanisms that are involved in any other learning.*

"(9) *Though criminal behavior is an expression of general needs and values, it is not explained by those general needs and values since non-criminal behavior is an expression of the same needs and values.*"

What we have, then, in these nine propositions is a theory showing how crime is rooted in the social organization of the community. A child grows up in a family, and a family's place of residence is largely determined by its income. The lower the rental value of the houses, the higher the delinquency rate. That does not mean that poverty itself produces crime. Such a view impugns the honesty of the poor, the majority of whom are, after all, law-abiding. One of the important insights yielded by Sutherland's theory is that even in an area where the delinquency rate is high, only some young people contribute to it. One active, outgoing, and athletic boy comes in contact with certain other boys in the neighborhood, learns delinquent behavior from them, and eventually becomes a criminal. Another equally active boy in the same neighborhood joins a scout troop and shuns delinquent associates. A third boy, somewhat introverted, spends most of his time at home and therefore also avoids "bad" company. Some of us may recall having been told by our parents or grandparents that "you're only as good as the company you keep." The insight conveyed by that popular adage is at the heart of the theory of differential association. That theory, as Sutherland notes, is most consistent with the available criminological evidence. For it explains

why the city crime rate is higher than the rural crime rate, why males are more delinquent than females, why the crime rate remains consistently high in deteriorated areas of the cities, why the juvenile delinquency rate in a foreign nativity [i.e., ethnic group] is high while the group lives in a deteriorated area and drops when that group moves out of that area, why second generation Italians do not have the high murder rate that their fathers had, why Japanese children in a deteriorated area of Seattle had a low delinquency rate even though in poverty, why crimes do not increase greatly in a period of depression.[11]

Criticisms of the Theory

There can be no doubt that the theory of differential association offers many intellectual advantages over nonsociological theories. Yet the form in which Sutherland stated his propositions left him open to several cogent criticisms. He has been taken to task for subscribing to a learning theory which is too mechanical. For example, in his description of the process by which a person learns criminal behavior, he seems to imply that *direct association* is necessary. He does not sufficiently allow for the possibility that an individual may *identify* with a role model without ever entering into direct interaction with him. However, it appears evident that delinquents may become a *reference group* (i.e., a group with which one identifies though one may not actually belong to it) for an individual who imaginatively seeks to emulate their conduct. As Daniel Glazer has put it:

The image of behavior as role-playing, borrowed from the theater, presents people as directing their actions on the basis of their conceptions of how others see them. The choice of another, from whose perspective we view our own behavior, is the process of identification. It may be with immediate others or with distant and perhaps abstractly generalized others of our reference groups. (The "amateur" criminal may identify himself with the highly professional "master"-criminal whom he has never met.)[12]

In those terms an individual actor may first learn criminal patterns from "afar" and only afterward come to be accepted by a criminal group. He may even remain a "loner." His learning of criminal patterns is thus accomplished by a process which Glazer terms *differential identification.* Glazer intended that concept to be a supplement to Sutherland's theory that would eliminate the implication of a mechanical learning process.

Another significant criticism is that Sutherland's theory fails to take two key factors into account. The first factor may be termed *opportunity.* It is clear that certain types of crime are at least partially a function of opportunity. Embezzlement, for instance, is a crime for which comparatively few blacks are convicted, for owing to the social-class position of the majority, they seldom have access to high positions of financial trust.

The second factor ostensibly overlooked by Sutherland may be termed *intensity of need.* "Need," it is argued, is a partial determinant of criminal behavior. That is presumably borne out by the fact that thefts are most frequently engaged in by members of the lower socioeconomic groups, who are in greater poverty. However, as Sutherland observed in response to such criticisms, the link between poverty and crime imputed here is problematic because the illegal appropriation of property and money also flourishes in the upper socioeconomic classes. Indeed, Sutherland was among the first to document the extent of such illegal practices in his study aptly entitled *White Collar Crime*—a subject to which we shall return in a later context. Furthermore, although it may be true that even the most respectable of persons would rather steal than starve, those are not the only options faced by most lower-class individuals who become criminals. There is a connection between poverty and crime, but for the most part it is not at the level of "steal or starve." To grasp the nature of that connection we must briefly consider a few more supplements to Sutherland's basic principles.

310

ANOMIE AND SUBCULTURES

In 1938 the distinguished American sociologist Robert K. Merton published an essay entitled "Social Structure and Anomie." In that essay he constructed a framework for the analysis of the social and cultural sources of deviant behavior. Elaborating the Durkheimian conception of anomie ["anomy"]—a situation in which cultural norms are unsuited to the actual state of the social system—Merton set out "to discover how some *social structures exert a definite pressure upon certain persons in the society to engage in non-conforming rather than conforming conduct.*"[13]

Merton begins by making an analytical distinction between two key aspects of a social structure. The first he calls *culturally defined goals.* That term refers to the major legitimate purposes, interests, and objectives which most members of a given society are taught to strive for. The second aspect he terms *institutional norms.* Those are the socially prescribed, legitimate means for pursuing the "culturally defined goals." The two analytical elements, *goals* and *means,* jointly contribute to the shape of prevailing social behavior.

"Conformity" describes the situation in which members of society adhere to both the goals and the means. It often happens, however, that a "disjointedness" develops between goals and means. Merton illustrates such disjointedness with the example of the United States, where the common cultural goal is monetary and material success; but where the legitimate means for pursuing those goals (e.g., inheritance of wealth, higher education, well-paying occupations, and access to profit-making business ventures) are not equally available to all members of society. Such "disjointedness" therefore suggests the following proposition: Given the prevalence of success-goals, high rates of deviance may be expected whenever the legitimate means of pursuing the goals are not equally available or not available at all to some groups and classes.

Thus Merton suggests that those who lack legitimate opportunities for reaching after coveted cultural goals often pursue them nevertheless —by illegitimate means. It is such persons and groups who become deviant and who engage in criminal practices. Members of lower socioeconomic groups who have less accessibility to legitimate means are especially subject to social pressures inclining them toward crime. Typically, those pressures are brought to bear in small groups which embody deviant norms.

311

THE EMERGENCE OF SUBCULTURES

In his book *Delinquent Boys* and in other studies, Albert K. Cohen extends Merton's analysis. Cohen agrees that deviance is most likely to occur where legitimate opportunities for achieving success goals are blocked. In the face of social disabilities, individuals will react against those institutions which have placed barriers in their way. Frequently they develop new goals and norms which "may permit or require behavior that violates the norms of conventional society; they may justify or demand deviant behavior."[14] That occurs in the context of a group with an emerging subculture. Describing the process, Cohen writes:

A number of people, each of whom functions as a reference object for the others, must jointly arrive at a new set of criteria and apply these criteria to one another. For this to happen, people with similar problems, . . . because they occupy similar positions in the social structure, must be able to locate one another and communicate with one another. They can then sound one another out, make tentative and exploratory moves in new directions, experience the feedback, and—if the feedback is encouraging—go on to elaborate what becomes a new and in some respects deviant subculture.[15]

Other students of delinquency have gone further, suggesting that there are elements of lower-class culture as a whole which contribute to the formation of delinquent subcultures. Walter B. Miller, basing his study on 21 corner groups in a slum district of a large eastern city, has described "Lower Class Culture as a Generating Milieu of Gang Delinquency."[16]

Miller's research team discerned certain "focal concerns" of lower-class culture: trouble, toughness, smartness, excitement, fate, and autonomy. Those concerns ostensibly distinguish the outlook of American lower classes from that of the middle classes. Getting into *trouble* and staying out of it is a major issue for lower-class adults and children of both sexes. For men trouble is frequently a matter of fighting, sexual exploits, and drinking; for women it is disadvantageous sexual experiences. The main concern over trouble revolves about the question of whether it is law-abiding or not. Thus a mother will evaluate the suitability of her daughter's boyfriend in terms of his trouble potential. If it is high she will discourage her daughter from pursuing the relationship. On the other hand, a young man's reputation for getting into trouble may be a source of prestige. That is particularly true in certain teenage and adult gangs which tend to engage in law-violating behavior. It is not trouble in and of itself that confers prestige in those circles but rather the willingness to court trouble, if necessary, in the pursuit of valued ends.

Toughness is expressed in physical prowess. It includes strength, endurance, and athletic skill and a general masculine demeanor characterized by a lack of sentimentality and by bravery in the face of a physical threat.

Smartness entails the ability to outfox others while remaining unduped oneself. Like trouble and toughness, smartness is not so much valued for itself as it is as a means for achieving money, material goods, and personal status. Such mental agility is manifest in a variety of street-corner activities such as gambling and verbal duels. One often demonstrates his smartness (and toughness) by means of an "ingenious aggressive repartée"—a semiritualized razzing, kidding, and teasing.

Excitement expresses itself in the search for thrills. The search is partially gratified in gambling and in the periodic "night on the town." Men make the rounds of various bars, drinking continually throughout the evening and trying to pick up women, which often leads to fights over women and other forms of "trouble." Indeed, trouble seems to be sought in those situations. Seeking risk, danger, and thrills in such fashion alternates with periods of relative inaction and passivity— "hanging out," doing "nothing," and "shooting the breeze."

Fate is another central theme of lower-class culture. The belief in fortune or luck prevails, and individuals are looked upon as the pawns of impersonal supernatural powers. If the cards are right, things will go your way. If luck is against you, pursuing a desired goal is futile. At the same time the lower-class male strongly asserts his personal *autonomy,* which expresses itself in resentment and resistance to external restraints on behavior and to unjust and coercive authority. Commonly expressed sentiments are: "No one's gonna push *me* around"; "I'm gonna tell him he can take the job and shove it"; and so forth.

The focal concerns of trouble, toughness, smartness, etc. figure prominently in the lower-class milieu. The social matrix in which they are generated is not the family but the one-sex peer group, which for teen-agers of both sexes appears to be the most significant group of lower-class society. Gaining membership in such groups and holding high status within them are widely sought after and are achieved by demonstrating one's toughness, smartness, autonomy, etc. Those qualities prove that one is no longer a "kid." "Adulthood" is proved not so much by assuming adult responsibility as by possessing the symbols of adult status—ready cash and a car—and by drinking, smoking, and gambling. 313

The lower-class teenager often works hard to prove his adulthood in such terms. Frequently the result is that a youth seeks avenues for proving his toughness, smartness, and autonomy with greater intensity than an adult and with less regard for the legitimacy of those avenues.

The foregoing discussion brings us closer to an understanding of why the commission of crimes is a customary feature of teenage and young-adult peer groups. Why do they engage in theft, assault, shoplifting, mugging, and the like? Certainly not because they are psychopaths or because they are physically or mentally "defective." On the contrary, they are physically and psychologically normal individuals, well aware of the illegal nature of their acts. They are primarily motivated, writes Miller, "by the attempt to achieve ends, states, or conditions which are valued, and to avoid those that are disvalued within their most meaningful cultural milieu, through those culturally available avenues which appear as the most feasible means of attaining those ends."[17]

TECHNIQUES OF NEUTRALIZATION: HOW DELINQUENTS RATIONALIZE THEIR LAW VIOLATIONS

Techniques of neutralization is a term which has been used to describe the process of reasoning engaged in by a teenager before, during, and after he has committed his delinquent act.[18] It is important to stress again that delinquents are *normal* individuals fully aware of their law-violating behavior. The causation involved in delinquency is primarily social; and a delinquent is more or less conscious of his motives. There is no need, therefore, to subscribe to the psychiatric notion that delinquency typically results from deep-seated and unconscious motives.

There is, however, an important question which requires more attention: How does a delinquent, or a person learning to become one, retain his self-respect? Winning a firm place for himself in a delinquent peer group is not sufficient. For no matter how distinctive and strong the delinquent subculture is, all of its members know that they are breaking the laws respected by the larger society. The answer is that a delinquent retains his self-respect by neutralizing in advance the disapproval he would otherwise experience both from himself and from respectable others. The typical delinquent develops a specific vocabulary of motives designed to rationalize his conduct. Through the use of one or more rationalizations he can "(a) remain committed to the conventional code but so qualify its moral imperatives that violations of them are not only 'acceptable' but 'right,' and (b) not seriously en-

314

danger his own self-conception."[19] Such rationalizations play a significant role in leading an individual to the point where, in Sutherland's
terms, he accepts "definitions favorable to the violation of law." As
we have already remarked, it is a well-known fact that most juvenile
delinquents do not go on to become adult criminals. The major reason
appears to be that most teenagers have strongly internalized the conventional norms of society. The delinquent individual violates those
norms only after convincing himself that his actions are somehow
justifiable.

Frank E. Hartung provides an illuminating discussion of several of
the rationalizations typically employed.

1. *Denial of Responsibility.* Hartung suggests that such a denial reflects the professional ideology of social workers, juvenile-court judges,
psychiatrists, and other persons in authority who deal with delinquents.
To justify his conduct a juvenile will often repeat in his own terms
what he has heard from those in authority. In effect, he attributes the
cause of his conduct to his underprivileged status and to his deteriorated
environment.

2. *Denial of Injury.* In defending his conduct, a juvenile will often
state, for example, that he has not really stolen the object, say an automobile, but merely "borrowed" it. Here again it appears that he is
echoing those in authority. The police often recognize the distinction
made by the delinquent, and the registered complaint is "unlawfully
driving away an automobile" rather than automobile theft.

3. *Denial of the Victim.* The victim "had it coming to him," the
offender insists. A crime is thereby transformed into a "just. punishment." One type of victim who may be "justly punished" is a homosexual. Other types are members of minority groups, "who have gotten
out of their place," "crooked" storekeepers, "unfair" teachers, and
"harsh" principals.

4. *Condemnation of the Condemners.* Often the offender attempts
to turn the tables on those who disapprove of his acts by describing
them as hypocrites motivated by spite. Hartung cites the experience of
Dr. Melitta Schmiderberg, who reports that "when I ask them [delin-
quent children] why they were sent [to training schools or to jail], they
tell me the judge didn't like them, or their lawyer was no good, or the
jury was rigged. Almost never do they say, 'I stole' or 'I shot a man.' "

5. *Appeal to Higher Loyalty.* In the final type of rationalization, the
delinquent defends his action by claiming that he did it out of loyalty
to his gang brothers. Thus Hartung shows that the delinquent often   315

derives his motives from the public domain and then employs them to rationalize his conduct both for himself and for others.

CRIME AND THE AMERICAN UNDERCLASS

We begin with some factual observations from the report by the President's Commission on Law Enforcement and Administration of Justice.

In a sense, social and economic conditions "cause" crime. Crime flourishes and always has flourished, in city slums, those neighborhoods where overcrowding, economic deprivation, social disruption and racial discrimination are endemic.

One of the most fully documented facts about crime is that the common serious crimes that worry people most—murder, forcible rape, robbery, aggravated assault, and burglary—happen most often in the slums of large cities. Study after study in city after city in all regions of the country have traced the variations in the rates for these crimes. The results, with monotonous regularity, show that the offenses, the victims, and the offenders are found most frequently in the poorest, and most deteriorated and socially disorganized areas of cities.[20]

In American cities crime rates tend to be highest in the city center and to decrease in relation to the distance from it. Throughout the period of massive immigration to the United States, city-slum crime rates remained high in spite of the succession of new ethnic groups moving in to displace the old. There were some notable exceptions: In particular, people of oriental ancestry, owing to their intense social and cultural solidarity, showed a unique capacity to insulate their youth from criminal influences. Nevertheless, these facts suggest that the crime rate remained high because the deteriorated conditions persisted and because a similar proportion of the youth from each successive ethnic group learned criminal patterns from their ethnic predecessors.

The same pattern of high crime rates holds true for blacks and other recent immigrants to the slum areas of American cities. Yet there is reason to expect even higher crime rates among the recent immigrants. As we have seen in our previous discussions of the underclass, major changes have occurred in the American economy, radically reducing the demand for unskilled labor. Since the end of World War II, skill and educational requirements for jobs have been steadily rising. The racial discrimination to which black Americans are subjected in employment, education, and housing is much harder to overcome than discrimination based on language and ethnic background. Working one's way out of the black ghetto is increasingly difficult. In the words of the President's

Commission report, "It could be predicted that this frustration of the aspirations that originally led Negroes and other minority groups to seek out the city would ultimately lead to more crime. Such evidence as exists suggests this is true."

According to the FBI's Uniform Crime Report for 1975, more than 25 percent of the persons arrested that year for all types of crime were black; and more than 47 percent arrested for violent crimes were black. The typical arrested black is a young male. Almost 57 percent of all persons arrested in 1975 were under 25 years of age, and more than 84 percent were male.[21]

According to a 1975 Law Enforcement Assistance Administration (LEAA) survey of *local* jail inmates, 42 percent were black, the majority of them (almost 65 percent) having failed to complete high school. Forty-six percent of the black inmates had been earning less than $2,000 a year when arrested, and another 12 percent less than $3,000 a year. The 1974 LEAA survey of *state* prison inmates yielded similar findings: About 47 percent of all prisoners were black, and of those at least 64 percent had not completed high school. Fifty-two percent of the incarcerated blacks were under 25 years of age, and 75 percent were under thirty.[22]

The typical victim of crime is also black. Blacks are more likely than whites to be victims of assault, robbery, rape, and burglary. Black households, from the very poor to the highest socioeconomic levels, have a higher burglary rate than do white homes. However, certain types of families are most likely to become crime victims. Especially vulnerable are families with several children, headed by women, black or Puerto Rican, and on public assistance. The poorest families, those with an income below $3,000 a year, are most often preyed upon.

The high arrest rate in the ghetto may be at least in part a function of racial discrimination. Typically, a police officer has a considerable measure of discretion whether or not to arrest an individual and thereby to initiate the criminal process. Numerous studies have shown that as compared with the proportion of the national population, an unusually high percentage of blacks are arrested and charged with violations.[23] Police are more likely to arrest a black than a white youth in discretionary situations; they are also more likely to stop and question blacks on the street and to arrest them after questioning. The same pattern holds for other disadvantaged ethnic groups. In their study of the relation of the police to Mexican-Americans in Denver, Colorado, Bayley and Mendelsohn found that the form of contact between the police and an individual was determined by the latter's ethnic-group membership.[24]   317

After arrest the black individual continues to be subject to unequal treatment. A study of the judicial system has been made based on data collected by the American Bar Foundation. Employing state court dockets for 1962, Stuart Nagel scrutinized thousands of cases from 194 counties in 50 states.[25] His study concentrated on larceny, the most frequently reported crime against property, and assault, the most frequently reported crime against persons. Nagel found that the rudimentary safeguards of the judicial system such as preliminary hearings, bail, and the right to counsel, grand jury, and trial by jury all revealed irregularities in relation to blacks and the poor. Bail, for instance, offers the accused a chance to remain free while preparing his defense. However, about 75 percent of the indigent cases failed to raise bail and stayed locked up. In contrast, 79 percent of the nonindigent assault cases and 69 percent of the nonindigent larceny cases succeeded in raising bail. Not only that, the percentage of nonindigent cases found guilty was consistently lower. The poor defendant, especially if he was black, was less likely to be granted probation or a suspended sentence. In state larceny cases 74 percent of the blacks found guilty were imprisoned, as compared with 49 percent of the whites. In federal larceny cases the percentages were 54 percent and 40 percent, respectively. Such studies clearly show that being poor or black or both is a definite disadvantage in the criminal-justice system.

Black Women and Crime

The data for black women yield similar conclusions, but it appears that the proportion of females among prison inmates is on the rise.[26] Between 1971 and 1974 the rate of increase for women incarcerated in state (15.6 percent) and federal (25.2 percent) prisons surpassed that of men, 12.2 percent and 2 percent, respectively. Between 1960 and 1970 there were increases in female arrests for robbery (277 percent), embezzlement (303 percent), and burglary (168 percent). These increases were also significantly higher than those for men. Female recidivism has likewise risen dramatically: From 1967 to 1973 the number of women with one or more commitments to prison rose by 21 percent.

Prison inmates are increasingly black. Although black Americans constitute only 11 percent of the national population, they make up 35 percent of the federal and 47 percent of the state prison population. Among those inmates black women are greatly overrepresented. In 1970 they comprised almost half (48.8 percent) of all women 18 years

318

of age and older in state prisons, and in 1974 more than half of the female inmates in federal prisons were black.

In some states the proportion of incarcerated black women exceeds 50 percent. In the Women's Detention Center (WDC) in the District of Columbia, for example, almost two thirds of the women were black in 1974. In the same year three quarters of all first bookings in that institution were black; and 83 percent of the women who returned to the WDC from their first court hearings were black. In Alabama as of 1973, 75 percent of the women in state prisons were black.

Black female offenders are almost invariably poor. In 1974 two thirds of the inmates of the Federal Correctional Institution (FCI) in Alderson, West Virginia, earned less than $3,000 a year prior to incarceration, and an additional 20.4 percent earned less than $5,000. Thirty percent of all women inmates were on welfare before incarceration. A high percentage of the black women had been employed as household domestics and in other low-skill, low-paying occupations. Moreover, the female offender is very likely to have dependent children. A majority (66.1 percent) of the FCI inmates in Alderson had dependent children, as did an even larger majority (80 percent) in Pennsylvania jails.

The unemployment rate for black youths, male and female, has reached alarming proportions. Little wonder, then, that the crimes for which black women are most frequently incarcerated are economically related. A 1973 Federal Bureau of Prisons study reported that of the 392 inmates in the "all other" category (i.e., mainly black), 196 were convicted of theft, forgery, or larceny. The next largest number of convictions was for drug offenses, which were often economically motivated. As Wyrick and Owens have observed after surveying the pertinent evidence, black female offenders "usually go to jail because they are broke and they often return to jail because they are still broke."[27]

The Street-Corner Context of Male Offenses

The street corner, the pool hall, the bar, the doorstep—those are the main ghetto settings in which homicides and assaults occur.[28] Most often offenses are committed by street-corner men and boys against ghetto friends, acquaintances, and strangers. The violence is rarely planned far in advance or based on old animosities; rather, it is most frequently triggered by minor altercations. The unemployed and underemployed young black male moves in a street-corner circle in which one's honor may be easily affronted—as in the course of a street rap or verbal duel, especially while drinking. Since the lower-class culture accepts the principles of toughness and excitement, violence is likely to follow verbal abuse: "He's a man and I'm a man, and I don't take no shit like that."[29] It is otherwise in middle-class culture, where an altercation between two acquaintances in a fashionable bar is likely to be peacefully resolved. But in street-corner society, if one party seems to be gaining the upper hand with his smooth rap, there is a great temptation for the less skilled party "to adopt a violent strategy in which you force others to give you what you fail to win by verbal and other symbolic means."[30] Physical confrontations are frequently bloody since black youths often carry knives and other cheap lethal weapons. Weapons possession is a correlate of toughness in the ghetto male's subculture. Those symbols of manliness are less expensive and more accessible than the automobile—the middle-class youth's extension of his masculinity—but they raise the likelihood that a relatively minor quarrel will eventuate in injury and even death. Citing such an incident Lee Rainwater writes:

Well, we were in the show and he was with his girl and there was a lot of crowd around, you know, just around. And I asked his girl for a piece of candy and he jumps up and smacks me in the face with a cap. Well, I could whip him. . . . I take the cap and hit him in the face. He grabbed me around the throat again so I pushed him against the wall again and I hit him about three or four times and he said, "Now I'm going to kill you," and he stuck his hand in his pocket. Well I went in mine and I was just a little faster than he was and I cut his throat.[31]

However, a large proportion of the homicides and assaults also take place within the ghetto household. Often the violence is between the economically marginal street-corner male and his legal or common-law wife. Economic pressure is an underlying condition. "The black street-corner marriage" says Lynn A. Curtis, "is burdened by far more conflict-building economic pressure than the black mainstream or black middle-class marriage. The situation perhaps most cited in participant and non-participant studies has the wife criticizing the husband's economic performance, particularly when it is juxtaposed against spending patterns related to contracultural lifestyles."[32] The wife's criticisms and the husband's response soon spiral upward, ending in violence and bloodshed. Ulf Hannerz relates the following incident to illustrate a typical conflict-precipitating exchange:

Randy Ballard . . . was obviously a little high. As he came in he asked his wife who was ironing in the kitchen what there was to eat.

"Ain't nothing for you to eat now," she replied curtly. "The kids finished it and there wasn't much anyway. I figured you wouldn't be in."

"Ain't that a bitch," Randy said. "I give her money for food, and I don't get nothing myself. Nothing in my own house!"

"Listen, nigger," his wife said, "you didn't hardly give me nothing, not compared to what you ought to have given me. And if you hadn't spent the money you kept drinking out there with those hoodlum friends of yours, I could have had more and you could have had your dinner. So hush your mouth."

"You hush your mouth," Randy shouted. "You ain't talking to me like that."

"I sure do when you come home drunk and think you can act any goddam way around here when you ain't got a nickel to spend on your own family. I'm sick and tired of taking all that crap."

"I'm a man," Randy said. "I ain't gonna take that from nobody. You better watch out."[33]

Here it is evident that the husband has fallen short of the wife's expectations. She resents his heavy drinking not only because he thus fritters away sorely needed funds, but also because it represents his strong commitment to street-corner culture.

What we have, then, is a chain of causal connections leading from the economy into the home. To begin with, the black ghetto youth or young adult encounters few economic opportunities that are both legitimate and decent. Footloose and economically marginal, he is attracted to the street-corner subculture. There he acquires certain focal concerns or values that lead him into "trouble." As his commitment to street-corner society grows, he learns that there are alternative avenues to social goals, and that the illegal avenue frequently provides greater and faster rewards for a smaller investment of time and energy. Actually, however, material rewards remain quite minimal. Most street-corner men stay poor—they cannot drink and support their families at the same time—and it is only a few street-corner "boss-men" who achieve notable material success. Hence the typical black street-corner male fails to fulfill the provider role and engenders the resentment of his wife.

The economically related man-wife quarrels and conflicts often lead one or both partners to seek comfort in others. The result is further tension, conflict, and often, separation. The street-corner husband's betrayal is partially motivated by the subcultural emphasis on sexual prowess and adventurism, while the wife's betrayal, when it occurs, "might better be interpreted as an assertion of independence. . . . [and] a search for comfort after failing to find it within the marriage, and a response to hurts inflicted by the husband."[34]

321

It is therefore clear that crime in the ghetto is rooted in blocked economic opportunities and in racial discrimination. Those are the conditions that bring ever-new recruits into the quasi-criminal and criminal street-corner subculture. The raw recruits as well as the journeymen and masters to whom they apprentice themselves are normal human beings. They learn criminal behavior as a part of the street-corner way of life, which itself is a way of coping with the reality of ghetto circumstances.

The sociological theory of criminal behavior which thus takes account of the structure of socioeconomic opportunity applies with equal force to organized crime. We have already observed that the theory and practice of criminal conduct has been learned by the successive ethnic immigrants to American slums. Daniel Bell, among others, has described the "queer ladder of social mobility" out of the slums which had organized crime as the first few rungs. The first to ascend those rungs were the Irish. They dominated the criminal organizations as well as the big-city political machines. Their control over the political machinery of large cities brought them a directing influence over construction, trucking, the waterfront, and public utilities—hence wealth, power, and respectability. In the 1920s the Irish were succeeded by the Jews, who were conspicuous in gambling and other spheres of the underworld for a decade. Then came the Italians. Now it appears that the Italians are being superseded. In the mammoth ghettos of Harlem and Paterson, New Jersey, the Italians are being pushed out of organized crime and are being replaced by the most recent immigrants: blacks and Puerto Ricans.

The beginnings of the pattern of Italian displacement by blacks is documented by Francis A. J. Ianni in his book *Black Mafia.* Ianni's study tends to confirm the sociological theory of crime, particularly the process of learning criminal behavior as described by Sutherland and his followers. A criminal organization emerges out of networks, the binding links of which are *childhood friendship.* Thus Ianni writes:

In every case of childhood friendship that grew into an adult criminal partnership, the individuals involved were of the same ethnic or racial group and usually were approximately the same age. It seems almost unnecessary to point out that this is not the result of any innate criminality in any of the ethnic groups but rather results from the fact that street society, where youngsters meet, is based on residential patterns, which tend to follow racial and ethnic lines as well as socio-economic ones.[35]

Another type of link develops "when *an experienced criminal in the neighborhood sees that a young boy (or gang of young boys) possesses talent and recruits him into organized criminal ventures.* This is a most common method of recruitment into organized crime and represents the first step in criminal apprenticeship." It is the pattern of association and role models that count. Ianni quotes the words of one of his black field assistants to that effect: "The ones you see are the ones that interest you. If it had been doctors and lawyers who drove up and parked in front of the bars in their 'catylacks,' I'd be a doctor today. But it wasn't; it was the men who were into things, the pimps, the hustlers and numbers guys [in Central Harlem]."[36] Lacking positive and legitimate associations and examples, many youths follow and emulate the men who are "into things" instead.

WHITE-COLLAR CRIMES

We see, then, why crime thrives on conditions of economic deprivation and racial discrimination. Our sociological analysis helps to explain why a disproportionately large number of poor young black males commit crimes like homicide, assault, and robbery. But crime also flourishes in conditions of affluence, where there is a powerful desire for material wealth and numerous opportunities for acquiring it illegally. If young poor black and white males figure prominently in crimes of violence in the street and at home, then older white upper- and middle-class males have cornered the market on white-collar crime.

The term *white-collar crime* was coined by Edwin H. Sutherland. He was among the first to note how misleading both the popular and the official conceptions of crime were. Official statistics were exclusively concerned with such crimes as murder, assault, burglary, robbery, larceny, sex offenses, and public intoxication. Hence "crime," as conventionally understood, had a high incidence in the lower socioeconomic classes. Sutherland, however, challenged the conventional understanding on the ground that it was biased in two important respects.

First, there is the rather obvious fact that well-to-do persons can escape arrest and conviction by employing skilled attorneys and by otherwise influencing the administration of justice in their favor. Second, there is the powerful bias in the legal code and in the system of criminal justice itself. Laws are applied in a special way to offenders from the upper socioeconomic classes:

Persons who violate laws regarding restraint of trade, advertising, pure food and drugs are not arrested by uniformed policemen, are not often tried in criminal courts, and are not committed to prisons; their illegal behavior 323 generally receives the attention of administrative commissions and of courts operating under civil or equity jurisdiction. For this reason such violations of law are not included in the criminal statistics nor are individual cases brought to the attention of the scholars who write the theories of criminal behavior. The sample of criminal behavior on which the theories are founded is biased as to socio-economic status, since it excludes these business and professional men. This bias is quite as certain as it would be if the scholars selected only redhaired criminals for study and reached the conclusion that redness of hair was the cause of crime.[37]

It was Sutherland's aim to eliminate that bias. He thus set out to demonstrate that upper-class individuals engage in much criminal behavior, the principal form of which is white-collar crime, "a crime committed by a person of respectability and high social status in the course of his occupation. Consequently, it excludes many crimes of the upper class, such as most of their cases of murder, adultery, and intoxication, since these are not customarily a part of their occupational procedures."[38] Sutherland devotes the bulk of his study to showing the large volume of upper-class crime in such forms as restraint of trade; rebates and discriminatory pricing; infringement on patents, trademarks, and copyrights; misrepresentation in advertising; unfair labor practices; financial manipulations and swindles; and war profiteering.

In the years since Sutherland published his pathbreaking study, white-collar crime has become a widely publicized fact. The prevalence of embezzlement, manipulation of accounts and stock records, collusion between employers and labor leaders at the expense of rank-and-file members, price fixing, and numerous other law violations is now a matter of common knowledge.

One of the best-known cases in recent years was a price-fixing conspiracy involving 29 electrical corporations. It was perhaps the biggest case in the history of the Sherman Anti-Trust Act. The companies ranged in size from the small shop of the Joslyn Manufacturing and Supply Company to billion-dollar giants like General Electric and Westinghouse. The executives of those companies were indicted in 1960 for "having conspired to fix prices, rig bids, and divide markets on electrical equipment valued at $1,750,000,000 annually. The twenty indictments, under which they were now to be sentenced, charged they had conspired on everything from tiny $2 insulators to multimillion dollar turbine generators and had persisted in the conspiracies for as long as eight years."[39]

In his 20 years on the bench, the presiding judge had earned a reputation for moderation and tolerance. But no sooner had the trial begun than it became clear that he was shocked by the conspiracy. He acknowledged that the Justice Department had insufficient evidence to convict those in the highest echelons of the corporations, but he went on to observe that "one would be most naive indeed to believe that these violations of the law, so long persisted in, affecting so large a segment of the industry and finally involving so many millions upon millions of dollars, were facts unknown to those responsible for the corporation and its conduct."[40] By the second day the trial was over. Seven jail sentences and 24 suspended sentences were handed down for vice-presidents and other officers ultimately responsible for corporation policy. General Electric was fined $437,500 and Westinghouse $372,500. In all, $1,924,500 worth of fines were levied.

Corporate executives and other businessmen often make decisions which place their actions in the gray area between law violation and extremely immoral and irresponsible behavior. Examples of such behavior have come to light quite recently. In congressional testimony in the spring of 1981, Vice Admiral Earl B. Fowler described in detail the problems encountered in the construction of the Trident submarine.[41] The Trident, approved by the Nixon administration in the early seventies, is an extremely large vessel, longer than the Washington Monument is tall. It was designed to carry 24 missiles and eventually to replace the Polaris and Poseidon, each of which carries 16. Nine Tridents were authorized, and the first was to begin its operation by April 1979. As of the time of Vice Admiral Fowler's testimony in March 1981, the first Trident, the *Ohio,* had yet to be completed. Fowler explained why, detailing the troubles that plagued its construction in the shipyards of Electric Boat, a subsidiary of General Dynamics in Connecticut. To begin with, navy inspectors discovered that Electric Boat had used a substandard quality of steel, with low crack resistance, in as many as 126,000 places in the Trident. Then the navy found several incomplete welds on a different type of submarine, also under construction by Electric Boat. This led to an inquiry which revealed that the company had no record of ever having inspected tens of thousands of welds on the Trident and other vessels. Reinspection on a vast scale was begun. On the Trident 36,149 welds originally required inspection, yet the company was unable to provide proof of having checked more than one quarter. When 2,772 of the still accessible welds were reexamined, one third were found to be defective.

As a result of these and other "unforeseen" circumstances, Fowler predicted that the *Ohio* would not begin its operations before the end of 1981—that is, two and a half years behind the contract schedule and   325 a full four years behind Electric Boat's original estimate that the vessel would be delivered by December 1977. The project called for seven Tridents to be at sea by the end of August 1983. Fowler said that only two would be ready by that date, with a third a month from delivery.

Naturally, costs soared as work fell behind schedule. Electric Boat originally estimated that the construction of the first Trident would require 14.6 million manhours. Its 1981 estimate was 23 million man hours, an increase of 60 percent. Even before the latest trouble over the quality of steel and the inspection of welds, which the navy calls "slippage," the estimated cost of the first vessel went up by 35 percent (from $1.47 billion in 1982 dollars to $2.053 billion). With the slippage, the costs are sure to be higher.

Admiral Hyman Rickover, the father of the nuclear submarine, addressed the key issues raised by such behavior in a prepared statement presented to the Joint Economic Committee of Congress in January 1982.

A preoccupation with the so-called bottom line of profit and loss statements, coupled with a lust for expansion, is creating an environment in which fewer businessmen honor traditional values; where responsibility is increasingly dissociated from power . . . ; where attention and effort is [sic] directed mostly to short-term considerations, regardless of longer-range consequences.

Political and economic power is increasingly being concentrated among a few large corporations and their officers—power they can apply against society, government and individuals. Through their control of vast resources these large corporations have become, in effect, another branch of government. They often exercise the power of government, but without the checks and balances inherent in our democratic system. . . .

Under pressure to meet assigned corporate profit objectives, subordinates sometimes overstep the bounds of propriety—even the law. The corporate officials who generate these pressures, however, are hidden behind the remote corporate screen, and are rarely, if ever, held accountable for the results.[42]

Admiral Rickover then proceeded to support his allegations by drawing upon his experiences in an area he knows best:

In recent years, several major navy shipbuilders when faced with large projected cost overruns resorted to making large claims against the navy. These large claims were greatly inflated and based on how much extra the contractor wanted rather than how much he was actually owed. Ignoring their own responsibility for poor contract performance, they generated claims which attributed all the problems to government actions and demanded hundreds of millions of dollars in extra payments—enough to recover all their cost overruns and yield the desired profit.

Sometimes the claims were many times the desired objective so that the company could appear to be accommodating the navy by settling for a fraction of the claimed amounts.[43]

When Rickover and other navy officials carefully scrutinized such claims, they found "numerous instances of apparent fraud."[44] In the 1970s the navy referred the claims of four large shipbuilders to the Justice Department for investigation. After nearly a decade this is the disposition of those cases: (1) Litton was indicted four years ago for fraud, but the Justice Department has taken no action to try the case. (2) The Justice Department conducted a lengthy investigation of Lockheed claims but did not issue an indictment. By now the statute of limitations has expired. (3) After investigating General Dynamics for four years, the Department of Justice recently announced they could find no evidence of criminal intent, although the claims were almost five times what the navy actually owed. (4) The Newport News investigation was recently dealt a serious blow when the Justice Department split up the investigating team and assigned the leading investigators other work. This happened shortly after they had reported their findings in the Newport case and had asked the department for more help to track down other promising leads.[45]

The "electrical conspiracy" discussed earlier and the conduct of these naval contractors serve as good examples of white-collar crime— law violation by respectable individuals from upper socioeconomic circles. In Sutherland's own sample of large corporations, he found that their officials violated laws with great frequency. No evidence he encountered anywhere in his study even vaguely suggested that the white-collar offender is "pathological" in some sense. Sutherland therefore concluded that his theory of differential association can explain middle- and upper-class crime as well as it explains the criminal conduct of the lower classes. However, some of his critics have protested that the middle- and upper-class offender, who has no history of criminal association, is an exception to the theory of differential association.

Should white-collar crime be treated as an exception? A white-collar offender may have had no previous criminal record, friends, or associates. But as we have already noted in connection with lower-class crime, direct contact with others who are already criminal is not always necessary for the development of criminality in a given individual. The same is true of the participants in the electrical price-fixing conspiracy. The motives of the executives involved were shaped by the general success goals of wealth and power and by the question of 327 whether the legitimate means of achieving those goals were sufficient. At some point those individuals decided that the legitimate means were inadequate. They went ahead with the conspiracy owing, in Sutherland's terms, to "an excess of definitions favorable to violation of law over definitions unfavorable to violation of law." The conspirators, like lower-class offenders, probably employed "techniques of neutralization" in an effort to rationalize their actions and safeguard their self-respect. While we know nothing about those techniques in this particular case, our general knowledge of white-collar crime gives us no reason to suppose that it constitutes an exception to the theory of differential association. It therefore seems clear that the best available principles for an understanding of crime are those set down by Durkheim and Sutherland and skillfully elaborated by their followers.

Crime is only one form of social deviance, albeit a major one. Several other forms would be considered in this chapter were it not for limitations of space. Nevertheless, we shall consider just one more form to illustrate further the sociological perspectives. The sociological conception of deviance, as we have seen, rests on this fact: Every society comprises individuals and groups who engage in behavior which is judged as offensive by a majority of society's members. Homosexuality, in those terms, may be comprehended as a form of social deviance.

A NOTE ON HOMOSEXUALITY AS SOCIAL DEVIANCE

The term *homosexuality* refers to sexual relations with persons of one's own sex. Such relations are widely regarded as evidence of a pathology, an attitude that no doubt originated in the dominant theological outlook of Western society, which has always looked upon homosexuality as a sin or evil. The psychological literature has also viewed homosexuality as a kind of sickness, often portraying it either as an illness rooted in physiological or mental disorders or as a form of social maladjustment. Homosexuals are, purportedly, unhappy people, the products of maternally dominant family backgrounds or otherwise defective social relations.

Much of what people say about homosexuals is erroneous, a form of stereotyping: All homosexual men are effeminate; all homosexual women are masculine; homosexuals are fond of sexual relations with children and are especially prone to violence. All those allegations, as we shall see, have been refuted by careful empirical studies.

328   In recent years homosexuals have established a number of organiza-

tions such as the Gay Liberation Front; One, Inc.; Daughters of Bilitis; SIR; and the Mattachine Society. The development of these organizations reflects the fact that homosexuals have become a self-aware deviant minority. The perspective they have advanced is that homosexuality, far from being a psychological disorder or hereditary defect, is rather an alternative lifestyle, as valid as that of "straight society." From that standpoint homosexuals are an oppressed minority group with striking similarities to other minority groups.

In the United States homosexual activity is a criminal offense in some 30 states, even between consenting adults in private. And even where such activity is not legally proscribed, homosexual men in particular are exposed to considerable police harassment. Police frequently display the conventionally hostile attitude toward homosexuality, arresting those they suspect of homosexual interests on such grounds as loitering, solicitation, or disorderly conduct. Homosexuals have organized themselves in order to bring such forms of social repression to an end.

Most people probably regard heterosexuality as the natural and normal condition of humanity. However, as we noted in chapter 3, there is no estrous cycle in humans, and no instinct governs human sexuality. What humans have instead is a wide range of erotic impulses which may be directed toward themselves, toward members of either or both sexes, and even toward animals and objects. The outcome in the case of every human being is contingent upon his mode of socialization.

Therefore, what is "natural" and "normal" in human beings is a remarkable diversity in all things, including sexuality. If we once again return to Durkheim's sociological framework, we can quite clearly see what it is that makes homosexuality "deviant." It is deviant neither because it is somehow injurious to society nor because it is pathological. Homosexuality is regarded as deviant simply because it offends the sentiments of the dominant majority.

The evidence at hand concerning human sexuality indicates that there exists a homosexual-heterosexual continuum. All gradations can exist, from apparently exclusive homosexuality, without any conscious capacity for arousal by heterosexual stimuli, to apparently exclusive heterosexuality. In the latter case, however, there may be transient homosexual inclinations as in adolescence. According to the psychoanalytic school, all humans pass through a homosexual phase, and available evidence indeed strongly suggests that a transient homosexual phase of development is very common.

In his renowned study *The Sexual Behavior of the Human Male,* Dr.   329

Alfred C. Kinsey formulated a homosexual-heterosexual continuum on a seven-point scale. The rating of six was assigned for sexual arousal and activity with other males only; three for arousals and acts equally with either sex; zero for exclusive heterosexuality; and intermediate ratings accordingly. The recognition of the existence of such a continuum is extremely important, for it leads to the conclusion that homosexuals cannot be reasonably separated from the rest of humanity.

Dr. Kinsey's study concluded that in the United States, 4 percent of adult white males are exclusively homosexual throughout their lives following the onset of adolescence. Kinsey also found evidence indicating that 10 percent of the white male population are more or less exclusively homosexual for at least three years between the ages of 16 and 65; and that 37 percent of the total male population have at least some overt homosexual experience, to the point of orgasm, between adolescence and old age. No comparable studies have as yet been made in Western Europe, but it is noteworthy that medical witnesses for the Wolfenden Committee expressed the opinion that similar figures would be established in Britain if comparable inquiries were made.[46]

Diversity and Deviance

The most recent study of human sexuality to have come out of the Institute for Sex Research (the "Kinsey Institute") is entitled *Homosexualities: A Study of Diversity Among Men and Women.* The authors, Alan P. Bell and Martin S. Weinberg, have marshaled considerable evidence to show that "there is no such thing as *the* homosexual (or *the* heterosexual, for that matter) and that statements of any kind which are made about human beings on the basis of their sexual orientation must always be qualified."[47] *Diversity* is the focus of this study —the significant ways in which homosexual persons differ from one another.

Employing the Kinsey scale, Bell and Weinberg had their respondents classify themselves, first with regard to their sexual behaviors and second with regard to their sexual feelings. In addition, respondents identifying themselves as homosexuals were asked whether they had had any sexual contact with persons of the opposite sex during the past year; whether they had experienced an orgasm in that context; whether their erotic dreams and masturbatory fantasies involved a heterosexual element; and whether they had been aroused by a member of the oppo-

330

site sex. Finally, those with both homosexual and heterosexual experiences were asked to compare the two.

The results were evidence that both the men and the women significantly differed from one another in the degree of their homosexuality. A strong heterosexual element was evident in their feelings and conduct; and many respondents, both male and female, had a history of sexual contact with persons of the opposite sex and were aware of a continuing potential for such contact. Respondents also showed considerable diversity with respect to the relative overtness of their homosexuality and the levels and frequency of their sexual activity.

The diversity of homosexual experience is brought into relief by means of a typology comprising five categories for men and women.

1. *Close-coupled:* Persons in a quasi-marriage. Among men, the close-coupled had fewer sexual problems and fewer partners than did the other respondents. They did little "cruising" (searching for a sexual partner).

2. *Open-coupled:* Persons also involved in a "marital" relationship, though less binding. They had several sexual partners and a few sexual problems, and they cruised more often than the close-coupled.

3. *Functionals:* Persons who were "single," not coupled. The other criteria for inclusion in this category were *high* scores on number of sexual partners and level of sexual activity and *low* scores on regret over homosexuality and number of sexual problems.

4. *Dysfunctionals:* Also single persons, they were chiefly distinguished from the functionals by their *high* scores on regret over homosexuality and number of sexual problems.

5. *Asexuals:* Also single persons who had *low* scores on level of sexual activity, number of partners, and amount of cruising. They had definite sexual problems (e.g., great difficulty in finding a partner) and more regret over homosexuality.

The majority of the male respondents were either coupled or functional while a distinct minority fell into the dysfunctional and asexual categories. That suggested to the authors that a large proportion of homosexual men manage their way of life satisfactorily, and that only a small proportion have conspicuous difficulties. Among the men there were many more open-coupled than close-coupled, whereas their female counterparts were more often close- than open-coupled. Women were more likely to be involved in a quasi-marriage marked by a high degree of stability (mutual fidelity).

331

The value of the typology became especially evident to the authors in connection with the question of psychological adjustment. The degree of such adjustment was probed by questions concerning general health, psychosomatic symptoms, happiness, exuberance, self-acceptance, loneliness, worry, depression, tension, paranoia, suicidal feelings, and professional help. If several diverse types of homosexual patterns had not been delineated, the authors candidly acknowledge, they might have concluded that homosexual adults in general fall short of the psychological adjustment of heterosexual men and women. However, by employing the typology one learns that

among both males and females, it is primarily the Dysfunctionals and the Asexuals who appeared to be less well off psychologically than those in the heterosexual groups.

Among the men, the Close-Coupleds could not be distinguished from the heterosexuals on various measures of psychological adjustments and actually scored higher on the two happiness measures. The Functionals, as well, hardly differed at all from the heterosexual men in terms of their psychological adjustment. Among the women, the need for delineating the homosexual adults on the basis of "type" is similarly striking. Like their male counterparts, the Close-Coupleds and the Functionals looked much like the heterosexuals psychologically. In fact, the Close-Coupled lesbians reported less loneliness than the heterosexual women, while the Functionals appeared to be more exuberant.[48]

The Bell and Weinberg study teaches us that the term *homosexual* tells us nothing about all the other dimensions of that person's character, values, and way of life, just as knowing that someone is a "heterosexual" conveys equally little about the whole person. Homosexuals differ among themselves with regard to all the important facets of social life and not least with regard to their sexuality. Though they are closer than most people to one pole of the homosexual-heterosexual continuum, they vary significantly in all the important dimensions of the sexual experience: level of activity, types of partnerships, techniques, levels of interest, problems, and acceptance of their homosexuality. That is why the authors have employed the plural noun *homosexualities* as the title of their study.

Homosexualities have existed in all societies throughout history; they are in evidence in all social classes and in all callings and occupations. What is beyond dispute, according to the best evidence available, is that deviating from the dominant heterosexual orientation of a society is not necessarily pathological. Indeed, even the "dominant" orientation includes a remarkable variety of sexual patterns, a variety that is obscured by the omnibus term *heterosexual*.

Increasingly, modern society has come to accept the principle that the intimate behavior of two consenting adults in private is not the law's business. In the United States that principle has already been recognized in 19 states, where laws against homosexuality have been abolished. In Britain, the authors of the Wolfenden Report have recommended the decriminalization of "homosexual behavior between consenting adults in private." And Bell and Weinberg have observed that

as long as these laws remain on the books, homosexuals will remain vulnerable to many kinds of social oppression. Besides the continual threat of arrest, they sometimes suffer extortion and blackmail by the police, by persons posing as police, by hostile sexual partners, or by a disenchanted friend. Since such persons threaten to expose their victim's homosexuality if their demands are not met, more often than not the victims will not turn to the police for help. . . . Understandably, they fear publicity at least as much as whatever legal penalties might ensue. In a homoerotophobic society such as ours, many persons feel free to rob or assault homosexuals, knowing that their victims are not likely to press charges and even feeling that they are getting just what they deserve.[49]

In light of the foregoing analysis, there is no good reason for perpetuating the legal and social repression of the homosexual minority.

13

Cities Then and Now:
The Urbanizing Process

What is a "city" and how did the earliest cities distinguish themselves from noncities? Some scholars hold that the city first emerged as a religiocultural center. Others view the city as a social form that arose as social classes replaced kinship as the chief principle of organization. Still others stress trade as a city's essential feature. Finally, scholars have focused on political and military power as the city's defining characteristic. It is clear that no single criterion will suffice to define a city. Indeed, even the earliest urban forms differed so significantly from one another that types of cities may be discerned.

One may begin to grasp the concept of "city" in its earliest forms by defining it as a community whose members live close together in a unified complex of buildings, often enclosed by a wall. But that definition fails to distinguish a city from certain types of villages, religious communities, and military camps. So one must add that a city is a community in which a significant portion of the population pursues its everyday activities inside the city, in nonagricultural occupations. However, since monasteries and factories surrounded by workmen's dwellings would also fall under that definition, one more feature must be added. A city exercises influence or control over the area outside it. Like all formal definitions, this one too is intellectually unsatisfying so long as it is unaccompanied by concrete illustrations. Yet it helps us to distinguish the earliest preurban types of human settlement from those that qualify as cities.

What is presupposed in all human settlements is a *food-producing* economy. Food-producing settlements are comparatively recent phenomena in human history, having emerged for the first time some 10,000 years ago, at the beginning of what archeologists call the Neolithic Age. Prior to that time human beings lived as food gatherers in very small groups constantly on the move. Of course the *Neolithic Revolution,* as V. Gordon Childe has termed it, was not something sudden. But it was a revolutionary change in that humans now began to increase their food supply by cultivating plants and domesticating animals. With the increase of the food supply and the storage of surpluses, the population increased. The food-producing revolution released humans from dependence on the natural food supply and thus permitted more stable habitation. That occurred for the first time in the Near East, where many archeological sites attest to the emergence of the earliest agricultural villages or towns. Two are presented here to illustrate the connection between agriculture and forms of human settlement.

The first is the site of biblical Jericho, the modern Tell es-Sultan, lying north of the Dead Sea on the eastern slope of the mountains separating the Jordan Valley from the Mediterranean coast. It is there that Kathleen Kenyon has uncovered the remains of an agricultural community dating back to the eighth millennium B.C. The community had existed before the local Mesolithic population, the Natufians, began to make pottery. The archeological evidence includes successive collapses and rebuildings of houses constructed with earth floors, mud-brick walls, and roofs of wood or branches. The houses were a translation into a solid medium of the impermanent huts of the nomadic population. The flint and bone industry of the original settlers was akin to that of the Natufian Mesolithic hunters of Mount Carmel. Hence Kenyon and other scholars have concluded that the site in question provides evidence "of the transition from man as a hunter to man as a member of a settled community."[1]

Neolithic Jericho extended over some ten acres and was enclosed by a solid, free-standing wall. It may have sheltered some 2,000 people. A population of that size could not have supported itself with wild grain and wild animals, and it is therefore highly probable that by the seventh millennium B.C., a successful agricultural system had been developed. Although Neolithic Jericho had a plentiful spring, the water supply was quite limited in its natural state. As the population expanded and a larger area of fields was required, the settlers learned how to extend and distribute the spring waters through a system of   335 irrigation ditches. The irrigation system, as Kenyon has explained, has important implications.

> The successful practice of irrigation involves an elaborate control system. A system of main channels feeds subsidiary channels watering the fields when the necessary sluice-gate is closed. Therefore the channels must be planned, the length of time each farmer may take water by closing the sluice-gates must be established, and there must be some sanction to be used against those who contravene the regulations. The implications therefore are that there must be some central communal organization and the beginnings of a code of laws which the organization enforces.[2]

From such evidence Kenyon has concluded that the complex organization required by irrigation transformed Neolithic Jericho from a peasant village into a primitive city, the first of its kind.

Objections have been raised, however, that Neolithic Jericho lacked at least two elements which according to our definition characterize a city. No significant portion of the settlement's inhabitants engaged in nonagricultural occupations; and the settlement itself exercised no apparent influence or control over the surrounding area. It has therefore been suggested that Jericho might more accurately be described as a highly developed agricultural village.

Another site of one of the earliest human settlements is a village called Jarmo. It is located on the foothills running along the Tigris-Euphrates plain to the east, near the border between Iraq and Iran (Persia). Jarmo and other sites in the same general region have been excavated by Robert J. Braidwood. Carbon-14 tests indicate that farming communities existed in the region by about 6750 B.C. Eight levels of continuous occupation at Jarmo strongly suggest that it was a permanent site and not a temporary station for nomads. The settlement contained at least 21 well-constructed mud-brick houses, each with several rooms. The house foundations were stone, though the floors were pisé (pounded clay). Basins, built-in ovens, and chimneys were common, as were storage pits in the floors where agricultural produce was preserved. The settlers used various Neolithic tools, some of which they made out of obsidian, a volcanic glass which can be chipped to a sharp, strong edge. The presence of obsidian implies trade with very distant areas since the nearest source for that material is Lake Van, far to the north in Armenia.

Jarmo supported some 500 settlers, most probably through agriculture. Agriculture is inferred, first, from the figurines of animals and fertility goddesses which were common; and, second, from the fact that 90 percent of the animal bones found at Jarmo are of domesticated beasts such as sheep, pigs, goats, and cattle. The remains of two types of wheat and field pea were found as well. The presence of querns, mortars, and storage pits also points unmistakably to an advanced level of agriculture.

However, food production in Jarmo was still supplemented by food gathering. Braidwood thus concluded that organized agriculture began in the uplands favored with wild wheat and that with the growth of population promoted by food production, the people moved down into the Tigris-Euphrates valley, carrying their agricultural skills with them. That would have occurred around 6,000 B.C. But in Jarmo as in Jericho, nonagricultural occupations were not in evidence; and both settlements had failed to extend their influence beyond that necessary for their subsistence.[3]

What we have, then, in those most ancient sites, is evidence for the Neolithic Revolution. This, as we have seen, was the era in which humans developed the tools, techniques, and forms of social organization which made sedentary communities and surpluses of food possible. The way was thus prepared for Childe's second revolution, the *urban.*

THE EARLIEST TRUE CITIES

In his remarkably thorough study, Mason Hammond has summarized the evidence concerning the question of where in the ancient world the true city first emerged.[4] His answer is Sumer in Mesopotamia, in the region of the Tigris and Euphrates. Sumer's urban history began sometime around 3200 B.C. Settlements were now closely built up and densely inhabited; they had an organized government for the exploitation of the surrounding agricultural territory; they developed a non-agricultural division of labor in crafts, religion, and administration; and they engaged in commerce and war and thus extended their influence far beyond their own boundaries.

Associated with the *Urban Revolution* is the introduction of copper and bronze—the so-called *Bronze Age.* Systematic irrigation on a scale that dwarfed that of Jericho provided the impulse for a highly complex social and political organization. At the same time the wheel, the plough, metalworking, and writing appeared. With the advance of urbanization, kinship increasingly gave way to wealth and power as the principles of organization. Furthermore, the urbanizing process continued throughout all the subsequent periods of Mesopotamian history.

Cities also appeared around 2500 B.C. in the Indus Valley. Since there were contacts between Sumer and that region, it is possible that the Indus Valley cities were inspired by the Mesopotamian example. In India urbanization was again associated with irrigation and flood control, specialization in nonagricultural crafts, and commerce. By 1500 B.C., however, those cities had been destroyed either by natural or social causes and left no inheritance to later India.

In Egypt the patterns were quite different. Thebes and Memphis, for example, never became the complex nerve centers characteristic of Mesopotamia. Irrigation in Egypt gave rise to a complex social and political organization as it did elsewhere, but Egyptian settlements were more in the nature of administrative centers of the pharaoh than autonomous urban centers. It is true that the inhabitants engaged in nonrural occupations of commerce and industry; but they were entirely controlled by the pharaoh's bureaucracy and not by the "city" itself. Hence Mason Hammond concludes that the city never fully emerged in Egypt until it was implanted in the Hellenistic period.

In Palestine, as we have seen, Jericho may have been a proto-city, but it left no enduring legacy. The northern Canaanites, on the other hand, developed settlements that had all the attributes of true cities. Canaanite cities exercised control over the surrounding agricultural population; and their most important economic activity was commerce. Byblos and Ugarit were coastal cities specializing in seaborne trade, while Alalakh was an inland city devoted to desert trade.

The famous Minoan civilization of Crete and its Mycenaean heir on mainland Greece also produced cities. In their earliest forms they resembled the quasi-cities of Egypt with whom they had commercial contacts; they were primarily palace communities. In time, however, they acquired the features of true urban centers.

In sum, in the ancient world prior to the rise of the Greek city-state, true cities had emerged in Sumer and in Canaan and to a lesser extent in the early Aegean civilization. Elsewhere in that era, there existed only proto-cities, quasi-cities, and abortive cities.

THE POLIS: THE GREEK CITY-STATE

The Acheans, who later became known as Ionians and Aeolians, were the forerunners of the people that came to be called Greek.[5] Seminomadic shepherds from the Balkan peninsula, their main unit of organization was the patriarchal clan called *patria* or *genos*. The members of a clan were descendants of the same ancestor and worshippers of the same god. Typically, two or more clans combined to form a *phratry,* a larger brotherhood, which in turn united itself with other *phratries* to form a *phyle* or tribe. When the *phratries* went out on hunting or war expeditions, they did so as units of a tribe. Each tribe had its own deity, organized fighting force, and *basileus,* a supreme king or chief whose authority superseded that of the clan and *phratry* chiefs. Finally, there were circumstances in which several tribes also formed alliances.

When the Acheans overran the mainland, they conquered the richest plains and the most powerful villages. Taking over the existing strongholds or constructing their own, they erected kings' palaces within the walls. Where space allowed, houses of chief officers and other dignitaries were attached to them. The tribal confederation was a religious as well as a political association. Hence the enclosed settlements always included a sanctuary for common worship. The settled tribal confederation and its institutions constituted the embryonic city-state, or *polis.*

The Greek "city" in earliest times was not simply an agglomeration of human beings together within the same walls. In fact, the first Greek cities were hardly places of habitation at all. The city was rather the sanctuary where the gods of the community resided; it was the fortifications which defended them and which they in turn sanctified; it was the political center of the confederation, the residence of the king, chiefs, and priests, the place where justice was administered. But the people lived outside the city in clan communities which divided the surrounding lands among themselves. Each clan had its own locale and shrine and was under the authority of the *pater* or clan chief. Only on certain ceremonial occasions of the confederation as a whole did the clan chiefs assemble in the city for common worship or for deliberations with the king. If it was a matter of preparing for war, the chiefs brought their families and servants, grouped them in *phratries,* and formed the army of the city under the king's command.

The term *polis* was at first reserved for the "lofty" part of the city. That was the *acropolis,* the fortified area containing the sanctuary and the palace. In the earliest Homeric poems the sacred and rich *polis* is distinguished from the *asty,* the lower town inhabited by craftsmen and traders. However, in the later books of the *Iliad* and in practically the entire *Odyssey,* a distinction is no longer made between the *polis* and the *asty.* The changing meaning of *asty* reflects the social changes which had taken place over several generations. The lower town gradually grew larger with the development of agriculture and com-   339 merce and soon acquired an importance equaling that of the upper city. Increasingly the men of property from the lower town mingled with the dignitaries of the *polis*. Accordingly, the *asty* was ultimately also encircled with towers, and the two words became synonymous.

In time the meaning of the term *polis* was further extended to include all the rural communities living under its protection. That was not a difficult step since all the people of the surrounding countryside recognized the authority of the same chief. Thus the word which began by denoting only the *acropolis* ended by indicating the entire city and its environs. The relatively large area inhabited by the people was given the name *demos,* a term ultimately transferred to the people themselves. The earliest Greek city thus expressed and enhanced the unity of the *demos* as a tribal confederation.

With the further development of agriculture and commerce, a new institution emerged in the *polis*. That was the *agora,* above all a public meeting area but also a marketplace. With the *agora* the *polis* became a true city. It now not only exercised influence over the rural populace; it comprised a significant body of people engaged in non-agricultural occupations. The *agora* was the place set aside for commerce. Here the people and the merchants congregated, buying and selling goods from overseas or from the interior of the country. But the *agora* was also a social center where men gathered to hear the latest news and to talk politics. It was expressly designed for full assemblies of the people, and it was there that public opinion was formed. Later in Greek history, with the rise of democratic cities, the ancient *agora* became too small for the huge popular assemblies. Thus in fifth-century Athens, for example, the citizens went out to the hill of Pnyx for their meetings.

In due course, hundreds of such city-states—each with its walls, shrine, palace, and marketplace—dotted the Greek landscape. So much so that the *polis* came to epitomize the Greek world. When Aristotle defined man as a "political being," i.e., as a "*polis* being," it was mainly the Greeks whom he had in mind.

However, if the *polis* was a city, it was also a "state"—a center of organized political power. Each *polis* was independent and sovereign. The boundaries dividing even two neighboring cities were practically insurmountable barriers. The neighbors were separated by calendars, money systems, weights and measures, interests, and history. The very origins of the city-state meant that it was a *patria*, a fatherland. The *patria* was the clan, the *phratry,* the tribe, the confederation. It was, in a word, all the kinship and political associations that were incorporated into the city-state. As soon as an individual stepped outside the boundaries of his city, he found himself in foreign and often in enemy country. The Greek was an intense patriot of his own *polis,* and hostility and war between neighboring city-states was the rule. The entire history of ancient Greece is replete with wars unleashed by the passions of city-state patriotism.

Numerous attempts were made to overcome the extreme particularism of the *polis* and to bring several cities into wider associations. The famous *amphictyonies,* or religious associations, were a step in that direction. However, they failed to achieve their purpose, always remaining hotbeds of intrigue and rivalry in which each city vied for advantages and supremacy. Since no city-state was willing to renounce any portion of its sovereignty, the *amphictyonies* never succeeded in transforming themselves into political confederations. Jealous particularism prevailed.

Social Stratification of the Polis

Already in the Homeric city there appeared at least four social classes: the nobles, the craftsmen, the hired men, and the slaves. The nobles possessed rich cornfields and vineyards as well as pastures where oxen, horses, and cattle grazed in large numbers. A noble's home was a palace, a large treasure house filled with precious utensils and ornaments, embroidered tapestries, jars of wine, and perfumed oils. The nobles traced their genealogies to the gods and enjoyed great power and prestige both in their own right and as members of the king's council.

However, the majority of the free men struggled to stay alive. Some had managed to clear a small patch and to bring a few vines and fruit trees to fruition. Others, less fortunate because they were installed in a scarcely cultivable mountain nook, found themselves eating roots more often than bread, or barley porridge. Still others were landless. The family or clan plot was not always sufficient for the support of all its members. Hence many landless men became craftsmen and artisans —carpenters, stoneworkers, leatherworkers, smiths, and potters. They lived and worked in the towns where the crafts concentrated around the marketplace.

But beneath the craftsmen was a large mass of people who were both landless and tradeless. It must be remembered that even the earliest *polis* was no egalitarian community. To be considered a citizen, one had first to belong to a well-off kinship group of repute, and the

341

poorer clans lacked the means of establishing themselves in the city. That meant that many men belonged to no *genos,* or brotherhood, and therefore had no place in the social structure of the city. Having no source of livelihood outside of their labor, and no clan brothers to fall back on for support, they hired themselves out for wages or sold themselves into servitude.

So the picture we have of the developing *polis* is one in which the wealthy and privileged clans increasingly monopolized power. The rest of the population was relegated to an inferior position. The artisans, working for the well-to-do "public" that frequented the stalls of the marketplace, were moderately well off. But the *thetes,* or hirelings, were scarcely distinguishable from the slaves. The condition of the rural population also worsened from day to day. Their small patches of land were dwarfed by the giant estates of the nobles who continually expanded their holdings by encroaching upon the communal pasturelands and by purchasing new territory. The majority of peasants lived in privation. In bad years they mortgaged their lands and borrowed grain from the neighboring lord for subsistence as well as for sowing. Debt bondage appeared as insolvent debtors were forced to sell themselves, their wives, and their children into servitude.

In the ninth century B.C., the city-state economy rested primarily on agriculture and stockbreeding, though there was also some bartering of goods obtained through piracy. By the end of the seventh century, however, a conspicuous economic change had occurred in the social life of the *polis.* A Greek merchant class appeared, moving about the entire Mediterranean looking for goods and customers. The volume of commerce and industry was unprecedented. Workshops and markets sprang up everywhere. Bartering in kind, now cumbersome and restrictive, was more and more replaced by money—electrum, gold, and silver. A form of "capitalism" spread throughout the Greek world.

Premodern Greek capitalism changed the social structure of Greek society. Nobles now sought to convert their crops into money and bullion, turning away from piracy, and opening large workshops with hirelings and slaves. A new wealthy merchant class appeared. Though their wealth was derived from commerce, they invested in land as well. At first the landed aristocrats disdained the new rich; but in time they recognized areas of common interest and formed political alliances with them.

The rule of such plutocracies together with the commercialization of the economy created new hardships for the peasantry. Paying dearly

342

for the goods manufactured in the towns, they obtained only low prices for their produce, owing to foreign competition. Usury, mortgage foreclosures, and debt servitude became more prevalent than ever before. The impoverished peasants, haunted and degraded by poverty though they were, found it difficult to frequent the *agora* and to express their resentment in political terms. The urban lower classes, however, now took an active interest in politics.

In every commercialized *polis* there existed a mass of "common people": small artisans, craftsmen, shopkeepers, hirelings, fishermen, and sailors. Herded together in the same deteriorated quarters of the towns and ports, the hirelings and other impoverished elements began to acquire a sense of common interest and a feeling of solidarity. In a historical train of events anticipating the revolutionary movements of the nineteenth and twentieth centuries, the army of discontented poor drew its leadership from the classes above them. The *polis* had now effectively split into two camps. Long and bitter conflicts broke out, accompanied by massacres, banishments, and confiscations. Revolutions and counterrevolutions continued from the seventh century B.C. right up to the time of the Roman conquest.

The results of those conflicts also anticipated the patterns of the modern era. The revolutionary movements of the *demos* gave rise not to genuine democracy but to what the Greeks called tyranny. In the face of the mounting opposition of the common people, the oligarchy of nobles and rich merchants sometimes ameliorated the material lot of the people. Typically, however, they refused to make concessions, thus opening the way to tyrannical rule.

The term *tyrant,* which probably originated in Asia Minor, meant "master" or "king." It was introduced into the Greek world as a term of opprobrium, to designate someone who ruled like a despot of the East. A tyrant held absolute power, not by right of lawful agreement between parties, but as a result of insurrection. It was the ruling oligarchy that gave the label "tyrant" to the leaders of the *demos* in an attempt to discredit them. The typical tyrant began his career as a demagogue, speaking out for the poor against the rich, for the common people against the oligarchy. As a rule, a demagogue who took up the cause of the urban poor was a discontented son of the privileged families. Often he already occupied some position of power such as a high office or a military command. By arming a band of followers and seizing a favorable opportunity, he succeeded in gaining control of the *acropolis.* Surrounded by a strong bodyguard, he then proceeded to consolidate   343 his regime by disarming his opponents. The most dangerous oligarchs were either banished or brought under control by taking hostages from them.

However, the problem remained of what to do about the urban poor. Here the tyrants were imaginative. In Corinth, for example, Periander prohibited the introduction of new slaves to ease the pressure upon wages. The tyrants were also great builders. Their projects kept the laboring people occupied and provided them with steady subsistence, ensuring social order. The tyrants thus gained popular support for their aqueducts, breakwaters, and other public works. They sponsored festivals and banquets for the people as well as games, contests, and theatrical performances.

But the tyrannical regimes did not long endure. Successive generations of rulers paid less and less attention to the needs and sentiments of the people. The tyrants, though brought to power by the people, became as harsh and oppressive as their predecessors. Their policies opened the way to the restoration of oligarchical and aristocratic regimes in virtually all the city-states of Greece.

Was Athens a democratic exception? As compared with many other city-states, the answer would probably have to be Yes. Greek writers inform us that both during and after the rule of the tyrants, the boundaries of personal freedom were widened for citizens. Debtors were forbidden by the Athenian constitution to pledge their persons for debts. No citizen could either be arrested for debt or reduced to servitude or slavery. The principle of individual responsibility was firmly established: No longer could a family be held responsible for the debts or crimes of its individual members. Political equality for all citizens existed as well. All had the same rights: All could speak in the assembly and vote; all could present themselves as candidates for the council and for other offices. Poverty did not prevent a man from participating in the public festivals, processions, games, theatrical presentations, and the like.

There were also significant attempts under Pericles to reduce the most glaring social inequalities. Thousands of unemployed and underemployed hirelings were settled abroad in Greek colonies and provided with land. To furnish work for the thousands who remained at home, the state turned employer. Immense projects were instituted for the construction of fleets, arsenals, markets, fortifications, port facilities, and monuments. A public-assistance program was available for those who could not work. War orphans were brought up at the expense of the state treasury; pensions were granted to wounded veterans and to disabled workmen. In peacetime the government ensured an adequate supply of cheap bread for all. Grain speculation was prohibited.

But all of those rights, we must remember, applied only to the citizenry. By our standards the Athenian regime under Pericles was not a democracy but rather a slaveholding aristocracy. Of course slavery was nowhere regarded in the ancient Greek world as fundamentally immoral or inhuman. The greatest of the Greek philosophers could not imagine a society without slaves. The number in Athens at the time equaled or even exceeded that of the free population. That means that all the menial labor and a substantial part of the industrial labor was done for the Athenian citizen by a subjugated population. If therefore the Athenian could enjoy the various rights of citizenship, it was because those rights were actually the privileges of a dominant class.

The second point to bear in mind is that Athens was not merely a city-state but an imperial power. Pericles continued the long-standing imperialist policy. The material and cultural benefits that the Athenian state conferred upon its citizenry were acquired at the expense of the subject states. Naturally, Athenian imperialism aroused the animosity of most of the Greek states, although Pericles sought to justify his policy with high-sounding phrases concerning Athens' destiny to be a leader and teacher of Greece. But the other city-states resented those pretensions as much as they valued their own autonomy. Under Pericles, Athens, in enlarging its empire, deprived the other states of their independence. Little wonder that they rebelled and attacked Athens. The Peloponnesian wars put an end once and for all to Athenian supremacy.

MEDIEVAL CITIES AND TOWNS

The history of Rome and cities under Roman rule offers several parallels with that of Greece.[6] It is highly probable that Rome was also founded by an association of clans and tribes. Clan membership was an essential qualification for Roman citizenship as it was for Greek. In Rome as in Greece, noble families encroached on the lands of the poorer clans and a class structure crystallized consisting of patricians, plebeians, impoverished peasants, and slaves. Under the Roman republic the patrician nobility allied itself with a new class of wealthy merchants. Together they swallowed up all the land of the peasants, 345 who were further ruined by compulsory service in the Roman legions. Dominating the countryside were *latifundia,* giant estates employing slave labor. The destruction of the smallholding peasantry meant that there no longer existed in the countryside a large rural population with a vital interest in opposing the invasions of the Germanic and other tribes. At the same time the growth of the urban poor and the fierce struggles between the patricians and the plebeians threw open the door to the emperors. Such were the underlying conditions which led to Rome's decline.

The disintegration of the empire was completed by the so-called barbarian invasions. Every frontier was penetrated by invaders, from Ireland's Celts to the Saharan tribes. The earliest large settlements were those of the Germans fleeing the Huns of the eastern steppe in the fourth century. The German influx continued through the sixth century. In the seventh century Islam appeared, and its incursions continued well into the eighth. By the ninth and tenth centuries, the major movements had ceased. The Scandinavians, the Magyar steppe-folk, and a final Moslem thrust further dismembered the empire.

Although the first intruders, the Goths and Franks, had sought to preserve Rome's cities, subsequent invasions were destructive of town life. The late Roman state was unable to arrest either the secession of the provinces or the disintegration of urban society. In the Mediterranean basin, owing largely to Islamic urbanism, town life survived Rome's decline and even flourished. In the Baltic area, to a much smaller extent, towns remained active trade centers. But on much of the European continent and in the British Isles as well, a new type of social, economic, and political system emerged: *feudalism.*

In its earliest stages feudalism was based upon a predominantly rural economy. Politically, it was a decentralized system made up of a multiplicity of domains. In each of those domains the feudal lord, originally a military chieftain, brought the cultivators under his control and compelled them to contribute a share of their produce to the lord's manor. Historically, feudal serfdom is associated with low levels of technique, simple tools, and a primitive division of labor. It is also associated with production for the immediate needs of the household or village community, and not for a wider market.

However, at the beginning of the twelfth century a revival of commerce in Western Europe was already affecting feudal society. Traders and trading communities appeared. The merchant's presence encouraged the tendency to barter surplus products and to produce for the market. Money insinuated itself into the manorial economy as lords sought money revenue to avail themselves of the many attractive items the trader had to offer. A market in loans evolved as did a market in land.

The growth of the exchange or market economy was accompanied by the establishment of towns or corporate bodies with varying degrees of economic and political independence. Towns acted as magnets to a rural population eager to escape the feudal regime. On the other hand, the lords themselves came under the urban influence. They borrowed from merchants, married their sons to merchants' daughters, and apprenticed their sons to urban crafts. As they became better off financially, they even purchased membership in urban guilds and engaged in trade.

The origin of the urban communities varied from town to town and from one country to another. A few of the medieval towns were remnants of old Roman cities which came to life again with the revival of trade. Other towns of the early period appear to have had a purely rural origin. Since they grew up within the feudal framework, they remained subject to feudal authority. The main qualification for citizenship was ownership of land, and only later did trade become a major occupation of the inhabitants. Still other towns appear to have sprung up as settlements of merchants' caravans. That process has been documented by the influential French historian Henri Pirenne. The earliest traders in the period in question were itinerant peddlers moving from fair to fair or from one feudal estate to another in caravans for their mutual protection. For a settlement they sometimes chose an old Roman town situated favorably at a junction of roads and at other times a feudal garrison or monastery. If the site appealed to them and there were advantages in remaining there permanently, they built a wall around the settlement. The wall not only gave them a military advantage; it also set them off as a community with a specific identity. As they attracted new settlers and as the walled towns grew in wealth and influence, they acquired privileges and protection from the king in return for money payments or loans. Such was the case with German and Italian merchants in England: Royal protection gave them a measure of freedom from the neighboring feudal lords. At a later stage the caravan became a guild, and the town inhabitants organized themselves to escape feudal authority altogether.

There is the question of what prompted the marked expansion of trade and, hence, the revitalization of urban life in the first place. Pirenne has also answered that question convincingly. It was the resurgence of maritime commerce in the Mediterranean which stimu-   347 lated the transcontinental trading caravans. Earlier, Mediterranean maritime commerce had been interrupted by the Islamic invasions and turned exclusively toward the East. However, beginning in the eleventh century with the Crusades, the old Western routes were reopened and even expanded.

In the larger continental cities there is evidence of another pattern. In addition to the merchants, there dwelt within the city a number of old aristocratic families who owned land in the city and its immediate environs. Often they dominated the urban government. They formed feudal-commercial associations using their feudal privileges to acquire exclusive rights in long-distance trade with the East. In some English towns too, there are traces of burghers or merchants of an aristocratic origin.

In most towns it was the craftsmen who, beside the burghers, constituted the major urban elements. The earliest period reveals only minimal differentiation between master and journeyman. No great disparity of income existed between them, and they toiled side by side in the workshop and shared the same table. A thrifty and hardworking journeyman could himself join the guild, open a workshop of his own, and enter retail trade. All the handicraftsmen of the early towns were at once small producers and traders. They owned their own tools and traded their products freely. Even outside handicrafts, the earliest traders, with a few exceptions, were little more than peddlers traveling between the town market and neighboring manors. They were hardly better off than the craftsmen, if at all, for their trade was strictly local, exchanging craftsmen's wares for peasants' produce.

Later, however, there did in fact emerge a better-off class of burghers engaged in a larger-scale and more lucrative trade. Mainly involved in foreign commerce, they acquired monopoly powers with the aid of their royal allies. They also thrived on local gluts and famines, exploiting the peasant and urban producer alike. Once capital began to accumulate, a further source of enrichment was usury. Peasants, craftsmen, knights, barons, and kings all came to rely on the rich merchants and money dealers. In time the wealthy merchants not only created trade monopolies in their own towns but effective interurban monopolies as well. They successfully maintained a margin between two sets of prices: those at which they bought local produce from the peasant or the craftsman, and those at which they sold it to the urban consumer. With the growth of their wealth and the expanding power of their

trading organizations, the rich burghers became an urban patriciate dominating town government and using their political power to further subordinate the craftsmen.

The craft guilds, in turn, imposed stricter controls over admission in the interest of limiting their number. In due course it became practically impossible for anyone who was outside the circle of privileged craft families and too poor to buy himself a place in the guild to set up a shop of his own. As a consequence a growing class of journeymen appeared. Though they were nominally members of the guild, they had no chance of advancement within it. They were among the forerunners of the industrial proletariat of later centuries.

THE BEGINNINGS OF CAPITALIST MANUFACTURING

The merchant soon became a capitalistic entrepreneur: He employed wage labor in the production of commodities for the market with the aim of realizing a profit. The town, or more correctly the burghers, thus further extended its sway over the countryside.

With the expansion of domestic and foreign markets, the guild system proved inadequate or too costly to meet the merchants' demand for commodities. The depressed state of the journeymen enabled the merchant to transform them into wage earners working directly for him. Weak and small guilds were also reorganized as small shops manufacturing goods for the entrepreneur. But even such supplementary forms of production failed to keep apace of the growing market. The capitalist, requiring additional sources of labor, quite naturally turned to the countryside. There the poorer peasants, caught in the economic squeeze between rising prices for everything they needed and declining incomes, were eager to obtain additional employment. It is true that they were relatively unskilled in the industrial arts, but that proved to be no obstacle since they were largely employed in new industries in which simplified techniques of production were introduced. The new system of production which arose in the countryside was known as the *putting-out* or *domestic* system because the work was done in the peasants' own cottages. Cottage industry appeared as early as the thirteenth century in English wool manufacturing but developed most rapidly between the fifteenth century and the middle of the eighteenth.

The worker in the putting-out system generally worked with his own tools, perhaps a loom or two. However, as production became more 349 complex and tools more expensive, they were furnished by the merchant. Whether or not he owned the instruments of production, the merchant-entrepreneur always supplied the raw material and completely owned the finished product. Thus a new system of production developed which differed from both the feudal and the guild systems. The worker in cottage industry was not bound to a master by guildlike regulations. His relationship to the entrepreneur was strictly formal: He produced commodities and the employer paid him wages. The only obligations the two parties had toward each other were contractual.

From the entrepreneur's standpoint, even the new system had serious shortcomings. It was difficult to supervise the labor of scattered workers; and there was a considerable waste of time in carrying materials back and forth through various stages of the productive process. The fact that the workers were scattered also meant that a cooperative division of labor could not be introduced. Under those circumstances it was difficult for the entrepreneur to cut costs and to increase production. Such considerations led to the rise of the *factory*—the bringing together of many workers in a central place under one roof, most often in barracks. The means of production were now completely in the hands of the entrepreneur who invested in tools, plant, raw materials, and labor. The only thing which the manufacturing shop needed in order to become a modern factory was large-scale, power-driven machinery.

Industry and Urbanization

Both the cottage industry and the manufactories were rural phenomena in spite of their having been under the control of the town burghers. With the Industrial Revolution, however, there was an unprecedented concentration of industry in the towns themselves or in their immediate suburbs. The key invention that precipitated the concentration of industry was James Watt's steam engine. It was patented in 1769 and applied in cotton manufacture some 15 years later. No less important was the invention of coal smelting in the iron industry and the application in 1788 of the steam engine to blast furnaces.

The introduction of Watt's engine as an industrial prime mover brought about a heavy concentration of industry and workers and removed the latter from their rural base. Steam power required coal, thus fostering industry near the mines or in places connected by canal and later by railroad. As Lewis Mumford has explained,

Steam worked most efficiently in big concentrated units, with the parts of the plant no more than a quarter of a mile from the power center: every spinning machine or loom had to tap power from the belts and shafts worked by the central steam engine. The more units in a given area, the more efficient was the source of power: hence the tendency toward giantism. Big factories, such as those developed in Manchester, New Hampshire, from the eighteen-twenties onward—repeated in New Bedford and Fall River—could utilize the latest instruments of power production, whereas the smaller factories were at a technical disadvantage. A single factory might employ two hundred and fifty hands. A dozen such factories, with all the accessory instruments and services, were already the nucleus of a considerable town.[7]

Urban concentration was further promoted after 1830 by the railroad, a new transportation system. As a cheaper means of transporting coal, it enabled industry to produce throughout the year without stoppages due to seasonal failures of water power. The characteristic pattern was, first, a massing of population in the coal areas where the new heavy industries located themselves; and, second, along the railroad lines, especially at great trunk lines, junction towns, and export terminals. Correspondingly there was a thinning out of the rural population. Political and commercial capitals also shared in industrial growth. They offered several advantages, notably proximity to centers of political power and to financial organizations, which controlled the flow of investments. A further advantage was the huge reservoir of cheap labor —the urban poor—that lay ready at hand. Thus the key elements of the new industrial town were the *factory,* the *railroad,* and the *slum.*

England, as we know, was the first country in which those developments occurred. In the 1830s Manchester was the most advanced industrial city in the world, producing textiles, cotton, iron, and copper for a global market. With its large port, Manchester was favorably placed to receive raw materials from abroad and to send its own products to all corners of the map. Three canals and a railroad made it possible to transport the city's manufactured goods throughout England. The largest coal mines in the country lay nearby, providing low-cost power for the machine-run industry. The city was therefore advantageously situated in all respects. By 1835 it had attained a population of some 300,000 and continued to grow at a prodigious rate.

In that year Alexis de Tocqueville visited Manchester. It is interesting to read the impressions which the most advanced industrial complex in the world made on the distinguished traveler from a comparatively unindustrialized society. Tocqueville saw large numbers of tenants typically crammed together in one house, with many occupying the damp, foul cellars. Dark pools of stagnant water covered the unpaved and poorly paved streets. Thousands of men, women and children were assembled in single factories. "Raise your head," wrote Tocqueville, 351 and all around you will see the immense palaces of industry. You will hear the roar of furnaces and the hissing of steam-engines. These vast structures prevent both light and air from penetrating the dwellings they dominate. . . . Here the wealth of a few, there the misery and poverty of the majority; here the productive forces organized for the profit of a single man. . . ; there the individual worker, more feeble, debilitated, and destitute than in a wilderness.[8]

Covering the expanse of the city was a thick, black cloud of smoke, which the sun failed to penetrate. "It is in this foul drain," Tocqueville continued,

that the largest stream of human industry originates and flows out to fertilize the whole world. From this filthy sewer comes gold. It is here that humanity attains its most advanced development and its most brutish; here civilization works its miracles and civilized man is turned almost into a savage.

Having been told by a Mr. Connel, one of the largest manufacturers of Manchester, that each of his 1,500 workers labored 69 hours a week, Tocqueville asked, "What kind of being will an individual necessarily become when he performs the same operation for twelve hours a day, every day of his life except Sunday?"[9]

MODERN CITIES

In his classic pioneering study of nineteenth-century urbanization rates, Adna Ferrin Weber demonstrated that the urban concentration of population was a universal tendency.[10] In the United States, for instance, the urban population increased eighty-seven-fold from 1800 to 1890, while the population as whole increased only twelvefold. Before 1820 the phenomenon of urban concentration was not yet evident in the United States. Early in the next decade, however, there opened the era of canals followed closely by the era of railroads. The urban population expanded accordingly. In the decade 1880–1890, the United States again experienced phenomenal industrial growth and the migration was cityward instead of westward. A similar correlation between urban growth and commerce and industry was found for Britain, continental Europe, Canada, Latin America, Asiatic and African countries, and others.

In 1800 no city in the world numbered a million people. London, the largest, had 959,310, while Paris had little more than a half million. By 1850 London exceeded two million, and Paris passed the one-million mark. Other cities also grew rapidly but fell far below those two leaders. But by 1900 there were no fewer than 11 metropolises with more than a million inhabitants, including Berlin, Chicago, New York, Philadelphia, Moscow, St. Petersburg, Vienna, Tokyo, and Calcutta. During the following 30 years, 27 additional cities joined the list, including metropolises on every continent. By the mid-twentieth century a multitude of new metropolitan areas emerged.[11]

Equally impressive was the rise of cities with a population exceeding a hundred thousand. The increasing magnitude of cities and their multiplication altered the balance between the urban and rural populations. Throughout earlier eras cities had been small islands in a vast agricultural sea, but now it was the agricultural areas that often appeared to be isolated green islands surrounded by asphalt, concrete, brick, and stone.

Certainly one important basis for the growth of cities and for the metropolitan agglomeration was the enormous increase of population during the nineteenth century. The peoples of Europe and America multiplied from about 200 million at the time of the Napoleonic wars to about 600 million at the outbreak of the First World War. Europeans and Americans, who accounted for about one-sixth of the earth's population in the late eighteenth century, rose to about a third of it in a little more than a century. Throughout the Middle Ages and indeed right up to the end of the eighteenth century, European cities depended almost entirely for their growth upon the influx of country people. Mortality was so high that the death rate annually equaled or exceeded the number of births. London, not atypical in that respect, had a birth rate which never exceeded its death rate for any extended period until the beginning of the nineteenth century. Thereafter London's rapid growth rate was in fact partly due to its ability to secure a natural increase of the population. The excess of births over deaths resulted from the application of the discoveries of medical science.

Nevertheless, it would be a mistake to regard population increase in and of itself as the chief reason for city growth. That was one of the important conclusions of Adna Ferrin Weber's statistical analyses. He argued that the essential reason for urban growth "must rather be sought in economic conditions." To illustrate that proposition he compared the demographic statistics of French cities for the decade 1881–1891 and

found that Lyons, Marseilles, and Bordeaux had fewer births than deaths; and yet surpassed the other large cities of France (Roubaix and Lille alone excepted) in their rate of growth.

The "mushroom" growth of American cities, which has been the subject of considerable comment, should occasion no surprise, since it is mainly due to the settlement of uncultivated territory. It is only when the growth of a city has proceeded more rapidly than the development of its contributory territory, that one needs to study other underlying causes. For then one is face to face with the problem of the concentration of population,—an increasing *proportion* of the population collected in cities.[12]

Weber went on to show that while the unprecedented population increase of the nineteenth century was one cause of the rapid urban growth, it was not the chief cause. In Portugal, for instance, the urban proportion of the population actually decreased in the last decades of the nineteenth century. The cities of France, on the other hand,

have been enjoying a rapid growth all the time that the population of the country as a whole has been virtually at a standstill, and those of Ireland have likewise grown while the population in general has declined.

It is now clear that the growth of cities must be studied as a part of the question of distribution of population, which is always dependent upon the economic organization of society.[13]

Some of the earliest and most systematic sociological studies of the modern city were made by members of the "Chicago School." That school is associated with the names of such outstanding urban sociologists as Robert E. Park, Ernest W. Burgess, R. D. McKenzie, and their distinguished students at the University of Chicago. Writing in the early decades of the twentieth century, Park and his associates focused attention on American cities and on Chicago in particular.

Perhaps the most notable contribution of the Chicago School was its application of the concept of "ecology" to the study of cities. Ecology, a concept borrowed from the biological sciences, concerns itself with the relations of organisms to their environments. Human communities, Park and his associates contended, may also be comprehended ecologically. A "city" is at once a physical and social organization by means of which human beings seek to control their environmental circumstances.

From a socioecological standpoint, urban populations tend to expand by extending themselves over wider areas. That process was described by the Chicago School in a chart consisting of several concentric circles.[14] The chart was conceived as an ideal-type construction which effectively conveyed the tendency of any city to expand from its central business district. The central district or "Loop" was encircled by zone II, an area of transition invaded by business and light manufacture. The third area was the zone of working people's homes. They had escaped from the most deteriorated areas but lived within easy access of their jobs. Next was the residential zone of higher-class and restricted single-family dwellings. Finally, there was the commuters' zone or suburban area a half-hour or hour's ride from the central business district. The concentric circles defining the several zones illustrated the tendency of each inner zone to expand its area by invading the next outer zone. Thus Park and his associates employed the ecological terms of *invasion* and *succession* to analyze the process of urban expansion.

The concepts of invasion and succession were also employed in the analysis of urban neighborhoods. The urban neighborhood is the most elementary form of residential organization in city life. Neighborhoods are informal associations that exhibit the distinctive social-class and ethnic character of their inhabitants. Each neighborhood is a locality with traditions and institutions of its own. Neighborhoods change as new ethnic groups move into them thus eventually displacing their predecessors. The large and famous neighborhood of Harlem in New York City, for example, began as a Dutch community, as the name suggests. Eventually the Dutch gave way to the Irish who in turn gave way to the Jews. By the early decades of the twentieth century, Harlem became predominantly black. In the history of Harlem as in the history of other urban neighborhoods, the invading group prompted the exodus of its predecessor. Much of American urban history has been characterized by such patterns of invasion and succession; and elements of the Chicago School's ecological model continue to apply to present-day urban processes.

FROM CITY TO SUBURB

In the early 1950s cost considerations prompted many industrial and other business firms to move out of the central city, where land was expensive and taxes were high. The superhighway system and other transportation facilities enabled companies to relocate in the suburbs. At the same time middle- and upper-class families were also leaving the cities and moving into a newly created suburbia. Following those developments, cities faced a progressive worsening of their traditional problems—poverty, crime, unemployment, and social tensions. They did so, moreover, with diminishing revenues and a decaying physical plant.

The 1920 census revealed that for the first time in American history, urban dwellers had become a majority of the population. Fifty years

355

later the census disclosed that more Americans resided in suburbs than in cities.[15] Of a population of 200 million, 76 million lived in the environs of cities but not inside them while only 64 million lived in the cities themselves. The massive movement to suburbia consisted of middle- and lower-middle-class families and to a lesser extent of families from the better-off strata of the industrial working class. They were impelled by the desire to get out of the deteriorating, problem-burdened cities. The movement was greatly facilitated by federal measures. Following the Second World War, the Federal Housing Administration (FHA) and the Veteran's Administration provided mortgage insurance and generous loans to families aspiring to new homes of their own. These programs fostered thousands of new real-estate developments. The federal guarantees of home loans, the postwar prosperity, and the easing of credit led to a construction boom that lasted nearly a decade. In 1950 alone almost a million single-family houses were built, the vast majority in the outskirts of cities and in suburban districts. The decade that followed witnessed a mushrooming of suburbs across the country.

What made that expansion possible was the postwar superhighway system designed for high-speed traffic. Automobile registrations rose from 25 million to over 40 million between 1945 and 1950 and reached nearly 60 million by the mid-1950s. As cities tried to accommodate to the proliferation of cars and trucks and the mounting costs of road construction, they turned to the federal government for assistance. In 1947 Congress authorized a 37,000-mile national-highway network and 3,000 miles of roads in or near 182 large cities. That network was further expanded in the 1956 Interstate Highway Act, which initiated an additional 42,000 miles of highway construction with links to every major American city. Real-estate developers now began to buy up huge tracts of land along the proposed interstate routes. Homes were built by the millions for the prospective suburbanites, and by 1970 the suburbs contained more housing units than the cities. Numerous businesses accompanied the new residential communities and established the sprawling shopping centers and free parking areas with which we are all familiar. Families could now meet all their material needs locally without having to return to the city's business districts.

Sociological studies have provided some insight into the motivations behind the residential suburban movement. Herbert Gans, for instance, in a study of Levittown, New Jersey, learned that most people have been attracted to suburbia by the opportunity to acquire more space. They desired single-family homes with recreation rooms, dens, large closets, backyards, and large green areas in the community where children could play safely. However, they also moved because they were repelled by such urban conditions as poor schools, deteriorating neighborhoods, and racial problems.

But disaffection with the cities was only one motivating factor. In the early 1950s, it had already become evident that families moved from one suburb to another with relative frequency. The typical Levittown family changed residences every two and a half years. In the late 1950s, 45 percent of Levittowners had previously resided in another suburb.[16] Suburbanites apparently did not think of their community as a permanent settlement for themselves and for future generations. They expected their children to live elsewhere, and they themselves hoped to move to a more "exclusive" suburb. However, only a fraction of the thousands of suburban communities now ringing American cities fall into the middle-class category. Many different types of suburbs exist: rich and poor, white and black, residential and industrial.

Nevertheless, there has been a pronounced racial facet to the suburban movement. Between 1950 and 1966, 70 percent of the increase in the nation's white population occurred in the suburbs. In the same period 86 percent of the growth in the black population took place in central cities. As we have noted in previous discussions of the black underclass, black Americans have in the last several decades become an urban people. By 1970 about 75 percent of blacks lived in cities as compared with about 64 percent of whites. In the North the urban proportion of blacks is markedly higher, some 93 percent. By the mid-1970s the black proportion of the population in six major cities—Washington, D.C., Newark, New Orleans, Baltimore, Atlanta, and Gary—was over 50 percent. Blacks also constituted very substantial minorities in Detroit, Cleveland, St. Louis, Philadelphia, Chicago, and Oakland. In addition there have been huge increases in the migration of other disadvantaged minorities to the cities. By 1970 there were more than 800,000 Puerto Ricans in New York City and more than 100,000 Chicanos in Los Angeles. The result has been an increasing ghettoization of blacks and other disadvantaged groups, and their separation from whites.

White Americans have left behind the least desirable jobs, the worst-quality housing, and the poorest health and educational facilities. Historically, white ethnic groups in the United States have been able to escape the ghetto slums, often within a generation after having arrived. Not so the blacks. Their ghettos have expanded, not diminished. Similar tendencies have been evident for Mexicans and Puerto Ricans, although 357 segregation is less marked for those groups than for black Americans.

In part the residential separation of black Americans may stem from the desire to live in predominantly black neighborhoods. Some members of all ethnic groups have preferred to live in their own cultural communities. The low income and educational levels of most blacks may also be a factor motivating whites to abandon the inner city. But such factors are insufficient to explain the almost total segregation of whites and blacks which is now characteristic of our metropolitan areas. To cite the report of the Kerner Commission: "This nearly universal pattern cannot be explained in terms of economic discrimination against all low-income groups. Analysis of 15 representative cities indicates that white upper- and middle-income households are far more segregated from Negro upper- and middle-income households than from white lower-income households."[17] It is clear that segregation is largely the result of white attitudes concerning "race" and color.

THE URBAN CRISIS

It is now widely recognized in the United States that the nation faces an urban crisis of unprecedented magnitude. The social and economic condition of the black and white underclass is at the core of that crisis. Because the scope of urban problems is national, a national urban policy is required to overcome them—or even to take the first few significant steps in that direction. Ostensibly the federal government has attempted to institute such a policy, but federal programs have not improved the lot of the underclass, and they may even have done more harm than good. That is borne out by the history of urban-renewal programs.

In 1949 President Truman signed the Wagner-Taft-Ellender Housing Act, setting a goal of 1.25 million new housing units a year for the following ten years. The main beneficiaries of that act were to be middle-class families who sought better housing in private markets with the aid of federal mortgage insurance and loans. But the act also provided for the construction within four years of 810,000 *public*-housing units for low-income groups. That project was to be supported by generous federal loans and subsidies. The omnibus bill combined several types of measures: slum clearance, public housing, and expanded mortgage insurance through the FHA. But the three elements were never coordinated.

358   Title I of the act called for urban redevelopment. The law stipulated

that redevelopment be chiefly residential, and it earmarked federal funds for the clearance of slums. However, the law contained a basic ambiguity. It allowed areas designated for redevelopment to be treated in either one of two ways: Before they were cleared, they had to be at least 50 percent residential; or, after they were cleared, all new construction had to consist of at least 50 percent residential units. The law was thus construed as providing for alternatives rather than two necessary steps. The result was that developers removed blighted residential slums and replaced them not with new housing but with factories, shopping complexes, luxury apartments, and parking lots—all of which tended to raise property values, assist private investment, and raise taxes. Hence urban renewal brought commercial revitalization to former slum areas in Pittsburgh, Boston, Cincinnati, Chicago, and other cities, but it failed to rehouse the displaced low-income slum dwellers.

The declared aim of the 1949 Housing Act was to provide "a decent home and a suitable living environment for every American family." In effect, however, the law benefited only high- and middle-income families. Very small sums were appropriated under the act for public housing; and it took twenty years, not four, to construct the 810,000 public-housing units which the authors of the bill had envisioned.

In 1954 and 1961 additional amendments to the 1949 act removed redevelopment even further from the solution of housing problems. Thus effectively ignoring the needs of the disadvantaged groups, federal policy renovated certain inner-city districts in the interest of middle- and high-income residents and private investors. Urban renewal, proceeding in that fashion, was particularly onerous for blacks whose districts were being renewed but who lacked the financial means to acquire new and decent housing. Rents for the new dwellings constructed on cleared land were prohibitively high. It has been estimated that 200,000 housing units a year were destroyed between 1950 and 1956, and 475,000 a year between 1957 and 1959.

Federal home-financing policies have also worked against the interests of the black and white poor. Through the FHA generous mortgage insurance has been made available to middle-class families, who have thereby found it easier to obtain bank mortgage loans and to flee the central city. Income-tax deductions for mortgage insurance and real-estate property taxes have likewise benefited the middle-class groups, subsidizing their suburban movement. The poor, on the other hand, displaced by urban renewal and enjoying none of those forms of assistance, were left behind in a worsened condition.

Highway construction has also been responsible for displacing poor 359 families from their homes. Expressways and interchanges were designed to cut through low-rent districts where, planners insisted, dilapidated structures had to be demolished anyway. Residents were uprooted and left to fend for themselves. Urban renewal and redevelopment has thus proceeded at the expense of the poor.

The flight to the suburbs of middle-class whites has left the central city with a steadily diminishing tax base. At the same time, the spiraling costs of municipal services have forced cities to raise their tax rates. Hence more businesses and homeowners have been prompted to leave the central city.

Thus "crisis" is not too strong a word with which to describe the plight of American cities. Given the national scope of the crisis, a national policy is required to deal with it. If such a policy is to succeed, the nation will have to reconsider two propositions that are legacies of American history. The first is the notion that the ownership and exploitation of land for private profit is the only way to organize urban life. An alternative view is that land is a national resource. That view does not necessarily imply that all land must be removed from private ownership and markets. It simply proposes that a proportion of the total land capable of development be set aside for planned urban growth; and that such land be acquired in advance of need. A type of land bank would thus be established. Using land as an instrument of national policy is certainly not new in American history: The federal government is already one of the largest landholders of the nation. Land has been set aside for parks, conservation areas, defense facilities, and other public purposes. In earlier eras land was regulated, sold, and even given away by the federal government to homesteaders and railroads. The authority to acquire land for public uses on the principle of eminent domain is well established.

The second proposition that will have to be reconsidered is the inviolability of local autonomy. Doubtless the principle of local autonomy has been a valuable traditional safeguard of freedom. Yet in the decades since the Second World War, the growth of freely incorporated municipalities has created an extreme situation. There now exist in every metropolitan area so many independent, competing, narrowly self-interested, and even hostile governmental jurisdictions that comprehensive planning and coordination has proved to be impossible. Here, as in the matter of land, a national policy is needed if the jurisdictional conflicts splintering metropolitan areas are to be overcome. Only a national policy, it would seem, can maximize opportunities for the

360

creation of economically sound and balanced communities consisting of residential, commercial, and industrial complexes.

The inner city might then enjoy genuine renewal and development. For it would ensure its inhabitants of both decent housing and a dignified means of paying for it through gainful employment.

14

Social Change, Modernization, and Development

The study of social change is concerned with the processes through which societies and cultures are transformed. Sociologists quite naturally view the relations of humans to their environment, as the universal source of social change. That fundamental premise requires clarification.

The human being, as we emphasized in the early chapters of this book, lives and acts in two environments: the human and the physical. The latter, consisting of climatic, geographic, and geological conditions, cannot be regarded as the major determinant of change, for those conditions evolve slowly as compared with social ones. So slowly in fact, that for all practical purposes the physical environment may be viewed as a "constant." And as Wilbert E. Moore has reminded us, "A constant cannot explain a variable in any system of logic."[1] Furthermore, the "physical environment" of human society is never purely physical. Topography and other physical conditions are profoundly altered by human activity through technology, social organization, and cultural values.

The same logic applies to human heredity. Biological changes are slow and imperceptible as compared with social changes. Given the superorganic nature of culture, "natural selection" in the human species is always partially "social selection." It follows that the source of social change is to be sought in human social organization and in its relation to the physical environment.

In the history of sociological analysis, several theories of social organization and change have been put forth. One of the most influential is functionalism. Functionalism, it will be recalled, treats groups and societies as self-sufficient systems with "system needs," "functional prerequisites," and "adaptive" and other capacities. As our criticisms in chapter 5 demonstrated, the problems and ambiguities associated with those concepts are numerous. Those criticisms will not be repeated here. For present purposes it will suffice to call attention to functionalism's major shortcoming, its *nonhistorical* approach to the study of society.

An understanding of the alterations a society undergoes requires a study of its history. That is a truth which the adherents of functionalism have yet to recognize. Perhaps their neglect of history has resulted from the modeling of their methods on those of the physical and biological sciences. But it cannot be stressed too strongly that the sociological imitation of the physical sciences carries with it a fatal flaw. In physiology, for instance, "history" is relatively unimportant because physiologists can rely on repeatable cycles. In human society, however, "cycles" are more metaphorical than real. To rely on cycles is to deprive oneself of the knowledge of emerging conditions and causes. The social scientist who studies a social organization without due regard for its specific history will never truly understand any given state of that organization or the forces operating to change it.

The neglect of history has been characteristic of functionalist and other theories which have conceived of societies as self-sufficient, self-equilibrating systems. However, it would have been unthinkable for any of the classical thinkers on whom we have relied in this book—notably Marx, Weber, and Durkheim—to ignore history. Durkheim, who was the most explicitly functionalist of the three, nonetheless emphasized the methodological importance of history. In his *Rules of the Sociological Method,* he called for an analysis of the functions of social facts, that is, the social ends they serve. But he also maintained that an adequate explanation of social facts demands a historical account of how the facts came to be what they are.[2]

Functionalism, like all other theories sharing its mechanistic, self-equilibrating view of social systems, tends to foreclose questions of change. Such theories exaggerate the unity, stability, and harmony of social organizations. Or as W. E. Moore has observed, "If discordant internal elements are brought into the analysis, the theoretical model will predict one direction of change, and one only—change that restores the system to a steady state."[3]

FUNCTIONALIST THEORIES OF SOCIAL EVOLUTION

Functionalist premises have also been applied to evolutionary theories. Talcott Parsons, a leading functionalist, introduced the notion of "evolutionary universals in society." An "evolutionary universal," he wrote, "is a complex of structures and associated processes the development of which so increases the long-run *adaptive capacity* of living systems in a given class that only systems that develop the complex can attain certain *higher* levels of general adaptive capacity."[4] Here we meet again the familiar but still ambiguous term *adaptive capacity*. Now, however, it is combined with the evaluative assumption of a "higher" developmental stage. The conceptual difficulties remain. "Adaptive capacity" to what conditions? How do we determine whether or not certain social structures heighten the adaptability of a social system?

In Darwin's theory of "natural selection," the failure of a species to adapt to its natural habitat ultimately results in its extinction. However, in Parsons's evolutionary scheme, the failure to adapt does not necessarily result in the extinction of a society. Parsons acknowledged that relatively disadvantaged societies do survive, but he did not proceed to enlighten us as to how, from an evolutionary standpoint, one distinguishes the advantaged from the disadvantaged systems.

For Parsons there were four basic prerequisites of a human society: religion, language, social organization, and technology. Those, he claimed, "may be regarded as an integrated set of evolutionary universals at even the earliest human level. No known society has existed without *all* four in relatively definite relations to each other. In fact their presence constitutes the very minimum that may be said to mark a society as truly human."[5] But that statement, we should note, is a definition rather than a scientific proposition. For Parsons nowhere pointed to "fossil" societies which have perished for lack of one or more of the alleged prerequisites. In a scientific evolutionary theory postulating prerequisites, it is necessary to show that certain empirical conditions contribute to a society's adaptive capacity, and that the absence of those conditions leads to its dissolution or disadvantage. But while Parsons admitted that social extinction is not the criterion of maladaptation, he did not provide other criteria. Hence his "evolutionary universals" are nothing more than a list of presumed prerequisites for an undemonstrated state called "adaptive capacity." The fundamental ambiguities of functionalism remain.

No functionalist has ever been able to show that a given society perished because it lacked one or more functional–evolutionary prerequisites. On the contrary, as George C. Homans has remarked, the historical records we have of now extinct societies reveal "that they possessed institutions fulfilling *all* the functions on all the usual lists of prerequisites. If they died it was not for lack of institutions but for lack of resistance to measles, or firewater, or gunfire, or something of the sort." Historically, Homans continued, the competition between societies has been "very different from the Darwinian struggle for existence. The weaker did not get eliminated but, in all but a few cases, absorbed. The victors were usually content to let the institutions of the weaker people alone so long as they themselves controlled the government and collected the taxes. It was impossible to demonstrate any general Darwinian mechanism that would eliminate [so-called] dysfunctional institutions." The Darwinian analogy "failed all along the line: the institutions of society did not mesh as closely as did the organs of a body, nor did societies compete in the way animals were supposed to do."[6]

CLASSICAL EVOLUTIONARY THEORIES

The nineteenth century was the heyday of social evolutionism. Edward Burnett Tylor, Lewis Henry Morgan, Auguste Comte, Herbert Spencer, and numerous others all subscribed to a doctrine according to which the human race progressed through successive stages. Doubtless we owe an immense intellectual debt to those extraordinary thinkers, for we have acquired from them a large body of substantive knowledge about a wide range of human societies and cultures. The question remains, however, whether their conception of social evolution is scientifically sound.

Most nineteenth-century evolutionary schemes had in common certain stated and unstated premises which may be summed up in the following statement: *Change is natural, directional, immanent, continuous, and derived from uniform causes.*[7] Those premises were shared by such otherwise diverse thinkers as Hegel, Saint-Simon (and his pupil Comte), Tocqueville, Spencer, Morgan, and Durkheim. For Hegel the developmental idea expressed itself in his view of the spirit of freedom, which grew from its modest beginnings in the ancient Orient until it attained its highest form in the Prussia of his day. In Saint-Simon and Comte the idea is evident in their law of three stages: Knowledge evolves from the religious through the metaphysical to the positive (or scientific) stage. In Tocqueville we find societies increasingly embody-

365

ing the spirit of equality, thus proceeding from aristocracy to democracy. For Spencer the direction of evolution was from the relatively homogeneous "military" society to the complex "industrial" one. Morgan perceived three main stages of societal development: "savagery," "barbarism," and "civilization." And Durkheim, finally, placed the subject of social solidarity into an evolutionary framework, arguing that society normally progresses from a *mechanical* to an *organic* stage of solidarity. Many more thinkers could be mentioned to illustrate the prevalence in the nineteenth century of evolutionary schemes. As Robert A. Nisbet has stressed in his illuminating monograph *Social Change and History,* all such schemes are drawn "from the metaphor of growth, from the analogy of change in society to change in the growth-processes of the individual organism."[8]

How did the idea of growth, development, and evolution become so pervasive in the nineteenth century? Historians have sought the answer to that question in the romantic-conservative reaction to the French Revolution. Conservatives throughout Europe deplored the consequences of that great social upheaval. They looked upon it as a disaster resulting from the folly of the revolutionaries who, intoxicated with Enlightenment ideas, attempted to reorder society according to mechanistically rational principles. In opposition to the eighteenth-century exaltation of the individual, conservatives elevated the *group,* the *community,* and the *nation* to paramountcy. Eighteenth-century thought had been dominated by mechanistic metaphors: The Newtonian universe was "spring and wire," and even the human being was likened to a machine. Nineteenth-century thought, in contrast, adopted organic metaphors. Why and how that came about is exemplified in the political philosophy of Edmund Burke who, along with many other thinkers, insisted that society was a living organism.

In advancing his organic conception of society, Burke was explicitly rejecting the abstract rationalist views of the philosophes. Those pre-Revolutionary French thinkers had held that there existed natural laws and natural rights which could be discovered by the mind. Laws made by humanity, they contended, should conform with the ideal natural laws as nearly as possible. In their application of that doctrine, Burke argued, the revolutionists had treated society like a machine, thinking they could simply pluck out the obsolete parts and replace them with new ones. They thus discarded old and established institutions which had developed through time and sought to replace them in accordance with an abstract formula. The individual was proclaimed more important than the nation, the part more important than the whole. The state, far from being conceived of as organically related to the rest of the social order, was reduced to a mere contractual relationship. That idea in particular had obvious revolutionary implications: If the state was a mere contract, then it could be dissolved as soon as the contracting parties decided that it no longer satisfied their interests—which is precisely what occurred in the French Revolution.

In his *Reflections on the French Revolution,* Burke presented a point-for-point rebuttal of the Enlightenment's mechanistic rationalism. The individual, he maintained, has no abstract rights. On the contrary, he has only those rights and privileges which prevail in a given community and which he acquires by virtue of having been born there. Rights and privileges develop slowly and organically; they are historical in character, not abstract. A community exists not merely in the present; it is an endless chain of generations, each generation inheriting from its predecessor and each constituting but one link. The revolutionaries therefore had no right to destroy customs and institutions which belonged not solely to them but to past and future generations. As for the state, it is no mere contract. It is a higher organic unity, an integral part of the national community. The state and the nation are not deliberate, calculated inventions but are the organic products of a prolonged process of growth.

Thus Burke formulated a developmental, organic view of society. He was only one among numerous nineteenth-century thinkers whose categories of analysis had become thoroughly dominated by evolutionary metaphors or by what J. B. Bury called "the idea of progress." That idea Bury identified as a distinctively modern product which emerged in its earliest forms in the seventeenth century and reached its fullest expression in the nineteenth.[9] The idea of progress stands in sharp contrast to the idea of cycles in Graeco-Roman antiquity.

With knowledge of the historical grounds of the evolutionary idea, we are better prepared to address the key question: Are nineteenth-century evolutionary theories to be regarded as objective and scientific accounts of social change? If, following Robert A. Nisbet, we define social change "as a succession of differences in time in a persisting [social] entity," then we have to answer that question in the negative. For the theories with which we are concerned have failed to demonstrate a series of developmental steps in the transformation of a single social entity. Typically, evolutionists selected their evidence for stages from diverse societies and from various historical periods. The great anthropologist Lewis Henry Morgan, for example, illustrated "savagery" with one society, "barbarism" with a second, and "civilization" 367 with a third. Selecting their data from divergent cultural areas and historical periods, the evolutionists have then arranged them in a series resembling the actual historical series in the West. Thus this school of thought has given us, Nisbet has convincingly noted, not a theory of the actual course of development of a single social entity, but rather a

"series" as in a movie film. It is the eye—or rather, in this instance, the disposition to believe—that creates the illusion of actual development, growth, or change. It is all much like a museum exhibit. (It might be observed in passing that the principles of museum arrangement of cultural artifacts have not been without considerable influence on the principles of cultural evolution.) The last one I saw was an exhibit of "the development of warfare." At the beginning were shown examples of primitive warmaking—spears, bows and arrows, and the like. At the far end of the exhibit were examples (constructed miniatures) of the latest and most awful forms of warfare. In between, constructed in fullest accord with the principles of logical continuity, was the whole spectrum or range of weapons that have been found or written about anywhere on the earth's surface at whatever time. All of this, observers were assured, represented the development of warfare. But the development of warfare where? Not, certainly, in the United States, or in Tasmania, or in China, or in Tierra del Fuego, *or in any other concrete, geographically identifiable, historically delimited, area. What "develops" is in fact no substantive, empirical entity but a hypostatized, constructed entity that is called "the art of war."* (Italics added.)[10]

That is a telling criticism. The social evolutionists, far from having proved the validity of their theories, have merely given expression to the dominant intellectual and cultural ideas of their time. The entire evolutionary theory and method rested on the prior acceptance of the idea of progressive development. Thus evolutionary theory has suffered from an inherent circularity which it has never overcome. Robert Nisbet and other scholars have included Marx among the evolutionists. But the view that Marx belongs in that category needs to be reexamined.

IS CLASSICAL MARXISM AN EVOLUTIONARY THEORY?

Not surprisingly, Karl Marx, for all of his originality, was also influenced by the dominant intellectual ideas of his day. Traces of evolutionism are evident throughout his writings and those of his colleague Frederick Engels. In their early co-authored book *The German Ideology,* they described several stages of ownership forms—tribal, ancient, feudal, and capitalist. In his *Critique of Political Economy* and in *Pre-Capitalist Economic Formations* (a part of his *Grundrisse*), Marx de-

368

lineated the Asiatic, ancient, feudal, and capitalist modes of production. In common with the evolutionists of their time, Marx and Engels constructed their developmental stages by selecting forms from various times and places. The impression that Marx and Engels viewed society as developing in stages is further strengthened by their enthusiastic reception of Lewis Henry Morgan's *Ancient Society* and by Engels's heavy reliance on that work in his *Origin of the Family, Private Property, and the State.*

Moreover, Marx and Engels frequently employed progressivist-evolutionary language and apparently saw some parallels between organic and social evolution. In his funeral oration over Marx's grave in 1883, Engels stated that "as Darwin discovered the law of evolution in organic nature, so Marx discovered the law of evolution in human history." In 1888, in his preface to an English edition of *The Communist Manifesto,* Engels prophesied that Marx's ideas are "destined to do for history what Darwin's theory has done for biology." Analogies between historical processes and evolutionary biology are scattered throughout *Capital.* In Marx's own preface to the second edition in 1873, he stated that capitalism is "a passing historical phase," and quoted from a Russian review of the first edition. The review praised Marx's work for demonstrating that the unfolding processes of economic life are analogous with biological evolution and for "disclosing the special laws that regulate the origin, existence, development, and death of a given social organism and its replacement by another and higher one." And Marx approvingly commented that the reviewer had accurately portrayed his "dialectical method." Marx's study of England as the most advanced capitalism of his time also suggests that for him the "less developed" societies were ultimately destined to mirror the conditions of the "more developed."

And yet, as several Marx scholars have convincingly argued, it would be a mistake to take all those organismic metaphors as anything more than rhetoric. Marx himself eventually recognized that his rhetoric had misled interpreters of his work. In a letter to Mikhailovsky, Marx rejected the latter's attempt to transform his sketch of the origins of capitalism in Western Europe into a suprahistorical theory "of the general path every people is fated to tread, whatever the historical circumstances in which it finds itself."[11] In both that letter and in a later one to Vera Zasulitch, Marx expressly excluded Russia from the picture of the genesis of capitalism that he had sketched in *Capital,* and he emphasized that the picture was intended for England and Western

Europe only.[12] A careful examination of Marx's writings tends to confirm that this was in fact his standpoint: The emphasis on the role of socioeconomic processes in social change is a *historically specific* proposition relating to the Western European origin of capitalism and not to societies in general. It has become increasingly clear in recent decades that Marx subscribed neither to evolutionism nor to any other suprahistorical doctrine. His social analysis was firmly rooted in the historical record, in concrete, empirical-historical evidence.

The major scientific aim of Marx's theory was to guide the exploration of the manifold and historically changing connections between the economy and all other facets of human society. His method enjoins the investigator to give due attention to the "mode of production," which includes the following four conditions:

1. the direct producers in their cooperative relations and the technological knowhow with which they carry on production;

2. the instruments and means of production;

3. the property and other power relations governing access to and control of the means of production and its products;

4. the natural base and the way it conditions the productive process.

The "mode of production" is the "foundation" of human society inasmuch as humanity's economic activity is everywhere essential and indispensable. But nowhere did Marx assert that the mode of production is the universally decisive factor in determining the various forms of society. It is strictly a matter for empirical investigation whether economics, politics, religion, or ideology will be decisive for change or nonchange in any particular case.

For Marx, then, there was no universal "prime mover" of history; there were no "iron laws," no universally necessary and inevitable stages. It is such a reading of Marx that has enabled some contemporary social scientists to recover the scientific kernel of the Marxian legacy. Its distinctive marriage of sociological concepts and historical analysis has proved particularly fruitful in the study of change and resistance to change.

There are, of course, many significant questions that might be studied under the rubric of social change. One question, however, appears particularly urgent: What are the historical processes that account for the economic modernization of some societies and for the comparative economic backwardness of others? That is the question to which we now turn.

MODERNIZATION AND DEVELOPMENT:
THE CLASSICAL CASES

Most countries of the Third World have thus far failed to experience the kind of capitalist-industrial development which has been characteristic of Western Europe and the United States. To understand why, we must begin by reviewing briefly the precapitalist conditions that prevailed in Europe.

The precapitalist economy of Europe was overwhelmingly agrarian. Power and authority rested firmly in the hands of a class of overlords who had bound the direct producers to the soil and exacted dues and services from them. That was feudalism, or serfdom, which varied in its form from one region to another. The feudal economy, as we noted earlier in other contexts, was associated with simple instruments of production and a low level of technique. Production was largely individual in character and the division of labor primitive. Each household and village community produced for its immediate needs and was self-sufficient. Economic historians have called such arrangements a *natural economy* to contrast them with the market economy that emerged later. Finally, the precapitalist economy was associated with a politically decentralized system in which each overlord exercised judicial and administrative functions in relation to the dependent population.

In Western Europe the feudal or quasi-feudal society began to show significant modifications in the twelfth century. The major catalyst was the growth of trade and with it traders and trading communities. The presence of the merchant encouraged the bartering of surplus products, or production for exchange. With exchange came money, which slowly undermined the self-sufficiency of the manorial economy. Merchants, artisans, and the growth of towns injected a powerful commercial impulse into the feudal order. Those factors, together with the monetization of the economy, led to an intensified feudal pressure upon the underlying agricultural population and to significant increases in agricultural output. Accompanying those developments was an extraordinary accumulation of wealth in the hands of merchants, commercialized lords, and well-to-do peasants and artisans. Thus the scope and speed of the accumulation of merchant capital played a major part in weakening the feudal structure.

In England, as we have seen in previous discussions, the commercial impulse behind the primary accumulation of capital was the rapid expansion of Flemish wool manufactures and hence the rising price of English wool. The new English nobles, increasingly dependent on   371 money, took advantage of those circumstances. They converted arable land into sheepwalks. The Enclosure movement had begun—the painful process of expropriating the agricultural population from the land. Peasant proprietors were expelled en masse, and a large class of landless proletarians came into being.

With the destruction of the peasant-proprietor class, revolutionary changes occurred in the structure of the English countryside. Landlords now leased portions of their giant estates to capitalist farmers who in turn employed rural proletarians in agricultural production for the market. The revolution in the countryside was both social and technical. Production was carried out on a larger scale with improved methods and a more complex division of labor. What had formerly been produced by peasants for their own use and consumption now became (1) capital in the hands of the lord and farmer, and (2) means of subsistence which the producer (now a wage earner) could only obtain by selling his labor power. With the proletarianization of the peasants, rural domestic manufacturing was also destroyed. Manufacturing was more and more separated from agriculture, and the new capitalists found a domestic market in the wage earners. That is the way the earliest form of agrarian and manufacturing capitalism came into being in England. The main promoters of capitalism were not only merchants but members of the landed upper classes and large and middle-sized yeomen. Commercial interest, private property, and economic freedom were as much the ideology of the enclosing landlords as of the bourgeoisie.

The promoters of capitalism found the controls imposed upon them by the old regime highly restrictive. Thus the underlying cause of the English civil war of the seventeenth century may be traced to the resulting tensions: The pressure of the commercial classes of both the countryside and towns to eliminate those restrictions and the efforts of the crown and sections of the titled nobility to preserve the old order. The war smashed royal absolutism, putting an end once and for all to the crown's interference in the affairs and property rights of the capitalist interests. The result was a "bourgeois revolution" not in the sense that capitalists took power, but rather in the sense that it brought about the victory of parliamentary democracy and capitalism. Although the aristocracy continued to rule politically well into the nineteenth century, it was a transformed aristocracy, one of money and commercial agriculture rather than birth. Given the commercial interests of the landed upper classes, they had no strong reason for opposing the advance of capitalist industry. Add to those conditions the national unity and political independence of the country, and it is not difficult to understand how England became the first capitalist-industrial society in the world.[13]

If we compare England with other Western European societies, we can see how the specific historical circumstances of each country came into play to produce distinctive economic and political patterns. In England, as we have seen, the process that began with the commercialization of the countryside led to the independence of the aristocracy from the crown and to the disappearance of the peasantry. France in the same period offers a point-for-point contrast. First, the nobility, far from becoming independent, was so severely weakened by French absolutism as to be reduced to a powerless appendage of the crown. Second, market stimuli led not to the commercialization of agriculture, but to an intensified exploitation of the peasants. Third, a land-hungry, smallholding peasantry remained characteristic of France, with revolutionary consequences not only in the eighteenth but well into the nineteenth centuries.

It is true that there was in France some fusion between bourgeois and noble. However, the nobility was never basically transformed as it was in England. The French landlord derived his profit not from the market but from squeezing the peasants for rents and dues. Commercial agriculture failed to take hold. Wine in France never came to play the revolutionary role that wool played in England. As Barrington Moore, Jr., has observed, "Viniculture cannot form the basis of a textile industry as can sheep raising. Nor can it provide for feeding the city population as does wheat growing."[14] The French aristocrat kept the peasant on the land and used feudal levers to extract more produce from him. The upshot is that though there were significant commercial influences in France and though they penetrated the countryside, they "did not undermine and destroy the feudal framework. If anything, they infused new life into old arrangements, though in a way that ultimately had disastrous consequences for the nobility."[15] Only following the French Revolution and several subsequent revolutions of the nineteenth century were the conditions created in France for the unimpeded development of a capitalist-industrial economy.

CAPITALISM, COLONIALISM, AND UNDERDEVELOPMENT

How did the great economic advances of England and France, the "first arrivals" to industrialization, affect the economies of other societies?   373

For all of their power, England and France could not prevent several other strong Western countries from embarking on industrialization. By the end of the nineteenth century, the United States, Germany, and to a lesser extent Russia had joined the ranks of the industrializing societies. In Asia there also appeared a conspicuous industrializer, Japan, a case we shall presently examine in detail.

However, contacts between the first arrivals and the various societies of the Third World produced quite different results. The economic growth of the latter was actually impeded. That was a consequence of Western penetration which even the most sophisticated observers of the time failed to anticipate. Marx, for example, not atypical in this respect, believed that the British construction of railways in India would inevitably lead to India's industrialization. The colony was ultimately destined to mirror the mother country. In an article entitled "The Future Results of British Rule in India," published in the *New York Daily Tribune* in 1853, Marx wrote:

I know that the English millocracy intend to endow India with railways with the exclusive view of extracting at diminished expenses the cotton and other raw materials for their manufactures. But when you have once introduced machinery into the locomotion of a country, which possesses iron and coals, you are unable to withhold it from its fabrication. You cannot maintain a network of railways over an immense country without introducing all those industrial processes necessary to meet the immediate and current wants of railway locomotion, and out of which there must grow the application of machinery to those branches of industry not immediately connected with railways. *The railway system will therefore become in India, truly the forerunner of modern industry.*[16]

And in *Capital* Marx expressed the same view in general terms: "The country that is more developed industrially only shows to the less developed the image of its own future."[17] In actuality, however, things developed quite differently from the way Marx predicted. To grasp the source of his error we need to take a look at British colonial policy. The classic case of India serves as a good example.

The British Rule in India

The British did in fact build a large railway network in India. They also constructed harbors and port facilities, established industries for the extraction of national resources, and organized markets for their imported commodities. The result was a substantial transformation of the colonial economy. The native handicraft industry of the great mass of peasants was destroyed and with it the self-sufficiency of the peasant

374

communities. Compelled to purchase the industrial commodities of the mother country, the colony became increasingly dependent upon it.

In some ways the destruction of the traditional methods of production in the colonial areas paralleled the process which had taken place a few generations earlier in the imperialist country itself. But there was one essential difference: Imperial domination not only undermined the colonies' original economies; it allowed only a partial and lopsided industrialization in them. Marx's expectations were never fulfilled either in India or in any other colonial area because it became a cardinal principle of every imperialist's policy to retard and prevent industrial development in the colonies, in order to ensure economic dependence on the mother country. Fully industrializing the colonies would have given rise to both an indigenous capitalist class and a large class of urban industrial workers. Experience had shown that even in their early stages of growth, those classes developed a national consciousness and became a major source of opposition to imperial domination. The perpetuation of imperial rule therefore dictated a policy of permitting colonial industrialization, in kind and degree, only insofar as it appeared compatible with imperial interests.

The imperialist presence, vastly outnumbered by the colonial population, sought allies among them. Typically, the most important of those allies were the landlords. British rule, wrote Jawaharlal Nehru,

consolidated itself by creating new classes and vested interests who were tied up with that rule and whose privileges depended on its continuance. There were the landowners and the princes, and there were a large number of subordinate members of the services in various departments of the government, from the patwari, the village headman, upward. . . . To all these methods must be added the deliberate policy, pursued throughout the period of British rule, of creating divisions among Indians, of encouraging one group at the cost of the other.[18]

By bolstering the privileges of the landlords, the imperialists gained the support of that rich and powerful class. The latter, now vitally interested in colonialism, played a large part in controlling the masses. Extreme poverty became the normal condition of the people. On the one hand, their misery was perpetuated by industrial stagnation, and on the other, by an imperialist-landlord alliance that made anything but the most superficial of agrarian reforms unthinkable.

Romesh Dutt, who had been a high-ranking civil servant in the British administration of India and a lecturer in Indian history at University College, London, summarized the impact of British rule this way:

It is, unfortunately, a fact, that in many ways, the sources of national wealth in India have been narrowed under the British rule. India in the eighteenth century was a great manufacturing as well as a great agricultural country, and the products of the Indian loom supplied the markets of Asia and of Europe. It is, unfortunately, true that the East Indian Company and the British parliament, following the selfish commercial policy of a hundred years ago, discouraged Indian manufactures in the early years of British rule in order to encourage the rising manufactures of England. Their fixed policy, pursued during the last decades of the eighteenth century and the first decades of the nineteenth, was to make India subservient to the industries of Great Britain, and to make the Indian people grow raw produce only, in order to supply material for the looms and manufactures of Great Britain. This policy was pursued with unwavering resolution and with fatal success; orders were sent out, to force Indian artisans to work in the Company's factories; commercial residents were legally vested with extensive powers over villages and communities of Indian weavers; prohibitive tariffs excluded Indian silk and cotton goods from England; English goods were admitted into India free of duty or on payment of a nominal duty. . . . The invention of the power-loom in Europe completed the decline of the Indian industries; and when in recent years the power-loom was set up in India, England once more acted towards India with unfair jealousy. An excise duty has been imposed on the production of cotton fabrics in India which . . . stifles the new steam-mills of India. Agriculture is now virtually the only remaining source of national wealth in India . . . but what the British Government . . . take as Land Tax at the present day sometimes approximates to the whole of the economic rent. . . . This . . . paralyses agriculture, prevents saving, and keeps the tiller of the soil in a state of poverty and indebtedness. . . . In India, the State virtually interferes with the accumulation of wealth from the soil, intercepts the incomes and gains of the tillers . . . leaving the cultivators permanently poor. . . . In India, the State has fostered no new industries and revived no old industries for the people. . . . In one shape or another, all that could be raised in India by an excessive taxation flowed to Europe, after paying for a starved [native] administration. . . . Verily the moisture of India blesses and fertilizes other lands.[19]

British rule thus broke down the foundations of the precolonial economy and stifled the growth of the industrial economy. By ruining India's handicraft and village economy, British policy created the infamous slums of India's cities which still today teem with millions of starving paupers.

What applies to India's colonial experience applies, *mutatis mutandis,* to all other colonial areas. We are here speaking of colonies of native peoples and not of Western settlers. The latter, created out of British and other emigrants, include the United States, Canada, Australia, and New Zealand. The United States overcame its colonial status in the late eighteenth century. However, American national unity was ensured only after the Civil War, when the major preconditions for modern large-scale industrialization were realized for the first time. Canada and the others, however, remained subordinate to the imperial power of Britain well into the twentieth century. In the case of Canada, for example, the legacy of imperial domination has had a lasting impact. Canada's gross national product is largely generated in foreign-owned firms; and major industries such as automobiles, machinery, chemicals, and electrical goods are dominated by foreign interests, mainly American. But the disadvantageous consequences of Canada's economic dependence, first on Britain and then on the United States, are simply not to be compared with the infinitely more disastrous experiences of India and other genuine colonies.

In documenting the tragic effects of British rule in India, we do not mean to suggest that in the absence of imperial domination India would have become some sort of paradise. But in light of the historical evidence this much seems beyond doubt: "Had the amount of economic surplus that Britain has torn from India been *invested in India,* India's economic development to date would have borne little similarity to the actual somber record."[20]

Japan as a Test Case

Japan is the only Asian country that industrialized itself in the late nineteenth century, thereby escaping the fate of all the other societies in the now underdeveloped world. To understand Japan's exceptional situation we must briefly review its history.

Early in the seventeenth century, Tokugawa Ieyosu established a form of centralized feudalism over much of Japan. His regime was a shogunate, a hereditary military dictatorship which did not remove the emperor but rather relegated him to a titular status only. Thus the shogun exercised actual power. The new regime was wholly committed to preserving the agrarian character of the economy. It was also xenophobic since the foreigners on Japanese soil were the source of the social changes which the new government feared and despised. The Spaniards were expelled in 1624 and the Portugese in 1638. After 1640 all foreigners were excluded except for a small trading station in Deshina (Nagasaki), where the Dutch and the Chinese were strictly controlled.

The object of the shogun's policy was to seal off Japan from foreign influences and to preserve the feudal agrarian structure intact. Parallel in several key respects to Western feudalism, Japanese social structure   377 consisted of *daimyo* (a landed nobility), *samurai* (a warrior stratum), and peasant cultivators. Rice, the staple crop, was the main source of wealth and power. The *samurai* owed allegiance to the lord in return for rice stipends. Originally they were not only warriors but tillers of the soil as well. But in 1587 the ruler of Japan, Hideyoshi, conducted the famous sword hunt "whereby he decreased the danger of popular revolt and also accentuated the class distinction between farmer and sword-bearing warrior."[21] Later, with the revolution in military organization due to the introduction of firearms, the *samurai* were gathered into castle towns, leaving the fields to be farmed by the peasantry.

The long years of peace under the Tokugawa shogunate sapped the martial ardor of the *samurai* and rendered them superfluous. Their stipends were increasingly cut off by the *daimyo,* and large numbers of them became frustrated *ronin,* or "wandering men." Some settled in the cities where they surreptitiously studied Western languages and science. Eventually they became the harbingers of Japan's opening to the West. Indeed, a great mass of *samurai,* filled with resentment and bitterness toward the regime, ultimately came to champion the Meiji Restoration of the late nineteenth century.

Not surprisingly, the shogunal government placed the *chonin* (merchant class) at the bottom of the social scale. Looked upon as a shifty, money-grubbing class, they were hedged in by the government with numerous restrictions. Nevertheless, their wealth continually increased and a money economy gradually supplanted the "natural economy" based on rice. The result was a growing dependence of both the *daimyo* and the *samurai* on the *chonin* and a fusion, through marriage, of both aristocratic classes with the merchants. Yet the restrictions were not without effect, so that the *chonin* lagged far behind the British and Dutch trading companies of the seventeenth and eighteenth centuries. Accordingly, the *chonin* also became wholehearted supporters of the Meiji Restoration.

As for the peasants, they were prohibited from leaving the countryside and from alienating their land. When the government failed to increase productivity through administrative squeezing, it began to encourage intensive agriculture, which meant that the peasant was more and more pressed by the need for money to purchase fertilizer, tools, and other goods. By manipulating prices and generally taking advantage of the poorer peasants, merchants and usurers increasingly gained control of the land and became the legal cultivators. A new, landowning, usurer class now added to the peasant's traditional feudal burden. Thus afflicted by debts, taxes, and *corvée,* the peasantry turned to both passive and active resistance. A massive flight to the cities ensued which the government could not check; and violent peasant rebellions became endemic. Peasant resistance and rebellions were a major factor in the strengthening of the antigovernment movement.

The leadership of the opposition to the shogunate was drawn primarily from the ranks of the lower *samurai.* As the feudal structure weakened and the *samurai* lost their traditional raison d'être, they saw no more worthy cause than to oppose the shogun. Other leading representatives of the opposition movement were the *kuge,* an ancient pre-aristocratic stratum specializing in humanistic studies. Having been reduced to penury by the government, they actively supported the opposition. Finally, four powerful *daimyo* commercial clans also led the antigovernment forces. They had continued to engage in clandestine trade with China throughout the Tokugawan period, and they greatly resented the government's policies. In sum, the *ronin,* the lower *samurai,* the *kuge,* and a few dissident *daimyo* clans formed the leadership of the government opposition.

Although the emperor lived in relative obscurity throughout the Tokugawan period, the shogun never challenged his ultimate right to reign. Thus when the dissident forces combined the slogan "Revere the Emperor" with "Expel the Barbarian," there was no way for the shogun to discredit the tactic. Soon the government's difficult situation was aggravated by the first successful Western penetrations of Japanese defenses. A national consciousness now began to emerge, including the revival of Shinto religion, a symbol of indigenous Japanese culture. Rebelling against the infatuation with everything Chinese, the leaders of the antigovernment movement turned westward. It was the *ronin* and the lower *samurai* in particular who avidly absorbed Western science and ideas via the Dutch language. Why they viewed Western knowledge as compatible with the developing Japanese consciousness has been convincingly suggested by Paul Baran: "The exceptional Japanese receptiveness to Western knowledge . . . was largely due to the fortunate circumstance that Western thought and Western technology were in Japan not directly associated with plunder, arson, and murder as they were in India, China, and other now underdeveloped countries."[22]

Furthermore, Japan, in contrast with China, had no entrenched Confucian bureaucracy. The shogunal administration was composed of *bushi,* a military caste. Thus as the West demonstrated the superiority of its military technology, the government leaders hastened to adopt it. They did so not only for nationalist reasons "but to maintain their own

379

prestige in a society that glorified military virtues."[23] Western military science could therefore be absorbed in Tokugawan Japan without threatening the existing hierarchy.

In the final decades of the Tokugawan regime, agrarian distress was severely aggravated by earthquakes, floods, and other natural disasters. Peasant revolts and riots, led by *ronin* and petty officials, increased in frequency. Thus faced with revolt at home, the government also found itself menaced by invasion from abroad.

The Opening of Japan

In the 1850s the United States, second only to Britain as a naval power, sought treaty rights guaranteeing shipping interests in the Far East. The United States turned to Japan because it was the only country still "available" and because American forces were in no position to challenge the established colonial powers in other areas. In 1853–1854 Commodore Perry "opened" Japan's door, and in 1858 the first commercial treaty was negotiated between Japan and a Western power. Those events served to intensify antiforeign and antigovernment feeling in Japan; and they accelerated the distintegration of the already weakened feudal economy through the substantial importation of foreign merchandise.

In the face of American penetration, the shogunate tightened Japan's defenses, imposing the financial burden on the agrarian population. That burden added to the distress of the peasantry and precipitated more frequent and violent revolts. Masses of impoverished peasants, beggars, and vagrants poured into the cities, creating general chaos. Those events also heightened the resentment of the *ronin* and sped up the process that resulted in the Meiji reforms and the rapid industrialization of Japan.

We are now in a position to address the central question: What was it that enabled Japan to industrialize and thus to avoid the fate of all other countries in the now undeveloped world? The answer is that *Japan never became a colony*. In that respect it was unique in Asia.

Japan's avoidance of colonialization may be attributed to several circumstances. It was never viewed by the Western powers as a potential market or source of cheap raw materials. Its economy was backward, its population poor, and its land almost devoid of natural resources. Even today, after more than 100 years of explorations, Japan has no natural wealth to speak of compared with other industrial societies. There is no oil, bauxite, or nonferrous metal and very little coal and iron. It is a remarkable fact that Japan's heavy industry was in the main built up with scrap metals imported from the industrial countries. Thus it is with good reason that the Western powers largely ignored Japan in the nineteenth century, giving their attention to the big prizes to be gained in India, China, and the Middle East.

By the time the United States arrived on the scene, Japan had acquired *strategic* importance as a staging area for further advances into China. However, the United States was itself a newly emerging industrial society and lacked the power to conquer Japan outright. Furthermore, the rivalry of the established imperialist powers had resulted in a system of checks and balances, which had much to do with preventing Britain from acquiring China for itself, as it had earlier done with India. With the various imperial powers thus watching one another closely, Japan was spared.

The final years of the Tokugawan regime witnessed a strengthening of the alliance between the *samurai,* the merchants, and the commercial *daimyo* clans. The open conflict with the shogun's forces was waged in 1864–1865. The antigovernment coalition won by employing for the first time *kiheitai,* shock troops drawn from nonmilitary classes such as peasants and townsmen and led by *samurai* and *ronin.* By 1868 the coalition was strong enough to depose the shogun, establish itself in power, and inaugurate the Meiji Restoration. (The word *Meiji* means "enlightened rule" and refers to the period of the reign of Emperor Mutsuhito [1868–1912]). In the ensuing years the following constitutional provisions were introduced:

1. legal equality for all classes;
2. the abolition of feudal rules and dress;
3. the disestablishment of Buddhism;
4. the institution of Western thought and technique;
5. the legalization of private property in land;
6. the political unification of the country.

Virtually all of those provisions served to dismantle the feudal framework and to promote the development of capitalism. The interest of national defense was a major factor in the extraordinary pace of Meiji industrial development. The new leaders had learned a vital lesson from the unequal treaties foisted upon China. Determined to save Japan from a similar fate, the Meiji regime poured huge subsidies into the development of heavy industry and military production. A close alliance was formed between the government and a few powerful private 381 oligarchies; and order was forcibly imposed on the peasantry, who bore the main burden of capital accumulation. In the decades that followed, Japan successfully took its place among the ranking industrial and military powers of the world.

A Further Test of the Thesis

In his study of social change in Indonesia, Clifford Geertz pauses to compare Java and Japan. There are of course many geographical, historical, and cultural differences between these countries, but they also share similarities. Both countries are heavily populated and both rest on a labor-intensive, small-farm, multicrop cultivation system centering on wet rice. In the mid-nineteenth century the similarities between the two were even greater. By the mid-twentieth century, however, it became apparent that as compared with Japan, Java was standing still. Japan had increased productivity per agricultural worker 236 percent; Java's productivity had scarcely increased at all. What accounts for the difference? This is Geertz's reply.

Where Japanese peasant agriculture came to be complementarily related to an expanding manufacturing system in indigenous hands, Javanese peasant agriculture came to be complementarily related to an expanding agro-industrial structure under foreign management. . . . In Japan, the industrial sector, once under way, then re-invigorated the peasant sector through the provision of cheap commercial fertilizer, more effective farm tools, support of technical education and extension work and, eventually after the First World War, simple mechanization, as well as by offering expanded markets for agricultural products of all sorts; in Java most of the invigorating effect of the flourishing agro-industrial sector was exercised upon Holland, and its impact upon the peasant sector was . . . enervating. The dynamic interaction between the two sectors which kept Japan moving and ultimately pushed her over the hump to sustained growth was absent in Java. Japan had and maintained, but Java had and lost, an integrated economy.[24]

After all is said and done, there is *one* crucial difference between Japan's economic history and Java's. Java until recently has been under Dutch colonial rule. In order to industrialize, it must therefore overcome its colonial legacy. Japan had no such legacy to overcome. As Geertz puts it, "The existence of colonial government was decisive because it meant that the growth potential inherent in the traditional Javanese economy . . . was harnessed not to Javanese (or Indonesian) development but to Dutch."[25]

The conclusion seems clear. The fact that Japan avoided colonization made all the difference. For Japan was thereby enabled to employ the economic surplus generated within its boundaries for its own economic development and not for someone else's. As a nationally independent power, Japan was able to protect its infant industries by means of tariff walls. In contrast, the experience of all colonies was "industrial infanticide."

And we may also draw a methodological conclusion: Comparative historical analysis is indispensable for the study of social change. No abstract sociological scheme, however sophisticated it may appear to be, can take the place of careful attention to history.

In the next and final chapter, we shall take a closer look at the so-called developed and less-developed countries. We shall see how intertwined their destinies are and, indeed, how the very survival of humanity, in this age of a nuclear "balance of terror," will depend on major progress toward international disarmament.

15

The Developed and the Developing Countries: A Program for Human Survival

The so-called developed countries of today are those that have created an industrial infrastructure which enables them to "take off" into self-sustained economic growth. In such societies traditional obstacles to growth have been overcome. Britain was the first and only country to experience the takeoff in the eighteenth century. All other major industrial countries—for example, France, the United States, Germany, Japan, Russia, and Canada—launched their respective takeoffs a century later. Britain's takeoff is commonly known as the *Industrial Revolution,* while that of the other advanced industrial states is often referred to as the *Second Industrial Revolution.*

The term *Industrial Revolution* was invented in the 1820s (by English and French thinkers) to describe the sudden and fundamental economic transformation that had begun in Britain in the 1780s. Today we appreciate that judged by its far-reaching consequences, the transformation was the most important event in world history—at least since the invention of agriculture and cities.[1] And yet by present-day standards the technical inventions of Britain's take off were quite modest: the flying shuttle, the spinning jenny, and the "mule" (a machine that twisted fiber into yarn). All such innovations were the work of artisans experimenting in their workshops. Even James Watt's rotary steam engine (1784), the most sophisticated machine of the age, was based on long-established principles of physics. It was not therefore the inventions themselves but rather a combination of favorable circumstances that enabled such modest technical innovations to bring about basic socioeconomic changes.

384

Among the most important of these favorable circumstances was the unique social structure of the British countryside. A small number of commercially minded landlords controlled the land, which they leased to capitalist-farmers who in turn employed landless laborers to cultivate it. The Enclosure movement, which had been going on for centuries, eliminated the British peasantry and replaced it with a large mass of rural proletarians. Britain's agriculture had become more highly commercialized and productive than that of any other country at the time. Agriculture was therefore ready to fulfill several necessary preconditions for industrialization: (1) it increased production and raised productivity sufficiently to feed the rapidly growing nonagricultural population; (2) it provided a large and expanding surplus of laborers to the emerging industries and towns; (3) it accumulated capital which could then be applied in the industrial sectors of the economy; (4) it created a large domestic market among the agricultural population; and (5) it produced a surplus for export which could be traded for needed capital imports. Owing to the advanced commercialization of its agriculture, Britain enjoyed two other favorable circumstances. The first is called *social overhead capital*. This refers to the highly expensive general facilities such as shipping and port installations, roads and waterways, that every industrializing society requires if it is to move ahead smoothly. The second favorable circumstance was the fact that the British political rulers, the commercialized lords, were quite favorably disposed to an economic system geared to profit.

Who were the pioneers of the first Industrial Revolution? They were the countless entrepreneurs and investors who recognized the exceptional rewards offered to the manufacturer of cotton and of textiles generally. A *world* market, not only a domestic one, had already existed for these staples. Small businessmen clearly perceived the possibilities for extraordinary profits if only they could expand their output quickly by means of cheap and simple innovations. The British cotton industry had originated as a by-product of overseas trade, which supplied its raw material, and in competition with the cotton goods or calicoes of India, whose markets the British manufacturers sought to capture with cheap imitations. Colonial trade had thus created the cotton industry and the industry in turn nourished slavery. For after the 1790s it was the slave plantations of the Southern United States that supplied the bulk of the raw cotton that met the soaring demands of the Lancashire mills. Thus "small men," who often started out with a few borrowed pounds, soon became powerful enough to prevail over the mercantile interests. The parvenus, these new, small capitalist manufacturers, undersold both the British East India Company and the Indian merchants. As a consequence, the Indian weavers' source of livelihood was destroyed and they were gradually ground to dust. The Indian subcontinent, systematically deindustrialized, became an insatiable market for British cotton. By 1840, for the first time since the dawn of histroy, Europe exported more to the East than it bought there.

For all of its successes, however, the new industrial-capitalist system soon began to exhibit certain flaws, notably the trade cycle of boom and slump. The economy experienced periodic crises resulting in declining production, unemployment, and bankruptcies. *Pre*capitalist crises were typically associated with natural catastrophes and war, which caused harvest failures and the destruction of food supplies. But the crises that became evident in the early nineteenth century were soon recognized as periodic phenomena rooted in trade and finance, in a shortage of profitable investment opportunities. "Slumps" occurred in 1825–1826, 1836–1837, 1839–1842, and 1846–1848. Such crises not only continued to occur but worsened.

Although cotton stimulated entrepreneurs to create the first Industrial Revolution, it was not cotton but rather *coal* that gave rise to the earliest basic capital-goods and other heavy industries in Britain. Up until the eighteenth century wood was the major fuel in Britain for both domestic and industrial uses. Smelting and the production of iron were done with charcoal, thus causing a drastic depletion of England's forests. In contrast with the expansion of the textile industry, the British iron industry had declined step by step until it gave the impression of having reached its end. In the face of deforestation a substitute for charcoal had to be found, as indeed it was in the coking of coal in 1735 and in the use of coke in blast-furnace operations in 1740. The mining of coal, however, encountered grave difficulties owing to perpetual flooding. The invention of the steam engine solved this problem. Crude demonstrations of the possibility of lifting water with fire had been made a century before the Industrial Revolution, but it was only toward the end of the eighteenth century that the steam engine reached the stage where it could produce the large quantities of coal necessary for modern industry. In 1800 Britain produced about 10 million tons of coal, or about 90 percent of the world output, with France, its nearest competitor, producing less than a million. It was coal that gave birth to the epochmaking invention of the era, the *railway*.

Not only did coal mining require large and powerful steam engines to prevent flooding; it also required means of transporting the huge quantities of coal from the coal face to the shaft and from pithead to destination. The railway was an obvious answer. The first of such lines was laid down in 1825, stretching from the coalfield of Durham to the coast. The railway became the true symbol of the Industrial Revolution, for it was this remarkable invention that created the unprecedented market for iron and steel, coal, heavy machinery, labor, and capital investment. In 1830 there were still only a few dozen miles of railroad in the entire world, mainly consisting of the line from Liverpool to Manchester. By 1840 there were more than 4,500 miles and by 1850 more than 23,500. Railway lines had by now been opened in several other countries: in the United States, in 1827, in France in 1828, in Germany and Belgium in 1835, and in Russia in 1837. Most of these lines were built, in large measure, by British capital, iron, machines and know-how.

THE SECOND INDUSTRIAL REVOLUTION

The *Second Industrial Revolution* refers to the remarkable technological changes of the late nineteenth century, which resulted in new sources of power and a marked rise in industrial productivity. Oil and electricity now joined coal; the gas engine and electric motor increasingly replaced steam as the sources of industrial power. Steel became a basic industrial material having received a new impetus from the introduction of the Bessemer process, the open-hearth furnace, and new ways of hardening steel with alloys. If the typical unit in Britain's industrial take off was the small family firm, the typical unit of the Second Industrial Revolution was the large corporation. The major reason for the growth in the size of enterprise was the application of the new technology in steel, aluminum, electricity, and chemicals in highly expensive, large-scale equipment. The enormous costs of fixed capital and other initial investments meant that the new industries became the preserve of giant corporations. Small businessmen faced a "natural barrier to entry."

The Second Industrial Revolution took place not in Britain but in the United States, Germany, and Japan—and to a lesser extent in Russia and Italy. These countries, the so-called latecomers, did not have to go through all the stages of Britain's industrial development; they began at the most advanced and modern stage, which gave them a decisive advantage. Since we cannot in this short space effectively discuss all the latecomers (the Japanese case was reviewed in the previous chapter), we shall focus attention on the United States.

With the secession of the South and the outbreak of the Civil War, control of the federal government passed out of the hands of the Southern legislators and their Northern allies. In their place came a new 387 group representing a coalition of Northeastern manufacturers and independent Midwestern farmers. The new political alliance wasted no time in enacting a legislative program that would provide optimal conditions for the growth of industry and the enrichment of its owners. Even during the Civil War the tariff was boosted to a level which afforded secure protection from foreign competition. The Morrill Act of 1861 was the beginning of a sharp upward climb in tariffs. It raised average rates from about 19 percent of value to 47 percent, more than double the rates of 1860. Following the Civil War the acts of 1883, 1890, 1894, and 1897 granted even more protection.[2]

American industry was also protected from domestic taxation. Profits were never highly taxed even during the war, and the moderate wartime income tax did not long survive the peace. Moreover, the federal government made generous appropriations for internal improvements. A Pacific railway was chartered, and the federal government granted vast areas of public land and a loan to the business interests that built it. With this stroke the market for American industrial goods was enormously enlarged. At the same time measures were taken to expand the agricultural sector through the Homestead Act, and to ensure a supply of cheap labor through the encouragement of massive immigration. It was this political program that created the favorable conditions for the explosive growth of American industry in the nineteenth century.

As large masses of immigrants streamed into the country, the domestic market expanded accordingly. The cultivation of the Western lands provided huge stocks of comparatively cheap food. The resulting development of manufacturing industry in the United States between 1860 and 1900 was fabulous. In 1860 products of American factories were valued at $1,800,000,000. That figure doubled by 1870 and trebled by 1880, becoming five times as large by 1890. Between 1860 and 1880 the production of pig iron increased by almost 5 times while the production of steel increased by 155 times. At the start of the Civil War the United States ranked fourth among the world's industrial powers. By 1894 it had surpassed all other countries and, in fact, produced more than Britain and Germany combined in that year.[3]

By 1900 the United States had constructed a huge railroad network and had built its major heavy industries and urban-industrial centers. In the following decades vast new industries were created, notably the automobile, airplane, and television industries. By the 1950s American manufacturing industry surpassed that of all other countries in the quantity of goods produced. If we compare the United States in that decade with Western Europe and the Soviet Union together with all of its Eastern European satellites, we find that American coal production was half as large but that it produced 2.6 times more oil, 1.2 times more electricity, 82 percent as much steel, and 86 percent as much pig iron. In other words, America's industrial plant in the 1950s was almost as large as that of all the industrial nations put together.[4]

Of course, we must remember that in the 1950s the Soviet Union and the other countries of Western and Eastern Europe were still recovering from the devastations of the Second World War whereas it was the good fortune of the United States never to have had a foreign bomb dropped on its soil. In the 1960s, however, the United States became embroiled in a protracted war in Indochina, which though it did not bring devastation to American shores, nevertheless proved to be quite costly both in human life and in other respects. Although it did not become immediately evident, the ten-year-long war that the United States waged in Vietnam and other areas of Indochina weakened the American economy considerably. The war entailed the channeling of huge resources, technical, material and financial, into military production, thus draining away those precious resources from the civilian sector of the economy. It was precisely in that decade, from the mid-1960s to mid-1970s, that other major industrial powers forged ahead and, much to the chagrin and surprise of Americans, caught up with and even surpassed the United States.

THE AMERICAN ECONOMY FALLS BEHIND

The United States, known until recently as the world's richest nation, had slipped by 1978 to sixth place in per capita income, $9,770. By 1979, according to the *World Bank Atlas,* the U.S. per capita income of $10,610 was only ninth-highest.[5] The leader was Kuwait with a per capita income of $20,250. But if we confine our attention to the advanced industrialized nations, we find the United States lagging behind Switzerland, Sweden, West Germany, Denmark, Norway, Belgium, and France in the late 1970s.

In March 1982 the Labor Department reported that 288,000 people had joined the jobless total in February. This brought the unemployment rate up to 8.8 percent of the work force, or a shade below the postwar high point of 9 percent. From the time that the unemployment rate began to rise markedly in July 1981, 1.8 million persons lost their jobs, raising the American unemployment total to 9.6 million. In February 1982 unemployment occurred in every category, rising to 22.3 percent among teenagers and to 42.3 percent among black teenagers—and this at a time when many job-creating programs had been disbanded as a result of the Reagan administration's deep cuts in government spending. (At the time of this writing well over 10 million people are unemployed in the United States.)

At the same time car sales fell to their lowest point in 20 years, and share prices as measured by the Dow Jones industrial index had declined to their lowest level in 22 months. Major airline companies and other industries have either failed (e.g., Braniff) or reported serious financial difficulties. One could go on and on, but inasmuch as this information can be found in the news media, there is no need. The point is that there are signs indicating that the American economy is in serious trouble.

Moreover, it has become evident that America's current economic difficulties are rooted in *structural* changes. There have been two pronounced upswings in recent American economic growth, from 1950 to 1953 and from 1962 to 1966. Both were accompanied by an even faster increase in manufacturing output. The 1970s, however, tell a different story. There occurred in that period an astonishingly rapid expansion in employment. From 1973 to 1979, nearly 13 million new nonagricultural jobs were created of which 11 million were in the private sector. The U.S. rate of growth in total employment in that period was more rapid than that of any other industrial country. It was more than three times as fast as Japan's rate, while employment stood still in France and the United Kingdom, and fell in West Germany. But there was something definitely peculiar about American growth in that period. Summing up the evidence, Emma Rothschild writes:

By 1979, 43 percent of all Americans employed in the private nonagricultural economy worked in services and retail trade. The two sectors together provided more than 70 percent of all new private jobs created from 1973 to the summer of 1980.

Even within these two vast sectors, the growth in employment was further concentrated. Three industries each provided more than a million new jobs during the 1973–79 period: "eating and drinking places," including fast-food restaurants; "health services," including private hospitals, nursing homes, and doctors' and dentists' offices; and "business services," including personnel supply services, data processing services, reproduction and mailing, and the quaintly named "services to buildings." . . .

The three "new" industries loom very large in total employment. . . . Thus the *increase* in employment in eating and drinking places since 1973 is greater than total employment in the automobile and steel industries combined. Total employment in the three industries is greater than the total employment in the entire range of basic productive industries: construction, all machinery, all electric and electronic equipment, motor vehicles, aircraft, ship building, all chemicals and products, and all scientific and other instruments.[6]

In a word, the United States has been transformed from a manufacturing to a service economy. The picture in our heads of the American economy should no longer be the smokestacks of Pittsburgh, Youngstown, and Detroit; it should be MacDonald hamburger! In international industrial classifications, "wholesale and retail trade, including restaurants and hotels" and "community, social and personnel services" together approximate the U.S. category of "trade and services" though it includes public employees and wholesale trade. If we compare the United States with the five major industrial countries, we find that the United States has by far the highest proportion of its employed population in that category. In fact, by 1978 the United States was the only major industrial nation in which more than half (51 percent) of the work force was employed in the "trade and services" sector.

There are, then, clear indications that where manufacturing industry is concerned, the United States is falling behind other capitalist countries. From 1972 to 1978, industrial productivity rose 1 percent in the United States but almost 4 percent in West Germany and over 5 percent in Japan. Other capitalist countries have moved ahead in methods and processes of production. While Japan, for example, now employs large oxygen furnaces and continuous casting, U.S. steel interests have invested vast sums of capital in obsolete open-hearth furnaces.

In response to the faltering state of the American economy, the Reagan administration has introduced policies designed to "liberate" free enterprise. Accordingly, social expenditures have been drastically reduced and tax policy has been restructured with the aim of encouraging savings and investment. In effect, the tax burden has been shifted from those who save (the rich) to those who consume (the poor). Rules and regulations that do not favor large business interests are being eliminated. So the question that naturally suggests itself is whether such a policy is likely to succeed in remedying America's economic ills. Will this policy reverse the trend in the United States toward deindustrialization?

In thinking critically about the administration's proposed solution, it is worth noting that no other capitalist country has achieved success by following such a route. Government absorbs slightly over 30 percent of the Gross National Product (GNP) in the United States but over 50 percent of the GNP in West Germany. Fifteen countries collect a larger fraction of their GNP in taxes than does the United States.[7] Other governments are not only larger but also play a more substantial role in shaping the economy. In West Germany union representatives are members, by law, of corporation boards. Japan has a central investment authority and a high degree of industrial planning. The French government owns the Renault automobile firm as well as other businesses. The West German government owns Volkswagen. As for rules and regulations, the American economy has the fewest, not the most. Furthermore, the nations that have surpassed the United States industrially have not done so by increasing inequality. If we look at the disparities in earnings between the top and bottom ten percent of the population, we find that the West Germans work hard with 36 percent less inequality than the United States and the Japanese work even harder with 50 percent less inequality.[8] Perhaps it is time for the United States, now no longer in the industrial forefront, to learn from the experience of others.

LEARNING FROM OTHER CAPITALIST COUNTRIES: JAPAN AND WEST GERMANY

In Japan large firms provide lifetime jobs, relative wages are almost totally dependent upon seniority rather than skill and merit, and income differentials are 50 percent smaller than in the United States. Yet the Japanese have the highest rate of productive growth. To understand why, we need to grasp the key features of the Japanese economic system. They are: (1) the enterprise-union concept, (2) lifetime employment, (3) consensual decision making, and (4) the collaboration of government and the private sector in formulating the nation's industrial policies and goals.

In Japan unions operate differently than they do in Western countries. The system of enterprise unions unites workers through their companies rather than their trade. Unions are much more involved in the decisions of the company. Most large firms have regular management-labor consultations to sort out problems and to prevent them from arising. Although talks fail occasionally and strikes do occur, they tend to be short. Many fewer days are lost through strikes than in the United States. Union-membership does not exclude workers from managerial positions. Japanese firms seek alternatives to laying off workers and do so only as a last resort. Japan's largest steelmaker, Nippon Steel Corporation, sends temporarily redundant employees to work at a company-owned ranch and fish hatchery until business picks up. Doubtless, the comparatively amicable relations between management and labor have promoted industrial growth just as the high level of growth has, in turn, smoothed labor-management relations. Official statistics place unemployment at about 2 percent.[9]

Japan's quality-control circles are by now quite famous. Meetings regularly take place to explore possibilities for improvement in production efficiency, safety, and quality. It is ironic that this idea originated in the United States but soon faded there. Promotions and raises come with seniority. Merit pay accounts for only 5 percent of the average employee's compensation. Not until an employee is in his mid-thirties or early forties can he expect to be singled out for promotion to a managerial or supervisory position. Even then the wage differential between manager and worker is small. On the average a middle-management employee with 15 years' seniority is paid only half again as much as the worker with 2 years' experience. Yet the Japanese manager appears to have considerably more responsibility than his American counterpart. Whereas the typical Japanese manager-supervisor in a large firm is responsible for some 200 workers, the average American manager is responsible for 10. Not only that, most Japanese managers have been union members. The fact that they have risen from the ranks of the workers adds to the stability of labor-management relations.

Still other aspects of the Japanese system contribute to a state of affairs in which workers, managers, and owners have a common interest in the prospering of the firm. Bonuses that depend on profits are negotiated annually, resulting in across-the-board awards equivalent to five or six months' salary. Robotization is supported by all parties concerned because mechanization does not result in the displacement of workers. From management's standpoint the value of mechanization lies not in the number of workers it can replace but in the improvement it can bring to the quality of workmanship. Managers strive to introduce robots in physically demanding and monotonous jobs because they recognize that workers' morale is higher when they work at more pleasant and challenging tasks requiring more thought and skill. Because factory work is inherently boring and repetitive, the quality-control circles constantly strive to alleviate tedium. There are frequent minor changes and occasional major changes to the routine in response to employee suggestions.

With regard to the Japanese pattern of "lifetime" jobs, we should note that it is not offered by small companies, which still employ more than half of all workers. Nor does it apply to women or temporary workers. Lifetime employment, then, is a privilege enjoyed only by male, full-time employees of large companies. Some students of the Japanese system have argued that lifetime employment may be an artifact of the 393 labor market, the result of economic forces. Those who have made this argument have questioned whether a distinctively Japanese pattern actually exists. After considering such arguments, however, most scholars agree that lifetime employment is a powerful ideal with considerable effect in Japanese social life.[10]

The final element of the Japanese system that merits serious consideration is the role of government in the economy. MITI (the Ministry of International Trade and Industry) is a governmental agency that collaborates with the private sector in planning the country's economic development. In fact MITI plays a directing role in this relationship; it is a planning authority.[11]

The foregoing discussion suggests that there is something to be learned from the Japanese experience. The point, of course, is not to copy but rather to consider whether there are worthwhile elements that can be adapted to American conditions. First, however, we need to look at one more case, the experience of West Germany with *Mitbestimmung,* that is, *codetermination.*

The idea of codetermination has its roots in the German Social Democratic movement and can be traced back to the nineteenth century.[12] It was not until 1951, however, that the Bundestag (German Parliament) passed a law stipulating that in the steel and coal industries labor was to be granted equal voting power with shareholders on each company's management board. This was the so-called Montan law according to which there are eleven-member management boards containing five members elected by shareholders and five by employees, including three by the union involved and two by the workers' council. The two sides elect a neutral outsider who holds the tie-breaking vote. The labor-relations director, who also sits on the management board, must have the approval of the labor board members.

In 1976 the unions were defeated in their attempt to extend the Montan law to other industries. According to a *Business Week* report of March 9, 1981 the Bundestag approved labor's participation on management boards in all companies with more than 2,000 employees but stopped well short of meeting labor's goal. The corporations and their political allies, fearing a loss of authority on key issues, rejected full worker parity and are now trying to undermine codetermination in the steel industry as well. In 1981 a law was proposed by the Bonn government in response to a move by Mannesmann, the industrial equipment giant, to reorganize itself so as to escape the full parity requirement of the Montan law. By transferring its steelworkers to the control of a separate subsidiary, the parent Mannesmann firm will no longer have

394

half of its sales in steel production. The parent firm will then be able to operate under the 1976 law, which gives shareholders a voting advantage over labor on the management boards. The steel-pipe subsidiary will still be under the 1951 Montan law. A compromise bill worked out by the government coalition partners, the Social Democratic party (SDP), and the Free Democrats (FD) requires that Manesmann and other firms making similar moves remain under the Montan law for six years. The unions wanted the Montan formula to remain in effect indefinitely. Union leaders fear that codetermination will disappear altogether unless the SDP can achieve an absolute majority in the Bundestag, which is not very likely in the prevailing economic and political climate. About three dozen giant firms are covered by the Montan law and many, no doubt, will do their utmost to follow the Mannesmann example with the expiration of the six-year period.

The battle over codetermination has grown bitter in the past several years. In 1978–1979, strikes lasting up to six weeks occurred in large parts of Germany's steel industry. The cause was union outrage over an attempt by the business interests to have even the watered-down 1976 codetermination law ruled unconstitutional in the high court. The business interests lost the case but the bitter feelings engendered have remained. As a result of the court challenge, the unions withdrew from Bonn's so-called concerted action meetings between top-level labor, business, and government officials. Until recently these meetings have helped create a consensus on economic policies that enabled the West German economy to grow rapidly. The consensus arrangement worked best from the mid-1960s until the late 1970s. But now that West Germany appears to be entering a period of slower growth, inflation, and unemployment, labor-management conflict is on the rise, and labor has been frustrated in its attempt to extend codetermination across the economy. Under the 1976 law, moreover, labor lost its power to veto the appointment of the labor-relations director, and the tie-breaking vote was placed in the hands of a shareholder nominee.

Ironically, the conflict over codetermination has sharpened in West Germany at a time when interest in the concept has picked up elsewhere. A committee of the European Parliament has been debating a measure that would make codetermination mandatory throughout the European Economic Community. At present, six other countries have legally mandated systems of codetermination: Sweden, the Netherlands, Norway, Denmark, Austria, and Luxembourg. Britain is in the process of considering it. Is it a mere coincidence that those countries that have instituted a measure of industrial democracy are among the most successful economies? In the United States, we should note, current economic difficulties have given rise to shared decision making at the plant level. The UAW and a few other unions have obtained seats on management boards as a quid pro quo for helping to rescue failing companies such as the Chrysler Corporation. It is too early to tell whether this will become a movement toward an American form of codetemination.

It is time to ask what might be learned from the Japanese and West German experiences. From Japan we learn that a long-term commitment to employees appears to be in the mutual interest of management and labor. We also learn that policy planning for industrial development has been an essential element of Japan's economic success. The planning authority includes representatives from government, business, and labor. From both the Japanese and the Western European experience we learn that consensual decision making by management and labor has been an integral part of their industrial systems. In this light we are moved to ask: Is economic planning essential for the United States if it hopes to reindustrialize? What would democratic planning entail?

DEMOCRATIC PLANNING FOR THE UNITED STATES?

Several congressional committees and at least one presidential commission have for several years been exploring the question of what national economic planning would look like in the United States. Wassily Leontieff, the father of input-output analysis, has provided an outline of the technical side of such planning.

The statistical index of a national economic plan may be visualized as a detailed, systematic annual survey of manufacturing, agriculture, transportation, and trade as well as federal, state, and local budgets. The survey would enable the planners to formulate a policy several years in advance but revise it annually in light of both past experience and new information. An economic plan is neither a forecast nor a rigid policy that is adhered to unswervingly. "The whole idea of planning," writes Leontieff, "assumes the possibility of choice between alernative feasible scenarios. Feasibility is the key word."[13] A national plan attempts to grasp the workings of the economy as a system of interdependent parts:

The trucking industry must be supplied by the oil refining sector with fuel, and, to be able to expand, it must be supplied by the automobile industry with trucks, in addition to the replacement of worn out equipment. To provide employment for additional workers, the automobile industry must not only be assured of an outlet for its products, but in the long run it must construct new plants and retool the old. In the process of doing so, it has

396

to receive from the construction industry more plant space, and from the machine-building industry additional equipment, not to speak of a greater flow of power, steel, and all other inputs.[14]

Traditional economic theory has supposed that by means of the competitive price mechanism, a capitalist economy will automatically bring about an equilibrium between supply and demand. But as often as not, a reliance on this so-called self-adjusting mechanism has resulted in periodic unemployment, idle productive facilities, and a misallocation of resources. As for conventional monetary and fiscal policies, they appear "to be no more successful in compensating for lack of systematic foresight than frantic pushing and pulling out the choke is able to correct the malfunctioning and stalling of a motor. Occasionally it works, but usually it does not."[15]

The complexity of present-day industrial capitalism is such, argues Leontieff, that we can no longer rely on either the traditional economic doctrine or on the Keynesian and Friedmanian theories. Planning is necessary. A *democratic* planning process would begin, however, not by enunciating some abstract goal but by presenting to the citizenry several possible future states of the economy, in concrete, nontechnical terms. The published volumes containing such alternative scenarios would resemble the *United States Statistical Abstracts* with sections devoted to industrial production, agriculture, trade, transportation, consumption, medical services, education, etc. on national, regional, and local levels. One advantage of a plan is that it would allow for an effective coordination between agencies that is presently imposible. Take, for example, our energy and environment policies, each of which

is controlled by a different department, not to speak of many smaller, often semi-autonomous agencies. Production of fuel and generation of energy is one of the principal sources of pollution. Any major move in the field of energy can be expected to have far-reaching effects on the environment, and vice versa. The energy-producing industry is immediately and directly affected by anti-pollution regulations. The obvious practical step to take to solve this problem is for both agencies to combine their data banks (their stocks of factual information) and to agree to base their policy decisions on a common model. This model should be capable of generating scenarios displaying jointly the energy and the environmental repercussions of any move that either one of the two agencies might contemplate making. Adversary policy debate could and should continue, but adversary fact finding would have become impossible, and policies that tend to cancel out or contradict each other would at least be shown up for what they are.[16]

Planning, then, would enable the American citizen to grasp the consequences of alternative policies.

Opponents of planning argue that a plethora of regulations will result from it. Such persons should be reminded that the American economy is already a regulated one: There are regulations for large farmers, for railroads, for the steel industry, for trucks, airlines, and so forth. Regulations in the United States are often designed to raise someone's income (through subsidies and tax policy) and therefore to lower someone else's. One group's gains are another's losses. As Lester Thurow has observed, "No one can say that a regulation is good or bad without a vision of what distribution of income should exist, and how the distribution ought to be created. In the abstract deregulation is popular; everyone is for it. In practice each of us opposes deregulation when it will lower our own income."[17] The real issue, then, is not regulation or no regulation, but rather what kind of regulations. The question is, How do we begin to revitalize the American economy? And how do we spread the costs that this will entail as equitably as possible?

The answer to these questions lies in the reversal of several dangerous trends that have become evident during the past decade. To begin with, there has been a substantial exodus from major Northeastern and Midwestern cities. Chicago has lost 12 percent of its population, Baltimore 14 percent, Cleveland 24 percent, and St. Louis 28 percent.[18] The proportion of taxpayers moving out was even larger. At the same time the population of certain Southern and Western cities has expanded commensurately: Houston gained 24 percent, San Diego 25 percent, Phoenix 33 percent.

In the same period the nation witnessed major American industries experiencing severe difficulties and failures: In 1979 U.S. Steel lost close to a half billion dollars; in 1980 the Ford Motor Company, Chrysler, and General Motors each lost between $1.5 and $2 billion, International Harvester almost $500 million, and Firestone $100 million. From Baltimore to St. Louis cities are deteriorating under the great strains caused by these industrial setbacks. Indeed, the traditionally powerful American industries have proved themselves woefully inadequate in the face of foreign competition and are unable to mobilize the vast amounts of capital required for modernization. As noted earlier, whatever growth the United States has experienced in the last decade or so has been in fast food places and other services.

The Reagan administration has not only failed to halt these trends but has devised policies to accelerate them. Inspired by certain fashionable economic notions, the administration regards these trends as the unavoidable consequences of the natural workings of the economic system in which there are winners and losers. The administration therefore has decided to back the rising businesses of the Sunbelt and to write off the declining industries (and cities) of the Northeast and Midwest. Critics of existing policies argue, however, that if these trends remain unchecked, they will have disastrous social and political consequences for the nation. Felix Rohatyn, a financier, maintains that the United States today

needs a second Industrial Revolution. The currently fashionable notion of backing the winners instead of the losers is as facile as it is shallow. The losers today are automotive, steel, glass, rubber, and other basic industries. That this nation can continue to function while writing off such industries to foreign competition strikes me as nonsense. Nor does it seem to me workable in the long run for a larger and larger proportion of our population to be diverted to such jobs as serving food and processing business paper while the industries that manufacture products for sale at home and abroad fall into a state of decline.

We cannot become a nation of short-order cooks and saleswomen, Xerox machine operators and messenger boys. These jobs are a weak basis for the economy. . . . To let other countries make things while we concentrate on services is debilitating both in substance and its symbolism. The argument that we are substituting brains for brawn is specious; brains without sinews are not good enough.[19]

Rohatyn goes on to argue that the restructuring of American industries is not only essential but cannot be accomplished without the government's support. He therefore urges the founding of a Reconstruction Finance Corporation (RFC), modeled after the famous agency of the New Deal period, though it was first created by Herbert Hoover in 1918. "The RFC of the 1930s," writes Rohatyn,

saved numerous banks, some cities, and many businesses, and prevented much larger dislocations from taking place. It financed synthetic rubber development during World War II and new aluminum capacity during the Korean War. It made money for the taxpayer. . . .

The proposed RFC should provide the kind of capital our older industries sorely lack: equity capital. In exchange for providing capital to industries that have a sound case for it, and the job security that would come with it, the relevant unions would be asked to make their contributions in the form of wage concessions and changes in work rules that would increase productivity. The lenders, the banks and insurance companies, could be asked to convert some loans to preferred stock and to join with the RFC in committing additional capital. . . . The RFC, like any other large equity investor, should have the right to insist on management changes in the board of directors if it deems them appropriate.[20]

Rohatyn is chairman of the Municipal Assistance Corporation (MAC), an RFC-type agency now functioning in New York City. Describing its successes, he writes that soon after it was created "as an independent agency by the state legislature, MAC was able to extract concessions from banks and unions, from the city and the state. It was then able to put together a financing package, including federal credit assistance, and to impose fundamental reforms which permitted the city to achieve a truly balanced budget in 1980, five years after near bankruptcy."[21] If the United States is to succeed in reversing its economic decline, Rohatyn concludes, it can do so only "by building a mixed economy, geared mostly to business enterprise, in which an active partnership between business, labor and government strikes the kind of bargains—whether on an energy policy, regional policy, or industrial policy—that an advanced industrial democracy requires to function and that in one form or another, have been made for years in Europe and Japan."[22]

The urgent task of this partnership between labor, business, and government is the restructuring of the American economy so that it provides employment for all who need it. As Rohatyn, again, has emphasized, our biggest problems are also our biggest opportunities: "Becoming self-sufficient in energy, rebuilding our basic industries and our oldest cities—there is work enough here for everyone in this country as far as the eye can see."[23]

This immense project must begin quite soon. In the United States, which prides itself on being a *work-ethic* society, restructuring the economy so that it provides full employment is not only an economic task but a moral duty. There have been too many periods in recent American history when large numbers of people were unable to find work. Unemployment has been endemic in peacetime: a depression from 1929 to 1940, and then recessions in 1949, 1954, 1957–1958, 1960–1961, 1969–1970, 1974–1975, and in the early 1980s. The conclusion seems inescapable: Since private enterprise has shown itself incapable of guaranteeing jobs for everyone who needs to work, government, and in particular the federal government, must play a leading role in instituting the necessary programs. Local and state governments cannot by themselves solve problems that are national in scope.

The United States is not the only developed country today experiencing serious economic difficulties. In the second quarter of 1980 industrial production dropped in each of the five largest industrial-capitalist economies.[24] And the unemployment rate appears to be on the rise in the European Economic Community. In the face of such developments, European leaders have expressed grave concern. Not atypical in this regard are the views of the West German chancellor, Helmut Schmidt:

400

The world economic crisis of today, almost 10 million unemployed in the US, two million unemployed in my country, over three million in the UK, is at least as great a strategic danger to the cohesion of the West as anything we have talked about so far, i.e., strategic-military questions. It is a strategic danger because it does spread social and political unrest in our countries, and it entails the danger of national economic protectionism against each other within the West.

For my taste, there is too much talk about so-called strategic questions in the military and political field, and too little talk and too little cooperation in the economic field. We have not seen a world economic recession of this degree since the 1930's. One could easily turn this into a general depression of the Western world.[25]

The cooperation that Chancellor Schmidt calls for will have to extend to the Third World countries as well. For as we shall see, the destinies of the developed and developing countries (also referred to as *under-developed countries*) of the world are intertwined.

THE DEVELOPING COUNTRIES

Every thoughtful and informed person recognizes that the coming decades may be fateful for humanity. The danger is real that in the year 2000 a large part of the world's population will still be living in poverty so extreme that it will result in massive starvation, social chaos, and brutal repression.

When we use the words *poor* and *poverty,* we must bear in mind how different the circumstances are in the developed and developing countries. In the advanced industrial societies that have reached high average levels of income, a *relative* poverty results when income is not equitably distributed, or when the economy is disrupted by recessions and depressions. In the Third World countries, in contrast, we are dealing with absolute poverty. Some 800 million people live in the low-income countries of sub-Saharan Africa and South Asia where the problem is not merely inequitable distribution. Their rate of growth in the past two decades—less than 3 percent per year—has been too slow to make much of a difference to the poor. Even if the total resources of these countries were equally divided, they would not serve to eliminate acute poverty on a massive scale. The low-income countries with a per capita GNP of less than $250 had a combined population of 1,215 million in 1976.[26] More than half this number live in absolute poverty and are concentrated in four large countries of Asia—India, Indonesia, Pakistan, and Bangladesh.

Poor health is still the fate of much of the Third World, and the death 401 rate is astonishingly high. In sub-Saharan Africa life expectancy is about 45 years. There are countries in Africa where one child in four fails to survive until its first birthday. Thirty to 40 million people in the Third World suffer from blindness, and many millions more are threatened with this affliction, which is caused by vitamin deficiencies and water-borne infections. No one knows for sure how many hungry and under-nourished people there are in the Third World, but estimates range from 600 million to 1 billion. Four out of five people living in the rural areas of the Third World have no access to safe water, the lack of which is a major cause of ill health and mortality. Women tread long distances to secure minimal water requirements for their families. In the absence of modern water technology and sanitation, there exist numerous water-borne, life-threatening diseases to which children in particular are sus-ceptible. Between 20 and 25 million children under five years of age die every year in the Third World countries, and a third of those deaths result from diarrhea contracted from polluted water.

In spite of these abominable conditions, the populations of the Third World countries continue to grow at a staggering rate. Every five days over 1 million people are added to the world's population, which is ex-pected to increase by almost 2 billion by the end of the century. The *increase* alone would therefore be greater than the world's total popula-tion during the first decade of the twentieth cenutry. Nine tenths of that increase will take place in the Third World. By the year 2000 the world's population is likely to have expanded from its 1982 level of 4.3 billion to 6 or 6.5 billion. It is true that family planning has enjoyed some suc-cess during the past several decades and that birth rates are declining in many of the developing countries. In some, such as the People's Repub-lic of China, birth rates have diminished so much faster than death rates that the rate of population growth has been cut in half. But the decline in fertility is not likely to make much difference to the world's total number during the next two decades. Even with a continuing decline in fertility the population of many developing countries is likely to double in the course of the next century. Population growth rates of 2 to 3 per-cent per annum are sufficient in those countries to produce a doubling of the population in 25 to 35 years.

Growing at an even faster rate than the total populations are the cities of the Third World, the largest of which are likely to exceed 30 million by the end of the twentieth century. Clearly there is cause for grave concern. If the population of the Third World is not soon stabilized, such extreme overcrowding is liable to bring about devastating social conflicts and other nightmarish consequences.

The vital importance of family planning can no longer be questioned. Vigorously pursued, family-planning programs have achieved considerable success. We have already mentioned China, which today acknowledges a population of one billion people. In the course of the 1970s it reduced its rate of growth from 2.3 percent to a little more than 1 percent. Its aim is to attain zero population growth by the year 2000. The first countries in Latin America to adopt family-planning policies were Chile, Colombia, and Costa Rica; they have reduced their birth rates by almost a third in the past 20 years. Similar successes have been achieved by such East Asian countries as Hong Kong, Singapore, and the Republic of Korea. The United Nations' fertility survey has analyzed data from 14 developing countries and found dramatic reductions in fertility in most of them. The marriage age for women is rising, which in itself tends to limit family size, and knowledge of family-planning methods has become widespread. In sum, birth rates are now falling significantly in most of the Third World countries, though Africa and the poorest Asian countries, Bangladesh and Pakistan, are notable exceptions. Over 60 countries, with 95 percent of the Third World's population, have adopted family-planning programs; but the need for material and educational assistance with these programs remains great. The United Nations Fund for Population Activities (UNFPA), for example, is only able to respond to two thirds of the requests it receives.

Family planning, however, cannot by itself solve the problems of the Third World. The acute poverty of those countries is not the result of rapid population growth in and of itself, but rather of rapid growth in a situation in which the economy is insufficiently developed to sustain so large a population. The economies of the Third World are *underdeveloped* and *lopsidedly* developed. What the Third World countries therefore most urgently need is sound economic development, that is, development that reflects their circumstances and meets their needs. We know from the experience of the advanced industrial societies that economic development itself provides the most favorable environment for stabilizing and reducing population growth—quite apart from the other advantages that economic development brings with it.

What does it mean to say that the Third World countries are afflicted with "underdeveloped" and "lopsidedly developed" economies? We have seen in our discussion of the first and second industrial revolutions that successful economic development required several preconditions. Britain, for example, the first industrializer, never could have laid the foundation for an industrial economy in the absence of (1) freedom from domination by foreign powers; (2) freedom of the industrializing

403

groups of the population from the constraints of the preindustrial order; and (3) an agricultural sector capable of producing food and raw-material surpluses for both the emerging industrial towns and for export. The same proposition applies no less forcefully to the participants in the Second Industrial Revolution. Japan, as we have seen in the previous chapter, was the only Asian country to have become a major industrial power in the early twentieth century. A necessary though not sufficient condition for her success was the avoidance of colonial domination. Japan enjoyed freedom from domination by foreign powers. The Meiji Restoration fulfilled the second precondition by dismantling the feudal obstacles to agricultural and industrial development. As for the United States, it fulfilled the first precondition with the War of Independence in 1776, but the second and third preconditions only after the Civil War. For there is good reason to suppose that the United States never would have become a major industrial power if the Southern secessionists had prevailed.

That there are, therefore, necessary preconditions for successful economic development appears to be a valid proposition. If we look at the Third World countries in light of this proposition, we can see that they have fulfilled the preconditions only partially or not at all. Many of the Third World countries have in fact achieved political independence but this has not always resulted in economic independence. Nor have many of the peoples of the Third World freed themselves from the extreme exploitation forced upon them by oppressive landed oligarchies. As a consequence, most of the Third World countries, far from possessing diversified agricultures and well-balanced economies, are highly dependent on one or a very few primary commodities which they produce for export.

Because Third World producers of agricultural and mineral commodities are so highly dependent upon the international market, they face price instability and, therefore, income instability. The greater the dependence, the more severe is the problem. Here are some examples of countries which in recent years have obtained almost all their export earnings from one commodity (leaving oil aside): Zambia, 94 percent from copper; Mauritius, 90 percent from sugar; Cuba, 84 percent from sugar; and Gambia, 85 percent from groundnuts and groundnut oil. The adverse consequences of high dependence on one commodity may be illustrated with the experience of Zambia in recent years.[27] World copper prices continued to rise from 1972, reaching a peak in April 1974 of $3,034 per ton; then the price plummeted to $1,290 before the end of that year. Prices of imports, however, continued to rise so that the volume Zambia could buy fell by 45 percent between 1974 and 1975, its Gross Domestic Product (GDP) falling by 15 percent. We may gain a better sense of the gravity of Zambia's situation by comparing it with the "oil shock" to the Western industrialized countries in 1974. The increase in their oil bill was equivalent to about 2.5 percent of their GDP. In numerical terms Zambia's shock was six times greater; but in human terms, for a country so poor to begin with, the shock was, of course, severer still.

Apart from price fluctuations, the developing countries are also concerned about price levels. Spokesmen for the single-commodity producers point to the long-term tendency for commodity prices to decline relative to the prices of the manufactured goods purchased from the industrialized countries. This tendency was quite evident form the mid-1950s to the early 1970s. The result was reduced export earnings and severely inhibiting effects on the efforts of the poor countries to import the capital goods necessary for their economic development.

The production and export of agricultural and mineral commodities could provide the basis for processing industries in the developing countries, which in turn would help advance and diversify their industrialization. But in this respect too, the poor countries are thwarted by the tariff and nontariff barriers imposed by the industrialized powers. Developing countries can export rice to the European Economic Community (EEC) free of duty, but they pay 13-percent tariff on processed rice and on rice products. The United States imposes a tariff of almost 15 percent on milled rice. Australia permits untreated wood to enter the country free but recently raised the duty on sawn timber from 7 to 14 percent. The EEC imposes a 4-percent tariff on crude palm oil but 12 percent on semirefined oil.

We see, then, that the poor countries are highly dependent for their economic development on international commodity prices. The industrialized countries, on the other hand, are highly dependent on commodity imports from the Third World and therefore have a strong interest in the security of supply. Apart from oil, these countries import coffee, tea, cocoa, natural rubber, jute, hard fibers, and a number of minerals such as nickel, copper, manganese, and tin. The dependence of the industrialized countries, West and East, on the importing of minerals and raw materials from the Third World will doubtless increase in the future. This means that in the long run the developed and the developing countries will have a substantial mutuality of interest. Let us take mining as an example. The poor countries cannot as a rule afford the expenditures required for mineral exploration. A major mine takes seven to

405

eight years from the time of discovery—and expenditures of a billion dollars—to the time when it is ready for production. Often such projects are not undertaken because the developing country believes it is not receiving fair remuneration for the exhaustion of a precious, finite resource. Historically, mineral exploitation in the developing countries has been dominated by international mining companies, which provided capital, technical knowledge, and marketing facilities while bearing the exploration risks and costs themselves. Most often there was a pronounced lack of advantage to the developing countries in the contracts negotiated with the companies. As a result, exploration and investment have markedly declined in recent years. The mining companies place the blame for this state of affairs on the Third World countries who have asserted sovereignty over their natural resources and are now reluctant to sign them away when they know too little about their extent and richness. The mining companies, on their part, are unwilling to invest in major ventures in the developing countries unless the terms are fully negotiated in advance. Clearly the consequences of this impasse can only redound to the disadvantage of both the industrialized and the developing countries. Both sides have an interest in continuing the exploration and exploitation of deposits in the developing countries; but the agreements between them will have to ensure to the host countries a full share of the benefits of mining, processing, and export.

Besides mining, other transnational or multinational corporations are closely involved with the Third World countries. The transnationals are major actors in the world's economy, controlling between a quarter and a third of total world production. In 1976 the total sales of their foreign affiliates were estimated at $830 billion, which is equal to the GNP of all the developing countries (excluding the oil-exporting ones). A small number of transnational corporations dominates the production, processing, or marketing of the following commodities: oil, bauxite, copper, iron ore, nickel, lead, zinc, tin, tobacco, bananas, and tea. The transnationals tend to invest in a small number of developing countries, mainly those offering favorable conditions such as political stability, tax incentives, large markets, cheap labor, and easy access to oil and other natural resources.

In recent years the transnationals have become a subject of controversy. Many developing countries have come to believe that the disadvantages and costs of foreign investment outweigh the benefits to them. Transnationals are criticized because much of the international trade they conduct goes on within their own organizations, between the parent firm and its subsidiaries and affiliates. According to some estimates, such "intrafirm trade" accounts for over 30 percent of all world trade. By this means corporations are able to shift profits from high- to low-tax countries and to get around price controls and customs duties. Transnationals have also been severely criticized for engaging in unethical political practices. It is almost certain, for instance, that the multinationals and the ubiquitous CIA were behind the coup d'etat that deposed the elected government of Allende in Chile. The major concern, then, is that the power of the multinationals is so vast that it can effectively evade the controls of nation-states and international organizations. The problems posed by these corporations are increasingly recognized by the industrialized countries as well since three quarters of foreign direct investment is made in those countries. But it is primarily the developing countries who, in an effort to obtain better terms, have begun to pursue more selective policies in admitting foreign investment. Here again, as in the relation of the mining companies to the host countries, we can see a potential mutuality of interest. Recent experience has shown that whenever the multinationals have made inequitable contracts with a host country, the contracts have run into trouble. Fair contracts are inherently more stable than unfair ones. It is therefore in the interest of the industrialized nations that the multinationals recognize the developing countries as sovereign states that have the right to exercise control over their own resources—and to derive maximal benefits from them.

Earlier we offered the proposition that one of the preconditions for industrialization is an agricultural sector capable of producing food and raw-material surpluses. Most countries of the Third World look upon industrialization as a central objective of their economic policy. Yet the sad truth is that 70 percent of the poor in the developing countries live in rural areas. Many of these countries have invested such disproportionate efforts in the development of the urban and industrial sectors that agriculture has been grievously neglected. In many cases this has caused food production to stagnate and fall. Countries that once were self-sufficient in food production now must use precious hard currency to import a large portion of their food supplies. It is now more and more recognized that a larger proportion of development funds must be directed to rural areas for investment in infrastructure, research, agricultural implements, seed, fertilizer, and pesticides. But experience has shown that unless such action is accompanied by *structural* changes, the impact on rural poverty will be negligible or even negative. The poorest countries are those with sharp disparities in land ownership: Landlords, a small minority of the population (5 to 10 percent) often own 60 or more percent of the usable land while the remainder of the population is crowded on small, fragmented plots or owns no land at all. In many countries of Asia, Latin America, and Central America much of the land is tilled on a tenant or sharecropping basis, with the landlord taking the lion's share of the total crop. Such agrarian structures are not only inefficient but unjust; they therefore generate social conflict and revolution. In recent years Americans have witnessed a cycle of such revolutions very close to home: in Nicaragua, El Salvador, and Guatemala. The former U.S. ambassador to the Dominican Republic (1962-1964), John Bartlow Martin, offers this explanation. The causes of the trouble, he writes,

> are simple. Indeed, they are one: injustice. In some places, four-hundred years of it. In some countries, out in the countryside, *campesinos* are poor even by Asian standards—they never see fifty dollars a year, yet their transistor radios give them a hint of what life could really be. For intellectuals, and for *campesinos* come to the city, injustice means military repression. But always the root cause is that too few people have too much land and money; too many have too little.
>
> Time and again we [the United States] have had a chance to join these revolutions. Time and again we have defended the indefensible status quo.[28]

As we reflect on the world's troubles, we can see that a substantial number of solutions depend on the recognition of the mutuality of interest between the developed and the developing countries. The industrialization of the developing countries, as it proceeds, will provide increasing opportunities for world trade, which need not conflict with the long-term interests of the developed nations. The fear still exists in the developed countries that imports from the Third World cause significant unemployment. However, a report of the EEC, summarizing numerous studies of the impact of Third World imports on North America, Western Europe, and Japan, has proved the fear to be groundless. It is with clothing and textiles that the Third World countries have most effectively penetrated the markets of the industrialized nations. In the mid-1970s their share of imports in the EEC, the United States, Canada, and Japan reached 7.2 percent. Yet the impact on unemployment was less than that of domestic technological change. The Third World needs fair contracts and good markets for both its raw materials and its manufactured products. It also needs advanced Western technology, which could provide the developed countries with big investment opportunities at a time when they are sorely needed. Yet the United States has not taken advantage of these opportunities. Notwithstanding the difficulties which the U.S. economy is presently experiencing, American companies are on the sidelines while the Japanese, in this respect as in others, lead the way.

Japan is now ahead of all other developed countries in transferring technology to the Third World. The major recipients of Japanese industrial technology are in Southeast Asia, the Pacific Basin, and Latin America, but the countries of Africa and the Middle East have also been increasing their trade with Japan. Forty-six percent of Japan's manufacturing exports now go to the developing countries as compared with 37 percent of U.S. manufacturing exports.[29] Between 1976 and 1978 over 50 percent of Japan's machinery exports went to the developing countries, compared to 30 percent for the United States. Japan's share of total exports from the developed countries to the Third World rose from 11 percent in 1962 to 25 percent in 1977.[30] In dollar volume Japan has surpassed the United States in exports to the developing countries of electrical power machinery, motor vehicle parts, metalworking machinery, and other types of sophisticated equipment. Japan is also ahead in the sale of entire plants; and for many Japanese firms the developing countries now constitute a larger market than does the United States.

The major reason for Japan's success in this area is the role of the Japanese government, which provides low-interest loans to the developing countries to finance large technological purchases and especially for the purchase of entire manufacturing plants. Sixty percent of Japan's Export-Import Bank loans are made in order to facilitate the sale of whole plants. The tax laws provide additional incentives for the transfer of technology and, finally, the government provides generous insurance against foreign losses. In contrast, U.S. government policy fails to encourage technological transfers to the Third World. As Robert B. Reich, the former director of policy planning for the Federal Trade Commission, explains:

US companies have lost ground to Japanese firms in the LDCs [less-developed countries] by being relatively unresponsive to their demands. US companies are often unwilling to give up control of significant parts of their businesses to the LDCs, to reinvest profits within them, to accept special local requirements, or to provide liberal financing. Moreover, US companies often fear that transfers of technology to LDCs, particularly in the form of whole plant sales, will create new competitors and threaten profitability.[31]

Instead of encouraging the transfer of productive technology to the developing countries, the U.S. government appears to be encouraging an arms race among them. For the United States is the largest supplier of armaments to Third World countries. It is true that the transfer of technology cannot solve the immediate problems of the developing countries, for whom foreign aid remains essential. But in the long run real economic growth in the developing countries will depend on their 409 productive relationships with the advanced industrial nations, just as the health of the industrialized economies will depend on the flow of credit (export financing) and technology to the Third World.

If there is a single formidable obstacle impeding the conquest of poverty in the Third World, it is the world's arm race. The world's annual military bill is approximately $450 billion. If only a portion of the unproductive arms spending were allocated for productive development in the Third World, the prospects would be a good deal more promising than they are today. Here are a few examples from the *North-South Report:*

1. The military expenditure of only half a day would suffice to finance the whole malaria eradication programme of the World Health Organization, and less would be needed to conquer river-blindness, which is still the scourge of millions.

2. A modern tank costs about one million dollars; that amount could improve storage facilities for 100,000 tons of rice and thus save 4,000 tons or more annually: one person can live on just over a pound of rice a day. The same sum of money could provide 1,000 classrooms for 30,000 children.

3. For the price of one jet fighter (20 million dollars) one could set up about 40,000 village pharmacies.

4. One-half of one percent of one year's world military expenditure would pay for all the farm equipment needed to increase food production and approach self-sufficiency in food-deficit low-income countries by 1990.[32]

And the authors of the report ask: "Could one be content to call something a 'new world economic order' if it did not include major progress towards disarmament?"[33] More and more people now recognize that there is only one way to answer this question if humanity is to survive.

ABOLISHING NUCLEAR WAR

All of humanity today lives in the shadow of the danger of a nuclear holocaust. In one of the Carter-Reagan television debates that took place prior to the presidential elections of 1980, Reagan was informed by the president that his daughter Amy's "greatest fear" was nuclear war. It appeared at the time that the nation laughed in derision at Carter's remark. Everywhere he went on the campaign trail he was greeted by signs sneering: "Amy for Defense Secretary." Today it is evident that the laughing has stopped and that many, many Americans are downright frightened. The nation appears to be shaking off the "psychic numbing" that has paralyzed it for so long. New peace groups have emerged, such as Physicians for Social Responsibilities and the Committee for National Security. The dangerous illusion of survivability in conditions of nuclear warfare is being shattered. It is evident in both Europe and the United States that the antinuclear movement has grown far beyond its traditional bases in religious pacifism and leftist politics.

In the United States there are unmistakable signs of growing opposition to nuclear war, not only at the grassroots level but in politically influential and elite circles as well. During most of the nation's history, criticism of American military policy has been rare. In the 1980s, however, the highest dignitaries of the Catholic Church, more than 40 bishops, are voicing opposition to the government's nuclear weapons policies. Forcefully stating their position, Archbishop John Quinn of San Francisco has proclaimed that "the teaching of the Church is clear: nuclear weapons and the arms race must be condemned as immoral."[34]

No less striking is the public stand taken by such eminent political personalities as George F. Kennan. Kennan is professor emeritus at the Institute for Advanced Study at Princeton. He was ambassador to the USSR in 1952 and to Yugoslavia from 1961 to 1963. But what is most interesting about George F. Kennan in connection with the subject at hand is the fact that he was one of the original architects of the Cold War. Kennan is Mr. X, the author in 1947 of a now-famous article in the semiofficial journal *Foreign Affairs*. In the 1947 article Kennan advocated a "patient but firm and vigilant containment of Russian expansive tendencies." This was to be accomplished by the "application of counterforce at a series of constantly shifting geographical and political points, corresponding to the shifts and manoeuvres of Soviet policy."[35] In 1947 the Soviet Union had not as yet exploded its first atom bomb, and Sputnik—demonstrating the possession of intercontinental ballistic thrust—lay almost a decade ahead. Hence Kennan could advise the United States to enter upon a policy of "firm containment, designed to confront the Russians with unalterable counterforce at every point where they show signs of encroaching upon the interests of a peaceful and stable world."[36]

In 1982, however, Kennan publicly condemned the fearful trap into which the accumulation of nuclear weapons is leading us. Can we escape from it? We can, says Kennan, but to do so we must begin by accepting the validity of two fundamental propositions:

411

The first is that there is no issue at stake in our political relation with the Soviet Union—no hope, no fear, nothing to which we aspire, nothing we would like to avoid—which could conceivably be worth a nuclear war. And the second is that there is no way in which nuclear weapons could conceivably be employed in combat that would not involve the possibility—and indeed the prohibitively high probability—of escalation into a general nuclear disaster.[37]

A start can be made in reversing the present suicidal course, says Kennan, by making

deep cuts in the long-range strategic missilery. There could be a complete denuclearization of Central and Northern Europe. There could be a complete ban on nuclear testing. At the very least, we could accept a temporary freeze on the further buildup of these fantastic arsenals. None of this would undermine anyone's security.

These alternatives, obviously, are not ones that we in the West could expect to realize all by ourselves. I am not suggesting any unilateral disarmament. Plainly, two—and eventually even more than two—will have to play at this game.

And even these alternatives would be only a beginning. But they would be a tremendous hopeful beginning. And what I am suggesting is that one should at least begin to explore them—and to explore them with a good will and a courage and an imagination the signs of which I fail, as yet, to detect on the part of those in Washington who have our destinies in their hands.[38]

The preoccupation with nuclear war, Kennan concludes, is morbid in the extreme. "There is no hope in it—only horror."[39]

How do we begin to rid ourselves of this horror? Let us consider this question by exploring the implications of the SS-20, the Soviet nuclear missile of intermediate range. The Soviets began to install these missiles in European Russia in 1977–1978, and their progressive deployment continued at least until March 1982. It was the deployment of the SS-20 which led to the controversial NATO decision to equip its nuclear forces with American Cruise missiles and Pershing-II ballistic missiles. Evidently the Soviet leaders had reached the decision in the early 1970s that in order to counteract the nuclear arsenals of West Germany and Britain plus the independent arsenal of France, they had to deploy some 300 SS-20s, as in fact they have. Presumably they now feel more secure than they did before; just as the West now feels insecure so long as the balance has not been repaired by means of the Pershing-II and Cruise missiles. The really important question is whether this security or insecurity perceived by both sides is real or imagined.

412   In response to the insecurity generated by the installation of the SS-

20s, the West now proposes that NATO must "match" them with medium-range missiles of our own (Pershing-II and Cruise) based in Europe. But Admiral Noel Gayler, former director of the National Security Agency and commander of all U.S. forces in the Pacific, has explained quite plainly why the NATO proposal is not as rational as it appears to be. Europe, he argues,

is under no special or unusual danger from SS-20s. There are plenty—very many more than plenty—of Russian intercontinental-range missiles, which can strike any target in Europe, simply by shortening the trajectory. It makes no difference to the target where a missile comes from—only where it lands.

European-based NATO missiles impart no special freedom of action or autonomy to the European allies. It is inconceivable that they could be used without American consent.[40]

A second proposal under consideration by NATO is the *neutron* bomb. It was at first supported by the Carter administration and then withdrawn; and though the Reagan administration has not openly promoted the proposal, it still has fervent and powerful supporters. The neutron bomb is a so-called tactical nuclear weapon designed for use by military forces. Their use, says Admiral Gayler, entails grave risks:

The enemy will certainly retaliate in kind, and he will up the ante. Rapid escalation to total nuclear war is a strong possibility, once the nuclear firebreak has been crossed.

Russian incursion will be fought on allied soil. Noncombatants will be killed in their hundreds and thousands, and these will be our friends and allies. Since our friends may well object to this outcome, the alliance may be fractured at the outset.

Even if we enjoy two miracles in a series—there is no escalation and the alliance holds together—we will be far worse off militarily after a tactical nuclear exchange than before. That is because we have the more critical and vulnerable targets: ports, airfields and lines of communication central to our defense.[41]

Are neutron bombs very different from standard nuclear weapons in practical effect? No, says Admiral Gayler, and they may be even more dangerous if they "lower the threshold to nuclear war. Are they really necessary to defeat tanks? No. There are better alternatives." Furthermore,

There is another difficulty of a different kind. Rightly, no President with all his marbles is likely to release nuclear weapons for use, except in the most extreme circumstances, if even then. The risks are simply too high to make the game worth the candle. The commander in the field therefore has an uncertain weapon on which he can never rely. Much better that he have effective means to victory that he can be confident will be available to him.[42]

Some officials continue to deny that nuclear war will constitute an unprecedented human catastrophe. But Admiral Gayler believes that

most of those who live near Washington will not survive a nuclear exchange. Nor will those who live in New York or Boston or Seattle or Chicago or any of our great cities. Nor those who live in Moscow of Kiev or Tbilisi or Vladivostok or remote Alma-Alta or anywhere else in either enormous country. . . .

Our basic need is truly to understand the nature of nuclear weapons. They are so enormous that, in a nuclear exchange, the first few weapons arriving do almost all the damage conceivable to the fabric of the country. So we have a military paradox: the power position of either side is not affected by the size of its nuclear forces.

In a similar way, technology has no great payoff. It makes no difference whether New York, say, is devastated by an old SS11 or a brand new SS19 MIRV warhead. In the absence of effective defenses, it makes no difference whether delivery is by a hypersonic maneuvering re-entry body, or a tired old Bear bomber. The results are the same.[43]

So what is to be done? Admiral Gayler fully supports George F. Kennan's proposals and goes beyond them: (1) immediate equal reduction of nuclear weapons of all categories, as proposed by George Kennan (this can become a progressive process); (2) renunciation of nuclear initiatives for any purpose whatever; (3) renunciation of any "counterforce" strategy against strategic nuclear forces; and (4) concurrent strong opposition to any form of nuclear weapons proliferation, by our clients or theirs. The first step must be our own reduction, or we have no credibility whatever. Are such agreements possible? Yes, because both the Soviet Union and the United States would gain by them.[44]

The most urgent task facing us all, therefore, is to ensure that such agreements are in fact reached. It is a difficult task, the realization of which will take years even under the best of circumstances. Nevertheless it must be immediately followed or preferably accompanied by negotiations to curb the arms race in conventional (nonnuclear) weapons, which account for 80 percent of the world's arms spending. All the wars since the Second World War have been fought with conventional weapons and have resulted in the death of ten million people. According to the Stockholm International Peace Research Institute, as of 1979, 70 percent of the arms imports to the Third World came from the United States ($5.8 billion) and the USSR ($4 billion), with France ($2 billion), the UK ($660 million), and Italy ($620 million) being the other leading suppliers.[45] Clearly, the major industrial powers, who are also the major arms-producing countries, have a special responsibility to reach arms-control agreements among themselves. Only then will we have taken an effective first step toward universal disarmament and the peaceful resolution of international conflicts.

Notes

CHAPTER TWO

1. Charles Darwin, *The Descent of Man* (London: John Murray, 1922), p. 193.

2. Ibid., pp. 192–93.

3. See M. F. Ashley Montagu, ed., *Culture and the Evolution of Man* New York: Oxford University Press, 1962). All the articles in this volume deal with the emergence of *Homo sapiens*.

4. *Descent of Man*, p. 96.

5. Karl Marx and Frederick Engels, *Selected Works*, 2 vols. (Moscow: Foreign Languages Publishing House, 1951), II, 74.

6. Ibid., p. 76.

7. Ibid., p. 77.

8. E. B. Tylor, *The Origins of Culture* (New York: Harper Torchbooks, 1958). First published in 1871.

9. This term was introduced by the renowned anthropologist Alfred L. Kroeber. See *The Nature of Culture* (Chicago: University of Chicago Press, 1952).

10. Cited by W. E. LeGros Clark in Morton H. Fried, ed., *Readings in Anthropology*, (New York: Thomas Y. Crowell, 1959), I, 45.

11. The present discussion is based on M. F. Ashley Montagu, ed., *Culture: Man's Adaptive Dimension* (New York: Oxford University Press, 1968).

12. Ibid., pp. 109–10.

13. Alfred L. Kroeber, *The Nature of Culture* (Chicago: University of Chicago Press, 1952), p. 27.

14. New York: Vintage Books, 1971.

CHAPTER THREE

1. See, for example, Konrad Lorenz, *On Aggression* (New York: Harcourt, Brace and World, 1966); Niko Tinbergen, *The Study of Instinct* (Oxford: The Clarendon Press, 1951); Desmond Morris, *The Naked Ape* (New York: McGraw-Hill, 1967); and by the same author, *The Human Zoo* (New York: McGraw-Hill, 1969).

2. See Robert Ardrey, *African Genesis* (New York: Atheneum, 1961) and *The Territorial Imperative* (New York: Atheneum, 1966); also Raymond Dart, *Adventures With the Missing Link* (New York: Harper & Row, 1959).

3. Thomas Hobbes, *Leviathan* (Middlesex, England: Penguin Books, 1968), p. 185 (first published in 1651); and Jean Jacques Rousseau, *The Social Contract and Discourses,* trans. by G. D. H. Cole (New York: Dutton, 1950), p. 222f.

4. V. C. Wynne-Edwards, "Ecology and the Evolution of Social Ethics," in J. W. S. Pringle, ed., *Biology and the Human Sciences* (Oxford: Clarendon Press, 1972), p. 59.

5. Adriaan Northland, "Aspects and Prospects of the Concept of Instinct," ibid., p. 207.

6. J. L. Cloudsley-Thompson, *Animal Conflict and Adaptation* (London: G. T. Foulis, 1965), p. 80.

7. Ueli Nagel and Hans Kummer, "Variations in Cercopithecoid Aggressive Behavior," in Ralph L. Halloway, ed., *Primate Aggression, Territoriality, and Xenophobia* (New York: Academic Press, 1974), pp. 159–84.

8. Charles Darwin, *On the Origin of Species by Means of Natural Selection* (London: John Murray, 1859), p. 62.

9. Charles Darwin, *The Descent of Man* (London: John Murray, 1871), p. 163.

10. Petr Kropotkin, *Mutual Aid: A Factor of Evolution* (Boston: Extending Horizons Books, n.d.), p. 6.

11. Ibid., pp. 64–65; italics in original.

12. Ardrey, *Territorial Imperative,* pp. 3, 29.

13. M. F. Ashley Montagu, *The Nature of Human Aggression* (New York: Oxford University Press, 1976), p. 237.

14. Ibid.

15. Ardrey, *Territorial Imperative,* p. 29.

16. See J. B. Watson, *Psychology from the Standpoint of a Behaviorist* (Philadelphia: Lippincott, 1919) and *Behaviorism* (New York: People's Institute, 1925).

17. B. F. Skinner, *Beyond Freedom and Dignity* (New York: Vintage Books, 1971), pp. 12–13.

18. Ibid., pp. 15, 25.

19. Ibid., p. 39.

20. Ibid., p. 302.

21. Ibid., p. 205.

22. See G. H. Mead, *Mind, Self and Society* (Chicago & London: University of Chicago Press, 1934).

23. This term is Harry Stack Sullivan's, not Mead's, though it is certainly implied in the latter's theory.

24. Mead, *Mind, Self and Society*, p. 174.

25. Ibid., p. 213.

26. Ibid., p. 356.

27. This case has been made by Dennis Wrong in a well-known article by that title. See the *American Sociological Review*, 26 (1963): 183–93.

28. Sigmund Freud, "Instincts and Their Vicissitudes," in James Strachey, ed. and trans., *Standard Edition of the Complete Psychoanalytical Works*, 24 vols. (London: Hogarth Press, 1953–1974), XIV, 118–19.

29. Ibid., pp. 121–22.

30. Ibid., pp. 177, 143.

31. Ibid., p. 147.

32. Sigmund Freud, *New Introductory Lectures on Psychoanalysis*, trans. by James Strachey (New York: W. W. Norton, 1933), p. 106.

33. In the present discussion, I rely on Herbert Marcuse's important study *Eros and Civilization* (New York: Vintage Books, 1961).

34. Ibid., p. 31.

35. Ibid., p. 32.

36. Ibid., p. 33.

CHAPTER FOUR

1. See G. H. Mead, *Mind, Self and Society* (Chicago and London: University of Chicago Press, 1934), p. 240f.

2. See Cooley's two major works, *Social Organization* and *Human Nature and the Social Order*, in one volume (Gencoe, Ill.: Free Press, 1956).

3. Erving Goffman, *The Presentation of Self in Everyday Life* (Garden City, N.Y.: Doubleday Anchor Books, 1959), p. 4. What follows is a concise summary of Goffman's perspective.

4. Ibid., p. 113. The quotation is from Simone de Beauvoir's *The Second Sex* (New York: Knopf, 1953), p. 543.

5. Goffman, *Presentation of Self*, p. 121.

6. See Erving Goffman, *Stigma* (Englewood Cliffs, N.J.: Prentice-Hall, 1963).

7. Ibid., p. 9.

8. Ibid., p. 127.

9. See Erving Goffman, *Strategic Interaction*, (Philadelphia: University of Pennsylvania Press, 1969), p. 9.

10. Ibid., p. 13.

11. Ibid., pp. 69, 79.

12. See *The Sociology of Georg Simmel*, trans. and ed. by Kurt H. Wolff (Glencoe, Ill.: Free Press, 1950), p. 122f.

13. Ibid., p. 135.

14. Ibid., p. 155.

15. Ibid., p. 163.

16. (New York: Harcourt, Brace & World, 1961.) See also the revised edition (New York: Harcourt, Brace, Jovanovich, 1974).

17. Ibid., 1st ed., p. 13.

18. Ibid., pp. 31–32.

19. Ibid., p. 56.

20. Ibid., p. 61.

21. Ibid.

22. See Homans's essay entitled "Fundamental Social Processes" in Neil Smelser, ed., *Sociology: An Introduction* (New York: John Wiley, 1967), p. 53.

23. Ibid., p. 55.

24. Ibid., p. 54.

25. Homans, *Social Behavior,* rev. ed., p. 76.

26. William F. Whyte, *Street Corner Society* (Chicago: University of Chicago Press, 1937); and Homans, ibid., pp. 264, 234.

27. Homans, Social Behavior, 1st ed., p. 235.

28. Ibid., pp. 235–36.

29. Ibid., p. 236.

30. Ibid., p. 236.

CHAPTER FIVE

1. Aristotle, *The Politics,* trans. by T. A. Sinclair (Middlesex, England: Penguin Books, 1962), pp. 191, 172.

2. Thomas Hobbes, *Leviathan* (Middlesex, England: Penguin Books, 1968), p. 150. First published in 1651.

3. Ibid., pp. 183–84.

4. Ibid., p. 186.

5. Ibid., pp. 223, 227.

6. Emile Durkheim, *The Division of Labor in Society,* trans. by George Simpson (New York: Free Press, 1964), p. 204. First published in 1893.

7. Ibid., p. 216.

8. Ibid., p. 56.

9. Ibid., p. 81.

10. Ibid., p. 173.

11. Ibid., p. 353.

12. Ibid., p. 374.

13. Ibid., p. 384.

14. Emile Durkheim, *Professional Ethics and Civic Morals,* trans. by Cornelia Brookfield (London: Routledge & Kegan Paul, 1957), p. 176.

15. Ibid., p. 203.

16. Ibid., p. 204.

17. Ibid., pp. 206–07.

18. Ibid., p. 207.

19. Durkheim, *Division of Labor,* pp. 382, 383, 386.

20. Durkheim, *Professional Ethics,* p. 213.

21. Marion J. Levy, Jr., *The Structure of Society* (Princeton: Princeton University Press, 1952), p. 113.

22. D. F. Aberle et al., "The Functional Prerequisites of a Society," *Ethics*, 60 (January 1950): 100–11.

23. Ibid.

24. F. M. Cancian, "Functional Analysis," in D. L. Sills, ed., *International Encyclopedia of the Social Sciences* (New York: Macmillan and the Free Press, 1968), VI, 29–43.

25. R. K. Merton, *Social Theory and Social Structure*, enlarged ed. (New York: Free Press, 1968), pp. 87–88; italics in original.

26. Ibid., p. 105.

27. Ibid., p. 106.

28. Karl Marx and Frederick Engels, *The German Ideology*, parts I & III (New York: International Publishers, 1960), p. 6.

29. Franz Mehring, *Karl Marx* (Ann Arbor: University of Michigan Press, 1962), p. 280.

30. Marx and Engels, *German Ideology*, pp. 6–7.

31. Karl Marx, *Economic and Philosophical Manuscripts of 1844* (Moscow: Foreign Languages Publishing House, 1961), p. 156.

32. Ibid., pp. 74–75.

33. Ibid., pp. 75–76.

34. Ibid., pp. 79, 111, 105.

35. Ibid., pp. 156, 141.

36. Marx and Engels, *German Ideology*, p. 19.

37. V. Gordon Childe, *What Happened in History* (Middlesex, England: Penguin Books, 1964), p. 30.

38. See Frederick Engels, *The Origin of the Family, Private Property and the State* (New York: International Publishers, 1942), p. 75f.

39. Lewis Henry Morgan, *Ancient Society* (Chicago: Kerr, 1877), pp. 85–86; cited by Engels, p. 79.

40. Engels, *Origin of the Family*, p. 155.

41. Marx, *Capital*, vol. 3, cited in Maurice Dobb, *Studies in the Development of Capitalism* (New York: International Publishers, 1947), p. 36.

42. Frederick Engels, *Anti-Duhring* (Moscow: Foreign Languages Publishing House, 1954), p. 377.

43. Karl Marx, *Capital*, 3 vols., trans. by Samuel Moore (Moscow: Foreign Languages Publishing House, 1954), I, 714.

44. Ibid., pp. 718–19; and More, cited by Marx in ibid., p. 720.

45. Ibid., p. 746.

46. Marx, *Economic and Philosophical Manuscripts*, trans. by T. B. Bottomore, in Erich Fromm, *Marx's Concept of Man* (New York: Frederick Ungar, 1961), p. 95.

47. Fromm, in ibid., p. 44.

48. Marx, *Capital*, I, 361.

49. Ibid., p. 422.

50. Ibid., p. 503.

51. Max Weber, *General Economic History*, trans. by Frank H. Knight (New York: Collier Books, 1961), pp. 208–09.

52. The present discussion is based on Weber's analysis in *Economy and Society,* 3 vols., ed. by Guenther Roth and Claus Wittich (New York: Bedminster Press, 1968), III, 956f.

53. Ibid., p. 973.

54. Ibid., p. 981.

55. Ibid., p. 983.

56. A corporate executive, retiring after 40 years in upper management, recently remarked: "I never gave an order in my life; I merely expressed my preferences."

57. Weber, *Economy and Society,* III, 989.

58. Ibid., p. 1049.

59. Alexis de Tocqueville, *Oeuvres Complètes,* vol. 2, part I, *L'Ancien Regime et la Revolution* (Paris: Gallimard, 1952), p. 109; the translation here is mine.

60. Ibid., pp. 130–31.

61. Weber, *Economy and Society,* III, 1149.

62. Cited in J. P. Mayer, *Max Weber and German Politics* (London: Faber and Faber, 1944), p. 127.

CHAPTER SIX

1. See Marx's letter to J. Weydemeyer, March 5, 1852, in *Selected Correspondence* (Moscow: Foreign Languages Publishing House, n.d.), p. 86.

2. David Ricardo, *Principles of Political Economy and Taxation* (London: Everyman's Library, 1965), p. 267.

3. Marx began his discussion of "classes" in the final chapter of the third volume of *Capital.* The manuscript breaks off, however, before the discussion is completed.

4. From Marx's article in the *New York Daily Tribune,* August 21, 1852.

5. Karl Marx and Frederick Engels, *Selected Works,* 2 vols. (Moscow: Foreign Languages Publishing House, 1950), I, 128–29.

6. Ibid., p. 303.

7. Karl Marx, *The Communist Manifesto,* in *Selected Works,* I, 39–40.

8. Max Weber, *General Economic History,* trans. by Frank H. Knight (New York: Collier Books, 1961), p. 208–09.

9. Max Weber, *Economy and Society,* 3 vols., ed. by Guenther Roth and Claus Wittich (New York: Bedminster Press, 1968), I, 305.

10. Ibid., II, 927.

11. Ibid.

12. Max Weber, "Politics as a Vocation," in H. Gerth and C. W. Mills, eds., *From Max Weber: Essays in Sociology* (New York: Oxford University Press, 1958), p. 80.

13. Fritz Machlup, *The Political Economy of Monopoly* (Baltimore: Johns Hopkins Press, 1952), p. 123.

14. Alfred S. Eichner, "Business Concentration and Its Significance," in Ivar Berg, ed., *The Business of America* (New York: Harcourt, Brace & World, 1968), p. 192.

15. G. D. H. Cole, *Studies in Class Structure* (London: Routledge and Kegan Paul, 1955), pp. 98–99.

16. Maurice Zeitlin, "Corporate Ownership and Control: The Large Corporation and the Capitalist Class," *American Journal of Sociology,* 79 (1974): 1073–1119. The quoted statement may be found on p. 1077.

17. Ibid., pp. 1081–82.

8. Ferdinand Lundberg, *America's Sixty Families* (New York: Citadel, 1946), pp. 506–08; cited by M. Zeitlin in "Corporate Ownership."

19. Zeitlin, "Corporate Ownership," p. 1085.

20. Sargent P. Florence, *Ownership, Control and Success of Large Companies* (London, 1961).

21. C. Wright Mills, *White Collar* (New York: Oxford University Press, 1956), p. 13.

22. *Final Report of the Select Committee on Small Business* to the House of Representatives, 87th Congress, House Report No. 2569, January 3, 1963, pp. 22–23.

23. See Anthony Giddens, *The Class Structure of the Advanced Societies* (London: Hutchinson University Library, 1973), p. 178.

24. Ibid., pp. 180–81.

25. David Lockwood, *The Blackcoated Worker* (London: George Allen & Unwin, 1958), p. 204f.

26. See Lewis Corey, *The Crisis of the Middle Class* (New York: Crown, 1935), p. 142.

27. See Solomon Barkin, "The Decline of the Labor Movement," in Andrew Hacker, ed., *The Corporation Take-Over* (New York: Harper & Row, 1964), p. 233f.

28. Ibid., p. 237.

29. These data are taken from *The Statistical Abstract of the United States, 1972* (New York: Grosset & Dunlap, 1972), p. 210f.; and from U.S. Bureau of the Census, *Current Population Reports,* Series p-23, No. 37 (Washington, D.C.: U.S. Government Printing Office, 1971), pp. 60–62, table 14.

30. Ferdynand Zweig, *The Worker in an Affluent Society* (London: Heinemann, 1961), p. lx.

31. John Goldthorpe et al., *The Affluent Worker in the Class Structure* (Cambridge: Cambridge University Press, 1969).

32. Serge Mallet, *La Nouvelle Classe Ouvriere* (Paris, 1963), cited in T. B. Bottomore, *Classes in Modern Society* (New York: Pantheon Books, 1966), pp. 31–32.

33. See Elton F. Jackson and Harry J. Crockett, Jr., "Occupational Mobility in the United States: A Point Estimate and Trend Comparison," in Jack L. Roach et al., eds., *Social Stratification in the United States* (Englewood Cliffs, N.J.: Prentice-Hall, 1969), p. 514f.

34. Giddens, *Class Structure,* p. 198.

35. *Fortune* (February 1940): 14, 20.

36. Neal Gross, "Social Class Identification in the Urban Community," *American Sociological Review,* 18 (1953): 401, table 2.

37. Richard Centers, *The Psychology of Social Classes* (Princeton: Princeton University Press, 1949), p. 85, table 20.

38. Solomon Barkin in Hacker, *Corporation Take-Over,* p. 225.

CHAPTER SEVEN

1. See Barrington Moore, Jr., *Social Origins of Dictatorship and Democracy* (Boston: Beacon Press, 1966), chapter 1.

2. "Some Principles of Stratification," 9 (1945): 242–49.

3. Ibid., p. 243.

4. Ibid.

5. Ibid., p. 244.

6. Ibid., p. 247.

7. Ibid.

8. Kingsley Davis, *Human Society* (New York: Macmillan, 1945), pp. 369–70.

9. *American Sociological Review,* 18 (1953): 387–94.

10. Ibid., p. 388.

11. Ibid., p. 389.

12. Ibid., p. 390.

13. Kingsley Davis, "Reply," *American Sociological Review,* 23 (1953): 395.

14. *American Sociological Review,* 23 (1953): 672.

15. Davis and Moore, "Some Principles of Stratification," p. 243.

16. Ralf Dahrendorf, *Class and Class Conflict in Industrial Society* (London: Routledge & Kegan Paul, 1963), p. 184.

17. Ibid., p. 213.

18. Ibid., pp. 247, 254.

19. Ibid., pp. 268, 276.

20. Ibid., p. 214.

21. T. B. Bottomore, *Classes in Modern Society* (New York: Pantheon Books, 1966), pp. 27–28.

22. Dahrendorf, *Class and Class Conflict,* 270–71.

23. Ibid., p. 276.

24. Ibid., p. 290.

25. Vilfredo Pareto, *The Mind and Society,* 4 vols., trans. by Arthur Livingstone (New York: Harcourt, Brace & World, 1935), IV, par. 2032.

26. Ibid., pars. 2054, 2055.

27. Ibid., par. 2179.

28. Ibid.

29. Gaetano Mosca, *The Ruling Class,* trans. by Hannah D. Kahn (New York: McGraw-Hill, 1965), p. 53.

30. Ibid., p. 30.

31. Ibid., pp. 55, 57.

32. Ibid., p. 61.

33. Ibid., p. 62.

34. Ibid., pp. 134, 139.

35. Cf. James H. Meisel, *The Myth of the Ruling Class* (Ann Arbor: University of Michigan Press, 1962).

CHAPTER EIGHT

1. See W. Boyd, *Genetics and the Races of Man* (Boston: Little, Brown, 1950); and his "The Contributions of Genetics to Anthropology," in A. Kroeber, *Anthropology Today* (Chicago: University of Chicago Press, 1953), pp. 488–506.

2. C. S. Coon, S. M. Garn, and J. B. Birdsell, *Races* (Springfield, Mass.: Thomas, 1950).

3. Theodosius Dobzhansky, *Mankind Evolving* (New Haven and London: Yale University Press, 1962), chs. 7, 8.

4. Ashley Montagu, *Man's Most Dangerous Myth*, 5th ed. (New York: Oxford University Press, 1974), p. 86f.

5. Dobzhansky, *Mankind Evolving*, p. 270.

6. Ibid., p. 271.

7. Cited in Montagu, *Man's Most Dangerous Myth*, p. 128.

8. Ibid., p. 186.

9. Ibid., pp. 186–87.

10. Andrew M. Greeley, *Ethnicity in the United States* (New York: John Wiley, 1974), p. 35.

11. This historical sketch is based on the following sources: Maldwyn Allen Jones, *American Immigration* (Chicago: University of Chicago Press, 1960); Carl Wittke, *We Who Built America*, 2nd ed. (Cleveland: The Press of Western Reserve University, 1964); Marcus Lee Hansen, *The Atlantic Migration, 1607–1860* (New York: Harper Torchbooks, 1961); and Oscar Handlin, *The Uprooted*, 2nd ed. (Boston: Little, Brown, 1973).

12. See Alexis de Tocqueville, *Oeuvres Complètes*, vol. 5 (Paris: Gallimard, 1958), p. 97; the translation here is mine.

13. Greeley, *Ethnicity in the United States*, p. 44.

14. Cited in Lerone Bennett, Jr., *Before the Mayflower*, rev. ed. (Baltimore: Penguin Books, 1968), p. 40. The present discussion is based on Bennett's work and on John Hope Franklin's *From Slavery to Freedom*, 3rd ed. (New York: Vintage Books, 1966).

15. Kenneth Stampp, *The Peculiar Institution* (New York: Vintage Books, 1956), pp. 21–22.

16. Quoted in Bennett, *Before the Mayflower*, p. 223.

17. Ibid., p. 233.

18. William J. Wilson, "The Changing Context of American Race Relations: Urban Blacks and Structural Shifts in the Economy," in Willem A. Veenhoven, ed., *Case Studies in Human Rights and Fundamental Freedoms* (The Hague: Martinus Nijhoff, 1976), V, 178. In the present analysis I rely heavily on Wilson's excellent summary of the main trends.

425

Notes

19. Ibid., p. 181.

20. Ibid., p. 184.

21. Ibid.

22. Ibid., p. 185.

23. Stanley L. Friedlander, *Unemployment in the Urban Core* (New York: Praeger, 1972); and Alice Handsaker Kidder, "Racial Difference in Job Search and Wages," *Monthly Labor Review* (July 1968): 24–36.

24. Wilson, "Changing Context of American Race Relations," p. 191.

25. See C. Wright Mills, Clarence Senior, and Rose Kohn Goldsen, *The Puerto Rican Journey* (New York: Harper & Row, 1950).

26. Nathan Glazer and Daniel Patrick Moynihan, *Beyond the Melting Pot*, 2nd ed. (Cambridge, Mass.: M.I.T. Press, 1970), p. 87. See Glazer's essay "The Puerto Ricans," pp. 86–142; see also Leonard Dinnerstein and David M. Reimers, *Ethnic Americans: A History of Immigration and Assimilation* (New York: Dodd, Mead, 1975).

27. See Sar A. Levitan et al., *Minorities in the United States: Problems, Progress, Prospects* (Washington, D.C.: Public Affairs Press, 1975), p. 73f.

28. Glazer, "Puerto Ricans," p. 112f.

29. See Rudolf O. de la Garza, "Mexican Americans in the United States: The Evolution of a Relationship," in Veenhoven, ed., *Case Studies,* V, 259–90.

30. For these and other historical details, see Wayne Moquin and Charles Van Doren, eds., *A Documentary History of the Mexican Americans* (New York: Praeger, 1971); Matt S. Meier and Feliciano Rivera, *The Chicanos: A History of Mexican Americans* (New York: Hill and Wang, 1972); and Carey McWilliams, *North From Mexico* (New York: Greenwood Press, 1968).

31. Garza, "Mexican Americans," p. 265.

32. Ibid., p. 266.

33. Dinnerstein and Reimers, *Ethnic Americans,* p. 105.

34. Cited in ibid., p. 106.

35. Levitan et al., *Minorities in the United States,* p. 52f.

36. Ibid., p. 54.

37. Ibid., p. 69.

38. Garza, "Mexican Americans, p. 287.

39. See Paul S. Martin, George I. Quimby, and Donald Collier, *Indians Before Columbus* (Chicago: University of Chicago Press, 1955), p. 16f.

40. Alfred L. Kroeber, "Demography of the American Indians," *American Anthropologist,* 36 (1934): 1–25; and Clark Wissler, *Indians of the United States* (New York: Doubleday Anchor Books, 1967 rev. ed. by Lucy W. Kluckhohn).

41. See David Bushnell, "The Treatment of the Indians in Plymouth Colony," *The New England Quarterly,* 26 (1953): 193–218.

42. Peter Farb, *Man's Rise to Civilization* (New York: Dutton, 1968), p. 75.

43. Cited by John R. Howard, ed., *Awakening Minorities* (Chicago: Aldine, Transaction Books, 1970), p. 21.

46. Levitan et al., *Minorities in the United States*, p. 84.

45. Ibid., p. 86.

CHAPTER NINE

1. George Peter Murdock, *Social Structure* (New York: Macmillan, 1949), p. 2.

2. See E. Kathleen Gough, "The Nayars and the Definition of Marriage," *Journal of the Royal Anthropological Institute*, 85 (1955): 45–80. See also Joan P. Mencher, "The Nayars of South Malabar," in Meyer F. Nimkoff, ed., *Comparative Family Systems* (Boston: Houghton Mifflin, 1965), pp. 163–91.

3. Melford E. Spiro, "Is the Family Universal?" *American Anthropologist*, 56 (1954): 839–46; and by the same author, *Kibbutz, Venture in Utopia* (Cambridge, Mass.: Harvard University Press, 1956) and *Children of the Kibbutz* (Cambridge, Mass.: Harvard University Press, 1958).

4. Robert H. Lowie, *Social Organization* (New York: Rinehart, 1956), p. 114. The present discussion is based on Lowie's summary of the available evidence.

5. Ibid., p. 117.

6. Peter Laslett, *The World We Have Lost* (New York: Charles Scribner's Sons, 1965). See too, by the same author, "Mean Household Size Since the Sixteenth Century," in Peter Laslett and Richard Wall, eds., *Household and Family in Past Time* (Cambridge, England: Cambridge University Press, 1972), pp. 125–58.

7. See Robert V. Wells, "Quaker Marriage Patterns in Colonial Perspective," *William and Mary Quarterly*, 29 (1972): 413–42; and his "Household Size and Composition in the British Colonies in America, 1675–1775," *Journal of Interdisciplinary History*, 4 (1974): 543–70.

8. John Demos, *A Little Commonwealth: Family Life in Plymouth Colony* (New York: Oxford University Press, 1970); and his "Demography and Psychology in the Historical Study of Family-Life: A Personal Report," in Laslett and Wall, *Household and Family in Past Time*, pp. 561–70.

9. Stuart M. Blumin, "Rip Van Winkle's Grandchildren: Family and Household in the Hudson Valley, 1800–1860," *Journal of Urban History*, 1 (1975): 293–315; and Edward T. Pryor, Jr., "Rhode Island Family Structure: 1875 and 1960," in Laslett and Wall, *Household and Family in Past Time*, pp. 571–89.

10. Jessie Bernard, *The Future of Marriage* (New York: World, 1972), p. 16.

11. Ibid., p. 21.

12. Ibid., p. 18.

13. Joseph Veroff and Sheila Feld, *Marriage and Work in America* (New York: Van Nostrand-Reinhold, 1970), p. 70.

14. Constantine Safilios-Rothschild, "Sex Discrimination: Theory and Research," in Willem A. Veenhoven, ed., *Case Studies on Human Rights and Fundamental Freedoms* (The Hague: Martinus Nijhoff, 1976), V, 164–65.

15. Susan Brownmiller, "Sisterhood Is Powerful," *New York Times Magazine*, March 15, 1970, p. 140.

16. Bernard, *Future of Marriage*, pp. 129, 130.

17. See U.S. Bureau of the Census, *Household and Family Characteristics*, Current Population Reports Series P-20, No. 366 (Washington, D.C.: U.S. Government Printing Office, September 1981); and U.S. Bureau of the Census, *Marital Status and Living Arrangements*, Current Population Reports Series P-20, No. 365 (Washington, D.C.: U.S. Government Printing Office, October 1981).

18. See Paul Glick and Arthur J. Norton, "Perspectives on the Recent Upturn in Divorce and Remarriage," *Demography*, 10 (1974): 301–14. See also Norton and Glick, "Marital Instability: Past, Present and Future," *Journal of Social Issues*, 32 (1976): 5–20.

19. Mary Jo Bane, *Here to Stay: American Families in the Twentieth Century* (New York: Basic Books, 1976), p. 32. In the present discussion, I rely on this author's persuasive analysis.

20. Ibid.

21. Ibid., p. 33.

22. Duane Denfeld and Michael Gordon, "The Sociology of Mate Swapping," in James R. Smith and Lynn G. Smith, eds., *Beyond Monogamy: Recent Studies of Sexual Alternatives in Marriage* (Baltimore and London: Johns Hopkins University Press, 1974), p. 75.

23. See the several articles collected in Smith and Smith, *Beyond Monogamy*.

24. Duane Denfeld, "Dropouts from Swinging: The Marriage Counselor as Informant," in ibid., pp. 260–67.

25. Ibid., p. 267.

CHAPTER TEN

1. William James, *The Varieties of Religious Experience* (New York: Collier Books, 1976), p. 116. First published in 1902.

2. Ibid., pp. 123, 139.

2. Ibid., pp. 123, 139.

3. Ibid., p. 397.

4. E. B. Tylor, *Religion in Primitive Culture* (New York: Harper Torchbooks, 1958), p. 8. First published in 1871.

5. Emile Durkheim, *The Elementary Forms of Religious Life*, trans. and ed. by J. W. Swain (London: George Allen & Unwin, 1964), p. 47.

6. Tylor, *Religion in Primitive Culture*, pp. 12–13.

7. Durkheim, *Elementary Forms*, p. 61.

8. Ibid., p. 133; italics in original.

9. Ibid., p. 189.

10. Ibid., p. 201f.

11. Ibid., p. 206.

12. Ibid., pp. 218–19, 221.

13. Ibid., pp. 226, 229.

14. Ibid., p. 421.

15. Bronislaw Malinowski, *Magic, Science and Religion* (Garden City, N.Y.: Doubleday Anchor Books, 1954), p. 58.

16. Ibid.

17. Ibid., p. 69.

18. James, *Varieties of Religious Experience*, p. 42.

19. Durkheim, *Elementary Forms*, pp. 225, 427.

20. Malinowski, *Magic, Science and Religion*, p. 28.

21. Ibid., pp. 28–29.

22. Sigmund Freud, *Totem and Taboo*, in A. A. Brill, ed., *The Basic Writings of Sigmund Freud* (New York: Modern Library, 1938), p. 915.

23. Sigmund Freud, *Moses and Monotheism*, trans. by Katherine Jones (New York: Alfred A. Knopf, 1949), p. 128.

24. Freud, *Totem and Taboo*, p. 917.

25. Ibid., p. 918.

26. Ibid., p. 919.

27. Ibid., p. 925, note 1.

28. Ibid., p. 925.

29. Ibid., p. 929.

30. Sigmund Freud, *The Future of an Illusion*, trans. by W. D. Robson-Scott (Garden City, N.Y.: Anchor Books, 1964), p. 47.

31. Ibid., pp. 67–68.

32. Durkheim, *Elementary Forms*, p. 427.

33. Ibid., p. 429.

34. Karl Marx, *Early Writings*, trans. and ed. by T. B. Bottomore (London: Watts, 1963), p. 44.

35. Ibid.

36. Durkheim, *Elementary Forms*, p. 52.

37. Gregory Baum, *Religion and Alienation* (New York: Paulist Press, 1975), p. 37.

38. Max Weber, *The Protestant Ethic and the Spirit of Capitalism*, trans. by Talcott Parsons (New York: Charles Scribner's Sons, 1958), p. 40.

39. Ibid., p. 41

40. Ibid., p. 110.

41. Ibid., pp. 110–11.

42. Ibid., p. 151.

43. Ibid., pp. 166, 172.

44. This subject is discussed in some detail in chapter 13 on urbanization.

45. Max Weber, *The Religion of China*, trans. and ed. by H. H. Gerth (Glencoe, Ill.: Free Press, 1962).

46. Max Weber, *The Religion of India*, trans. and ed. by H. H. Gerth and D. Martindale (Glencoe, Ill.: Free Press, 1962), p. 331.

47. Max Weber, *Ancient Judaism*, trans. and ed. by H. H. Gerth and D. Martindale (Glencoe, Ill.: Free Press, 1952), p. 4.

48. Ibid., pp. 119, 126.

49. Ibid., p. 167.

50. Freud, *Future of an Illusion,* pp. 73, 52, 92.

51. James, *Varieties of Religious Experience,* p. 393.

52. Durkheim, *Elementary Forms,* p. 427.

53. In a letter to Emmy Baumgarten, cited in J. P. Mayer, *Max Weber and German Politics* (London: Faber and Faber, 1956), p. 35.

54. Peter L. Berger, *The Sacred Canopy* (New York: Doubleday, 1967), pp. 107–8.

55. Brian Wilson, *Religion in Secular Society* (Middlesex, England: Pelican Books, 1969).

56. Baum, *Religion and Alienation,* p. 142.

57. Berger, *Sacred Canopy,* p. 108; Wilson, *Religion in Secular Society,* p. 122; and Baum, *Religion and Alienation,* p. 143.

58. Andrew Greeley, *The Denomination Society* (Glenview, Ill.: Scott, Foresman, 1972), p, 2.

59. James, *Varieties of Religious Experience,* p. 42; italics in original.

60. Robert N. Bellah, *Beyond Belief* (New York: Harper & Row, 1970), p. 170.

61. Ibid., p. 171.

62. Ibid., p. 168.

CHAPTER ELEVEN

1. These elements are implicit in the writings of several classical theorists. However, they have been expressly built into a framework for the analysis of social movements by Theodore Abel in *The Nazi Movement: Why Hitler Came to Power* (New York: Atherton Press, 1965), first published in 1938. I rely heavily on Abel's outstanding study in the following discussion of Nazism, and I also adapt his framework to the other movements considered in this chapter.

2. For historical details I have relied on the following sources: Theodore Abel, ibid.; William L. Shirer, *The Rise and Fall of the Third Reich* (New York: Simon and Schuster, 1960); Franz Newmann, *Behemoth* (New York: Harper Torchbooks, 1963); Richard Grunberger, *A Social History of the Third Reich* (London: Weidenfeld and Nicholson, 1971); Hermann Rauschning, *The Beast from the Abyss* (London: Heinemann, 1941); Roy Pascal, *The Nazi Dictatorship* (London: Routledge & Sons, 1934); William Ebenstein, *The Nazi State* (New York: Octagon Books, 1975); and Albert Speer, *Inside the Third Reich* (New York: Macmillan, 1970).

3. Shirer, *Third Reich,* p. 57.

4. Abel, *Nazi Movement,* p. 30.

5. Adolf Hitler, *Mein Kampf* (New York: Reynal and Hitchcock, 1940), pp. 242–44; cited in Abel, ibid.

6. Abel, ibid., p. 70.

7. Shirer, *Third Reich,* p. 118.

8. Ibid., pp. 158–59.

9. Quoted in Shirer, ibid., p. 227.

10. Abel, *Nazi Movement*, p. 145.

11. Max Weber, *Economy and Society*, 3 vols., ed. by Guenther Roth and Claus Wittich (New York: Bedminster Press, 1968), I, 241.

12. Ping-ti Ho, *Studies on Population of China* (Cambridge, Mass.: Harvard University Press, 1959), p. 298.

13. See Barrington Moore, Jr., *Social Origins of Dictatorship and Democracy* (Boston: Beacon Press, 1966), chapter 4.

14. Franklin W. Houn, *A Short History of Chinese Communism* (Englewood Cliffs, N.J.: Prentice-Hall, 1973), pp. 6–7.

15. Ibid., p. 8.

16. The present discussion of the rise of Chinese communism has been distilled from the following sources: Jerome Ch'en, *Mao and the Chinese Revolution* (London: Oxford University Press, 1965); Edward E. Rice, *Mao's Way* (Berkeley: University of California Press, 1972); and Houn, *Short History of Chinese Communism*.

17. Cited in Ch'en, *Mao and the Chinese Revolution*, pp. 199–200.

18. Edgar Snow, *The Other Side of the River* (New York: Random House, 1961), p. 76.

19. Mao Tse-tung, *Selected Works*, 4 vols. (Peking: Foreign Languages Press, 1961), IX, 35. Most scholars would agree that these figures are not exaggerated.

20. For historical details I have relied on the following sources: John Hope Franklin, *From Slavery to Freedom* (New York: Vintage Books, 1969); Lerone Bennett, Jr., *Before the Mayflower: A History of the Negro in America 1619–1964* rev. ed. (Baltimore: Penguin Books, 1968); Lewis M. Killian, *The Impossible Revolution?* (New York: Random House, 1968); and on personal notes taken while the events were current.

21. Cited in Bennett, *Before the Mayflower*, p. 312.

22. See Hanes Walton, Jr., *The Political Philosophy of Martin Luther King, Jr.* (Westport, Conn.: Greenwood, 1971); and Benjamin Muse, *The American Negro Revolution* (Bloomington: Indiana University Press, 1968).

23. Franklin, *Slavery to Freedom*, p. 623f.

24. Muse, *American Negro Revolution*, p. 206f.

25. Ibid., p. 230.

26. Cited in ibid., p. 236.

27. See Stokely Carmichael and Charles V. Hamilton, *Black Power* (New York: Vintage Books, 1967).

28. Ibid., p. 173.

29. Ibid., p. 295.

30. (New York: Bantam Books, 1968), p. 116f.

CHAPTER TWELVE

1. Emile Durkheim, *The Rules of the Sociological Method*, trans. by Sarah A. Solovay and John H. Mueller and ed. by George E. G. Catlin (New York: Free Press, 1964), pp. 55, 66.

2. Ibid., p. 69.

3. Ibid., pp. 71–72.

4. Emile Durkheim, *The Division of Labor in Society,* trans. by George Simpson (New York: Free Press, 1965), p. 81.

5. Ibid., p. 88.

6. Ibid., p. 102.

7. Ibid., p. 108.

8. For details about Lombroso and other key figures in the history of criminology, see Bernaldo de Quiros, *Modern Theories of Criminality* (Boston: Little, Brown, 1911); and Hermann Mannheim, ed., *Pioneers of Criminology* (Chicago: Quadrangle, 1960).

9. See, for example, Ernest A. Hooten, *The American Criminal: An Anthropological Study* (Cambridge, Mass.: Harvard University Press, 1939); and William H. Sheldon, *Varieties of Delinquent Youth* (New York: Harper & Bros., 1949). For a detailed critique of the methods employed in those studies, see George B. Vold, *Theoretical Criminology* (New York: Oxford University Press, 1958).

10. The following propositions are quoted from Sutherland's *Principles of Criminology,* 4th ed. (Philadelphia: Lippincott, 1947), pp. 5–9.

11. Karl Schuessler, ed., *Edwin H. Sutherland: On Analyzing Crime* (Chicago: University of Chicago Press, 1973), pp. 19–20.

12. Daniel Glazer, "Criminality Theories and Behavioral Images," in Donald R. Cressey and David A. Ward, eds., *Delinquency, Crime and Social Process* (New York: Harper & Row, 1969), p. 521. Glazer's article was originally published in the *American Journal of Sociology,* 61 (1956): 433–44.

13. Robert K. Merton, *Social Theory and Social Structure,* enlarged ed. (New York: Free Press, 1968), p. 186; italics in original.

14. Albert K. Cohen, *Deviance and Control* (Englewood Cliffs, N.J.: Prentice-Hall, 1966), p. 107.

15. Ibid.

16. This is the title of a well-known article by Miller published in the *Journal of Social Issues,* 14 (1958): 5–19. It was reprinted in Cressey and Ward, *Delinquency, Crime and Social Process,* p. 332f.

17. Ibid., p. 346.

18. See Gresham M. Sykes and David Matza, "Techniques of Neutralization: A Theory of Delinquency," *American Sociological Review,* 22 (1957): 664–70.

19. See Frank E. Hartung, "A Vocabulary of Motives for Law Violations," in Cressey and Ward, *Delinquency, Crime and Social Process,* pp. 456–57.

20. (Washington: Government Printing Office, 1971), chapter 2, "Crime in America."

21. U.S. Department of Justice, FBI, 1976.

22. U.S. Department of Justice, LEAA, 1976.

23. See Harry E. Barnes and Negley K. Teeters, *New Horizons in Criminology,* 2nd ed. (Englewood Cliffs, N.J.: Prentice-Hall, 1952); Donald L.

Halstead, *The Relationship of Selected Characteristics of Juveniles to Definitions of Delinquency* Ed.D. diss. University of Michigan, 1967); and Irving Piliavin and Scott Briar, "Police Encounters with Juveniles," *The American Journal of Sociology,* 70 (1964): 206–14.

24. David H. Bailey and Harold Mendelsohn, *Minorities and the Police* (New York: Free Press, 1969).

25. Stuart Nagel, "The Tipped Scales of Justice," *Transaction* (May/June 1966): 3–9.

26. See Eleanor Saunders Wyrick and Otis Holloway Owens, "Black Women: Income and Incarceration," in Charles E. Owens and Jimmy Bell, eds., *Blacks and Criminal Justice* (Lexington, Mass.: Lexington Books, 1977), pp. 85–92.

27. Ibid., p. 87.

28. In the present discussion, I rely on Lynn A. Curtis, *Violence, Race and Culture* (Lexington, Mass.: Lexington Books, 1975).

29. Cited in ibid., p. 50.

30. Lee Rainwater, *Behind Ghetto Walls: Black Families in a Federal Slum* (Chicago: Aldine, 1970), pp. 287–88.

31. Ibid.

32. Curtis, *Violence, Race and Culture,* p. 55.

33. Ulf Hannerz, *Soulside: Inquiries into Ghetto Culture and Community* (New York: Columbia University Press, 1969), p. 88.

34. Curtis, *Violence, Race and Culture,* p. 57.

35. Francis A. J. Ianni, *Black Mafia: Ethnic Succession in Organized Crime* (New York: Simon and Schuster, 1974), p. 282.

36. Ibid., pp. 283, 285.

37. Edwin H. Sutherland, *White Collar Crime* (New York: Holt, Rinehart and Winston, 1949), p. 9.

38. Ibid.

39. See Richard Austin Smith, "The Incredible Electrical Conspiracy," *Fortune* (April 1961): 132–80; (May 1961): 161–224. Reprinted in Cressey and Ward, *Delinquency, Crime and Social Process,* pp. 884–912.

40. Smith, "Incredible Conspiracy," in ibid., p. 885.

41. James Fallows, *National Defense* (New York: Random House, 1981), p. 15f.

42. Hyman Rickover, "Advice from Admiral Rickover," *New York Review of Books,* March 18, 1982, pp. 12–14.

43. Ibid., pp. 12–13.

44. *Ibid.,* p. 13.

45. *Ibid.,* p. 13.

46. The Committee on Homosexual Offenses and Prostitution, *The Wolfenden Report* (London: Her Majesty's Printing Office, 1962).

47. Alan P. Bell and Martin S. Weinberg, *Homosexualities: A Study of Diversity Among Men and Women* (New York: Simon and Schuster, 1978), p. 23.

48. Ibid., p. 216.

49. Ibid., p. 188.

CHAPTER THIRTEEN

1. Kathleen Kenyon, *Archeology in the Holy Land* (London: Ernest Benn, 1970), p. 43.

2. Ibid., pp. 45–46.

3. See Mason Hammond, *The City in the Ancient World* (Cambridge, Mass.: Harvard University Press, 1972), chapter 1.

4. Ibid., pp. 45f. and 148f.

5. The present discussion is based on the following works: Numa Denis Fustel de Coulanges, *The Ancient City: A Study of the Religion, Laws, and Institutions of Greece and Rome* (Boston: Lee and Shepard, 1894); G. Glotz, *The Greek City and Its Institutions* (New York: Barnes & Noble, 1965), first published in 1929; William Reginald Halliday, *The Growth of the City State* (Liverpool: University Press of Liverpool, 1923); and Warde Fowler, *The City-State of the Greeks and Romans* (London: Macmillan, 1918).

6. In the present discussion, I rely on Henri Pirenne, *Les Villes du Moyen Age* (Paris: Presses universitaires de France, 1971); John H. Mundy and Peter Riesenberg, *The Medieval Town* (Princeton: Van Nostrand, 1958); Max Weber, *General Economic History*, trans. by Frank H. Knight (New York: Collier Books, 1961); and Maurice Dobb, *Studies in the Development of Capitalism* (New York: International Publishers, 1947).

7. Lewis Mumford, *The City in History* (New York: Harcourt, Brace & World, 1961), p. 456.

8. Alexis de Tocqueville, *Oeuvres Complètes* (Paris: Gallimard, 1951), V, 81; the translation here is mine.

9. Ibid., p. 82.

10. See Adna Ferrin Weber, *The Growth of Cities in the Nineteenth Century* (Ithaca, N.Y.: Cornell University Press, 1965). First published in 1899.

11. Lewis Mumford, *City in History*, p. 726f.

12. Weber, *Growth of Cities*, pp. 156–57.

13. Ibid., p. 157.

14. Robert E. Park et al., *The City* (Chicago: University of Chicago Press, 1925), p. 50f.

15. Harlan Paul Douglass's *The Suburban Trend* (New York: Century, 1925) is one of the earliest systematic studies of the move from city to suburb. The most notable recent works would include: the editors of *Fortune* magazine, *The Exploding Metropolis* (Garden City, N.Y.: Doubleday, 1958); Jean Gottman, *Megalopolis: The Urbanized Northeastern Seaboard of the United States* (Cambridge, Mass.: MIT Press, 1961); Herbert J. Gans, *The Levittowners: Ways of Life and Politics in a New Suburban Community* (New York: Pantheon, 1967); Robert Wood, *Suburbia: Its People and Their Politics* (Boston: Houghton Mifflin, 1958); and Bennett M. Berger, *Working-Class Suburbs: A Study of Auto Workers in Suburbia* (Berkeley: University of California Press, 1968).

16. See Herbert J. Gans, *The Urban Villagers: Group and Class in the Life of Italian Americans* (New York: Free Press, 1962).

17. *Report of the National Advisory Commission on Civil Disorders* (The Kerner Commission) (New York: Bantam Books, 1968), p. 247.

CHAPTER FOURTEEN

1. Wilbert E. Moore, *Social Change* (Englewood Cliffs, N.J.: Prentice-Hall, 1963), p. 19.

2. Emile Durkheim, *The Rules of the Sociological Method,* trans. by Sarah A. Solovay and John H. Mueller (New York: Free Press, 1964), p. 110.

3. Moore, *Social Change,* p. 10.

4. Talcott Parsons, "Evolutionary Universals in Society," *American Sociological Review,* 29 (June 1964). Reprinted in Talcott Parsons, *Sociological Theory and Modern Society* (New York: Free Press, 1967), pp. 490–520. This particular quote may be found on p. 493; italics added.

5. Ibid., p. 495.

6. George C. Homans, *Sentiments and Activities* (London: Routledge & Kegan Paul, 1962), pp. 26–27.

7. The present discussion is based on Robert A. Nisbet's highly illuminating *Social Change and History* (London and New York: Oxford University Press, 1969), pp. 166–88.

8. Ibid., p. 166.

9. J. B. Bury, *The Idea of Progress: An Inquiry into Its Origin and Growth* (London: Macmillan, 1928).

10. Nisbet, *Social Change and History,* p. 197.

11. Karl Marx and Frederick Engels, *Selected Correspondence* (Moscow: Foreign Languages Publishing House, 1953), p. 379.

12. Ibid., p. 412.

13. See Barrington Moore, Jr., *Social Origins of Dictatorship and Democracy* (Boston: Beacon Press, 1966), Chapter 1.

14. Ibid., p. 47.

15. Ibid., p. 55.

16. This article, which appeared in the *New York Daily Tribune* on July 22, 1853, is cited in Henry M. Christman ed., *The American Journalism of Marx and Engels* (New York: New American Library, 1966), pp. 106–7.

17. Volume 1 (Chicago: Kerr, 1906), p. 13.

18. Jawaharlal Nehru, *The Discovery of India* (New York: J. Day, 1946), p. 304f.

19. Romesh Dutt, *The Economic History of India* (London: Routledge and Kegan Paul, 1901). The quoted excerpt is taken from the 7th edition, 1950, pp. viiif., and is cited in Paul A. Baran, *The Political Economy of Growth* (New York: Modern Reader Paperbacks, 1968), p. 147.

20. Baran, *Political Economy,* p. 148.

21. E. H. Norman, *Japan's Emergence as a Modern State* (New York: Institute of Pacific Relations, 1940), p. 16. In the present discussion, I rely on Norman's excellent monograph and on Moore, *Social Origins of Dictatorship,* chapter 5.

22. Baran, *Political Economy*, p. 160.

23. Norman, *Japan's Emergence*, p. 32.

24. Clifford Geertz, *Agricultural Involution: The Process of Ecological Change in Indonesia* (Berkeley: University of California Press, 1963). Quoted from Henry Bernstein, ed., *Underdevelopment and Development: The Third World Today* (New York: Penquin, 1976), pp. 47–48.

25. Ibid., p. 53.

CHAPTER FIFTEEN

1. The present discussion is based on the following sources: E. J. Hobsbawm, *The Age of Revolution: 1789–1848* (New York: Mentor, 1962); Max Weber, *General Economic History*, trans. by Frank H. Knight (New York: Collier Books, 1961); and W. W. Rostow, *The Stages of Economic Growth* (Cambridge: Cambridge University Press, 1964).

2. Barrington Moore, Jr., *Social Origins of Dictatorship and Democracy* (Boston: Beacon Press, 1966), p. 150.

3. Arthur M. Schlesinger, *Political and Social History of the United States, 1829–1925* (New York: Macmillan, 1925), pp. 280–81; and Louis M. Hacker, *The Triumph of American Capitalism* (New York: Simon and Schuster, 1940), p. 400f.

4. Engene N. Schneider, *Industrial Sociology* (New York: McGraw-Hill, 1957), p. 65.

5. *World Bank Atlas*, Washington, D.C.: World Bank, 1981.

6. Emma Rothschild, "Reagan and the Real America," *New York Review of Books*, February 5, 1981, p. 12.

7. Lester Thurow, *The Zero-Sum Society* (New York: Basic Books, 1980), p. 7f.

8. Ibid., p. 7.

9. Julia Elcock, "Japanese Labour-Management Relations," Toronto *Globe and Mail*, December 14, 1981.

10. See Rodney Clark, *The Japanese Company* (New Haven: Yale University Press, 1979), pp. 175–76.

11. Chalmers Johnson, *MITI and the Japanese Miracle* (Stanford, Calif.: Stanford University Press, 1982).

12. See D. W. F. Bendix, *Limits to Codetermination* (Pretoria: Sigma Press, n.d.); and Alfred L. Thimm, *The False Promise of Codetermination* (Lexington, Mass.: Lexington Books, 1980).

13. Wassily Leontieff and Herbert Stein, *The Economic System in an Age of Discontinuity: Long-Range Planning or Market Reliance* (New York: New York University Press, 1976), p. 30.

14. Ibid., pp. 30–31.

15. Ibid., p. 31.

16. Ibid., pp. 39–40.

17. Thurow, *Zero-Sum Society*, p. 123.

18. Felix Rohatyn, "Restructuring America," *New York Review of Books,* March 5, 1981, p. 16.

19. Ibid., p. 16.

20. Ibid., p. 18.

21. Ibid., p. 18.

22. Ibid., p. 20.

23. Ibid.

24. U.S. Department of Commerce, 1980, table II, F1.

25. Jonathan Power, "Chancellor Schmidt's Pacific Two-Step," *Manchester Guardian Weekly,* February 28, 1982, p. 9.

26. The data on which the present discussion is based may be found in Willy Brandt, ed., *North-South: A Program for Survival,* the report of the Independent Commission on International Development Issues (Cambridge, Mass.: MIT Press, 1980). See also *Poverty and Human Development,* a World Bank publication (New York: Oxford University Press, 1980).

27. See *North-South,* p. 145f.

28. John Bartlow Martin, "First Ask the Right Questions," *Manchester Guardian Weekly,* April 4, 1982, p. 15.

29. General Agreements on Tariffs and Trade (GATT), *International Trade,* 1978–1979 (1979), p. 8.

30. R. Mikesell and M. Farah, *U.S. Export Competitiveness in Manufactures in Third-World Markets* (Washington, D.C.: Georgetown Center for Strategic and International Studies, 1980), table II-C.

31. Robert B. Reich, "Japan in the Chips," *The New York Review of Books,* November 19, 1981, p. 50.

32. *North-South,* p. 14.

33. *Ibid.,* p. 14.

34. Quoted by Jay Mathews in "Catholic Hierarchy Questions Morality of Nuclear Weapons Policies," *Manchester Guardian Weekly,* January 3, 1982, p. 11.

35. George F. Kennan [Mr. X], "The Source of Soviet Conduct," *Foreign Affairs,* 25 (1947): 575–76.

36. Ibid., p. 581.

37. George F. Kennan, "On Nuclear War," *The New York Review of Books,* January 21, 1982, p. 8.

38. Ibid., p. 8.

39. Ibid., p. 12.

40. Noel Gayler, "Sticking by Our Guns," *Manchester Guardian Weekly,* July 12, 1981, p. 9.

41. Ibid., p. 9.

42. Ibid.

43. Ibid.

44. Ibid.

45. Cited in *North South,* p. 120.

Sources and Suggestions
for Further Reading

CHAPTER TWO

Boaz, Franz, *Race, Language and Culture,* New York: Free Press, 1940.
Childe, V. Gordon, *What Happened in History,* (Middlesex, England: Penguin Books, 1964.
———, *Man Makes Himself,* London: Watts, 1965.
Darwin, Charles, *The Descent of Man,* London: John Murray, 1922.
Dewey, John, *Human Nature and Conduct,* New York: Holt, 1922.
Kroeber, Alfred L., *The Nature of Culture,* Chicago: University of Chicago Press, 1952.
Linton, Ralph, *The Study of Man,* New York: Appleton-Century-Crofts, 1936.
Lowie, Robert H., *Social Organization,* New York: Rinehart, 1956.
Montagu, Ashley, ed. *Culture and the Evolution of Man,* New York: Oxford University Press, 1962.
———, *Culture: Man's Adaptive Dimension.* London: Oxford University Press, 1968.
Tylor, E. B., *Anthropology,* London: Macmillan, 1881.
———, *The Origins of Culture,* New York: Harper Torchbooks, 1958.

CHAPTER THREE

Allee, W. C., *Cooperation Among Animals.* New York: Henry Schuman, 1951.
Ardrey, Robert, *African Genesis,* New York: Atheneum, 1961.
———, *The Territorial Imperative,* New York: Antheneum, 1966.

Barnet, Richard J., *The Roots of War*, Baltimore: Penguin Books, 1973.

Bernstein, Irwin S., and Gordon, Thomas P., "The Function of Aggression in Primate Societies," *American Scientist*, 62 (1974): 304–11.

Cloudsley-Thompson, J. L., *Animal Conflict and Adaptation*, London: G. T. Foulis, 1965.

Dart, Raymond, *Adventures with the Missing Link*, New York: Harper & Row, 1959.

Darwin, Charles, *On the Origin of the Species*, London: John Murray, 1859.

———, *The Descent of Man*, London: John Murray, 1871.

Dewey, John, *Human Nature and Conduct*, New York: Modern Library, 1929.

Freud, Sigmund, *Civilization and Its Discontents*, London: Hogarth Press, 1930.

———, "Instincts and Their Vicissitudes," In *Standard Edition of the Complete Psychological Works*, Ed. by James Strachey. London: Hogarth Press, Institute of Psychology, 1955–1964, Vol. 6, pp. 117–40.

———, "Repression," In ibid., pp. 146–58.

———, *The Ego and the Id*, London: Hogarth Press, 1950.

———, *The Complete Introductory Lectures on Psychoanalysis*, Trans. and ed. by James Strachey, New York: W. W. Norton, 1966.

Fromm, Erich, *The Anatomy of Human Destructiveness*, New York: Holt, Rinehart and Winston, 1973.

Horney, Karen, *Collected Works*, New York: W. W. Norton, 1942.

Klopfer, Peter H., "Instincts and Chromosomes: What Is an 'Innate' Act?" *American Naturalist*, 103 (1969): 556–60.

Kropotkin, Petr, *Mutual Aid*, Boston: Extending Horizons Books, n.d.

Lehrman, Daniel S., "A Critique of Konrad Lorenz's Theory of Instinctive Behavior," *Quarterly Review of Biology*, 28 (1953): 337–63.

Lidz, Theodore, *The Person*, New York: Basic Books, 1976.

Lorenz, Konrad, *On Aggression*, New York: Harcourt, Brace & World, 1966.

Marcuse, Herbert, *Eros and Civilization*, New York: Vintage Books, 1955.

McDougall, William, *An Introduction to Social Psychology*, London: Methuen, 1908; 23rd ed., New York: Barnes & Noble, 1960.

Mead, G. H., *Mind, Self and Society*, Chicago and London: University of Chicago Press, 1934.

Montagu, Ashley, *The Nature of Human Aggression*, New York: Oxford University Press, 1970.

———, ed. *Culture and Human Development*, Englewood Cliffs, N.J.: Prentice-Hall, 1974.

Morris, Desmond, *The Naked Ape*, New York: McGraw-Hill, 1967.

———, *The Human Zoo*, New York: McGraw-Hill, 1969. Nagel, Ueli, and Kummer, Hans, "Variations in Cercopithecoid Aggressive Behavior," In Ralph L. Holloway, ed., *Primate Aggression, Territoriality, and Xenophobia*, New York: Academic Press, 1974, pp. 159–84.

Skinner, B. F., *Beyond Freedom and Dignity*, New York: Vintage Books, 1971.

Storr, Anthony, *Human Aggression*, New York: Atheneum, 1968.

————, *Human Destructiveness*, New York: Basic Books, 1972.

Sullivan, Harry Stack, *The Interpersonal Theory of Psychiatry*, New York: W. W. Norton, 1953.

————, *Clinical Studies in Psychiatry*, Vol. 2 of *Collected Works*, ed. by Helen Swick Perry et al. New York: W. W. Norton, 1956.

Tinbergen, Niko, *The Study of Instinct*, Oxford: Clarendon Press, 1951.

————, "On War and Peace in Animals and Man," *Science*, 160 (1968): 1411–18.

Watson, J. B., *Psychology, From the Standpoint of a Behaviorist*, Philadelphia: Lippincott, 1919.

Watson, J. B., *Behaviorism*, New York: The People's Institute, 1925.

Wilson, Edward O., *Sociobiology: The New Synthesis*, Cambridge: Harvard University Press, 1975.

Wynne-Edwards, V. C., "Ecology and the Evolution of Social Ethics," In J. W. S. Pringle, ed., *Biology and the Human Sciences*, Oxford: Clarendon Press, 1972, pp. 40–59.

CHAPTER FOUR

Cooley, Charles Horton, *Social Organization* and *Human Nature and the Social Order*, in one volume, Glencoe, Ill.: Free Press, 1956.

Goffman, Erving, *The Presentation of Self in Everyday Life*, Garden City, N.Y.: Doubleday Anchor Books, 1959.

————, *Stigma*, Englewood Cliffs, N.J.: Prentice-Hall, 1963.

————, *Strategic Interaction*, Philadelphia: University of Pennsylvania Press, 1969.

Hare, A. Paul, Borgatta, Edgar F., and Bales, Robert F., eds. *Small Groups*, New York: Alfred A. Knopf, 1955.

Homans, George C., *Social Behavior: Its Elementary Forms*, New York: Harcourt, Brace, Jovanovich, 1974.

Humphrey, George, and Argyle, Michael, *Social Psychology Through Experiment*, London: Methuen, 1962.

Mead, George Herbert, *Mind, Self and Society*, Chicago and London: University of Chicago Press, 1934.

Ofshe, Richard J., *Interpersonal Behavior in Small Groups*, Englewood Cliffs, N.J.: Prentice-Hall, 1973.

Slater, Philip E., *Microcosm*, New York: John Wiley, 1966.

Thibant, John W., and Kelly, Harold H., *The Social Psychology of Groups*, New York: John Wiley, 1959.

Tonnies, Ferdinand, *Community and Society* [*Gemeinschaft and Gesellschaft*], Trans. and ed. by Charles P. Loomis, East Lansing: Michigan State University Press, 1957.

Wolff, Kurt, ed. and trans. *The Sociology of Georg Simmel*, Glencoe, Ill.: Free Press, 1950.

CHAPTER FIVE

Aberle, D. F., et al. "The Functional Prerequisites of a Society," *Ethics,* 60 (1950): 100–111.

Aristotle, *The Politics,* Middlesex, England: Penguin Books, 1962.

Cancian, F. M., "Functional Analysis," In *International Encyclopedia of the Social Sciences,* ed. D. L. Sills, New York: Macmillan and the Free Press, 1968, Vol. 6, pp. 29–43.

Childe, V. Gordon, *What Happened in History,* Middlesex, England: Penguin Books, 1964.

Durkheim, Emile, *Professional Ethics and Civic Morals,* London: Routledge & Kegan Paul, 1957.

——, *The Division of Labor in Society,* New York: Free Press, 1964.

Engels, Frederick, *The Origin of the Family, Private Property and the State,* New York: International Publishers, 1942.

——, *Anti-Duhring,* Moscow: Foreign Languages Publishing House, 1954.

Fromm, Erich, *Marx's Concept of Man,* New York: Frederick Ungar, 1961.

Hobbes, Thomas, *Leviathan,* Middlesex, England: Penguin Books, 1968.

Levy, Marion J., Jr., *The Structure of Society,* Princeton: Princeton University Press, 1952.

Marx, Karl, *Capital,* Vol. 1. Moscow: Foreign Languages Publishing House, 1954.

——, *Economic and Philosophical Manuscripts of 1844,* Moscow: Foreign Languages Publishing House, 1961.

——, and Engels, Frederick, *The German Ideology,* New York: International Publishers, 1960.

Mayer, J. P., *Max Weber and German Politics,* London: Faber and Faber, 1944.

Mehring, Franz, *Karl Marx,* Ann Arbor: University of Michigan Press, 1962.

Merton, Robert K., *Social Theory and Social Structure,* Rev. ed. New York: Free Press, 1968.

Tocqueville, Alexis de, *Oeuvres Completes,* Vol. 2, part 1, *L'Ancien Regime et la Revolution,* Paris: Gallimard, 1952.

Weber, Max, *General Economic History,* New York: Collier Books, 1961.

——, *Economy and Society,* 3 vols., New York: Bedminster Press, 1968.

CHAPTER SIX

Anderson, Charles H., *The Political Economy of Social Class,* Englewood Cliffs, N.J.: Prentice-Hall, 1974.

Bendix, Reinhard and Lipset, Seymour Martin, *Class, Status and Power,* Glencoe, Ill.: Free Press, 1963.

Bottomore, T. B., *Classes in Modern Society,* New York: Pantheon Books, 1966.

————, and Rubel, Maximilien, *Karl Marx: Selected Writings in Sociology and Social Philosophy,* London: Watts, 1956.

Cole, G. D. H., *Studies in Class Structure,* London: Routledge & Kegan Paul, 1964.

Gerth, Hans, and Mills, C. Wright, eds. *From Max Weber: Essays in Sociology,* New York: Oxford University Press, 1958.

Giddens, Anthony, *The Class Structure of the Advanced Societies,* London: Hutchinson University Library, 1973.

Hacker, Andrew, ed. *The Corporation Take-Over,* New York: Harper & Row, 1964.

Lockwood, David, *The Blackcoated Worker,* London: George Allen & Unwin, 1958.

Mills, C. Wright, *White Collar,* New York: Oxford University Press, 1956.

Roach, Jack L., Gross, Llewellyn, and Gursslin, Orville, *Social Stratification in the United States,* Englewood Cliffs, N.J.: Prentice-Hall, 1969.

Sennett, Richard, and Cobb, Jonathan, *The Hidden Injuries of Class,* New York: Alfred A. Knopf, 1972.

Weber, Max, *General Economic History,* Trans. by Frank H. Knight, New York: Collier Books, 1961.

————, *Economy and Society,* 3 vols. Ed. by Guenther Rother and Claus Wittich, New York: Bedminster Press, 1968.

Zeitlin, Maurice, "Corporate Ownership and Control: The Large Corporation and the Capitalist Class," *American Journal of Sociology,* 79 (1974): 1073–1119.

CHAPTER SEVEN

Bottomore, T. B., *Elites and Society,* New York: Basic Books, 1964.

Dahrendorf, Ralf, *Class and Class Conflict in an Industrial Society,* London: Routledge & Kegan Paul, 1963.

Davis, Kingsley, and Moore, Wilbert E., "Some Principles of Stratification," *American Sociological Review,* 10 (1945): 242–49.

Meisel, James H., *The Myth of the Ruling Class: Gaetano Mosca and the Elite,* Ann Arbor: University of Michigan Press, 1962.

Moore, Barrington, Jr., *Social Origins of Dictatorship and Democracy,* Boston: Beacon Press, 1966.

Mosca, Gaetano, *The Ruling Class,* Trans. by Hannah D. Kahn, New York: McGraw-Hill, 1965.

Pareto, Vilfredo, *The Mind and Society,* 4 vols. Ed. by Arthur Livingstone, New York: Harcourt, Brace & World, 1935.

Tumin, Melvin M., "Some Principles of Stratification: A Critical Analysis," *American Sociological Review,* 18 (1953): 387–94.

CHAPTER EIGHT

Barron, Milton L., *Minorities in a Changing World,* New York: Alfred A. Knopf, 1967.

Bennett, Lerone, Jr., *Before the Mayflower: A History of the Negro in America, 1619–1964,* Rev. ed. Baltimore: Penguin Books, 1968.

Boyd, W., *Genetics and the Races of Man,* Boston: Little, Brown, 1950.

————, "The Contributions of Genetics to Anthropology," in A. Kroeber, *Anthropology Today,* Chicago: University of Chicago Press, 1953.

Brophy, William, and Brophie, Sophie, *The Indian: America's Unfinished Business,* Norman: University of Oklahoma Press, 1966.

Coon, C. S. et al., *Races,* Springfield, Ill.: Thomas, 1950.

Deloria, Vine, *Custer Died for Your Sins,* New York: Macmillan, 1969.

Dinnerstein, Leonard, and Jaher, Frederic Cople, eds. *The Aliens: A History of Ethnic Minorities in America,* New York: Appleton-Century-Crofts, 1970.

Dinnerstein, Leonard, and Reimers, David M., *Ethnic Americans: A History of Immigration and Assimilation,* New York: Dodd, Mead, 1975.

Dobzhansky, Theodosius, *Mankind Evolving,* New Haven and London: Yale University Press, 1962.

Elkholy, Abdo, A., *The Arab Moslems in the United States,* New Haven: College & University Press Publishers, 1966.

Farb, Peter, *Man's Rise to Civilization,* New York: E. P. Dutton, 1968.

Feinstein, Otto, *Ethnic Groups in the City,* Lexington, Mass.: Heath Lexington Books, 1971.

Foreman, Grant, *The Five Civilized Tribes,* Norman: University of Oklahoma Press, 1966.

Franklin, John Hope, *Reconstruction: After the Civil War,* Chicago: University of Chicago Press, 1961.

————, *From Slavery to Freedom: A History of Negro Americans,* 3rd ed. New York: Vintage Books, 1966.

Gerson, Louis L., *The Hyphenate in Recent American Politics and Diplomacy,* Lawrence: University of Kansas Press, 1964.

Glazer, Nathan, and Moynihan, Daniel Patrick, *Beyond the Melting Pot,* 2nd ed. Cambridge, Mass.: MIT Press, 1970, pp. 86–142.

Greeley, Andrew M., *Ethnicity in the United States,* New York: John Wiley, 1974.

Handlin, Oscar, *The Uprooted,* 2nd ed. Boston: Little, Brown, 1973.

Hansen, Marcus Lee, *The Atlantic Migration, 1607–1860,* New York: Harper Torchbooks, 1961.

Herzog, Stephen J., *Minority Group Politics,* New York: Holt, Rinehart and Winston, 1971.

Howard, John R., ed. *Awakening Minorities: American Indians, Mexican Americans, Puerto Ricans,* Hawthorne, N.Y.: Aldine, Trans-Action Books, 1970.

Jones, Maldwyn Allen, *American Immigration,* Chicago: University of Chicago Press, 1960.

Kitano, Harry H. L., *Race Relations,* Englewood Cliffs, N.J.: Prentice-Hall, 1974.

Kramer, Judith R., *The American Minority Community,* New York: Thomas Y. Crowell, 1970.

Levitan, Sar A. et al. *Minorities in the United States: Problems, Progress, Prospects,* Washington, D.C.: Public Affairs Press, 1975.

Montagu, Ashley, *Man's Most Dangerous Myth,* 5th ed., New York: Oxford University Press, 1974.

Pinkney, Alphonso, *Black Americans,* Englewood Cliffs, N.J.: Prentice-Hall, 1969.

Ryan, Joseph, ed. *White Ethnics: Their Life in Working-Class America,* Englewood Cliffs, N.J.: Prentice-Hall, 1973.

Smith, Drew L., *The Legacy of the Melting Pot,* North Quincy, Mass.: Christopher Publishing House, 1971.

Stampp, Kenneth M., *The Peculiar Institution: Slavery in the Ante-Bellum South,* New York: Vintage Books, 1956.

Vander Zanden, James W., *American Minority Relations,* 3rd ed. New York: Ronald Press, 1972.

Veenhoven, William A., ed. *Case Studies on Human Rights and Fundamental Freedoms: A World Survey,* 5 Vols. The Hague: Martinus Nijhoff, 1975.

Wittke, Carl, *We Who Built America,* 2nd ed. Cleveland: Press of Western Reserve University, 1964.

CHAPTER NINE

Bane, Mary Jo., *Here to Stay: American Families in the Twentieth Century,* New York: Basic Books, 1976.

Bernard, Jessie, *The Future of Marriage,* New York: World, 1972.

————, *The Future of Motherhood,* New York: Dial Press, 1974.

————, *Women, Wives, Mothers: Values and Options,* Chicago: Aldine, 1975.

Cavan, Ruth Shonle, *Marriage and Family in the Modern World,* New York: Thomas Y. Crowell, 1965.

Coser, Rose Laub, *The Family: Its Structure and Functions,* New York: St. Martin's Press, 1964.

Engels, Frederick, *The Origin of the Family, Private Property and the State,* New York: International Publishers, 1942.

Farber, Bernard, *Family: Organization and Interaction,* San Francisco: Chandler, 1964.

————, ed. *Kinship and Family Organization,* New York: John Wiley, 1966.

Goode, William J., *World Revolution and Family Patterns,* Glencoe, Ill.: Free Press, 1963.

Gordon, Michael, ed. *The Nuclear Family in Crisis,* New York and Evanston: Harper & Row, 1972.

Howell, Mary C., *Helping Ourselves: Families and the Human Network,* Boston: Beacon Press, 1975.

445

Kirkpatrick, Clifford, *The Family: As Process and Institution,* New York: Ronald Press, 1955.

Leslie, Gerald R., *The Family in Social Context,* New York: Oxford University Press, 1967.

Parsons, Elsie Clews, *The Family,* New York and London: G. P. Putnam's Sons, 1906.

Rossi, Alice S., et al., eds. *The Family,* New York: W. W. Norton, 1978.

Schulz, David A., *The Changing Family: Its Function and Future,* Englewood Cliffs, N.J.: Prentice-Hall, 1972.

Smith, James R., and Smith, Lynn G., eds. *Beyond Monogamy: Recent Studies of Sexual Alternatives in Marriage,* Baltimore and London: Johns Hopkins University Press, 1974.

Stein, Peter J., *Single,* Englewood Cliffs, N.J.: Prentice-Hall, 1976.

Yorburg, Betty, *The Changing Family,* New York and London: Columbia University Press, 1973.

CHAPTER TEN

Baum, Gregory, *Religion and Alienation,* New York: Paulist Press, 1975.

Bellah, Robert N. *Beyond Belief,* New York: Harper & Row, 1970.

Berger, Peter L., *The Sacred Canopy,* Garden City, N.Y.: Doubleday, 1967.

Dudley, Guilford, III, *Religion on Trial: Mircea Eliade and His Critics,* Philadelphia: Temple University Press, 1977.

Durkheim, Emile, *The Elementary Forms of Religious Life,* Trans. and ed. by J. W. Swain, London: George Allen & Unwin, 1964.

Eister, Allan W., *Changing Perspectives in the Scientific Study of Religion,* New York: John Wiley, 1974.

Eliade, Mircea, ed. *The History of Religions: Essays in Methodology,* Chicago: University of Chicago Press, 1966.

Fallding, Harold, *The Sociology of Religion,* Toronto: McGraw-Hill Ryerson, 1974.

Freud, Sigmund, *The Future of an Illusion,* Garden City, N.Y.: Anchor Books, 1964.

Hoult, Thomas Ford, *The Sociology of Religion,* New York: Dryden Press, 1958.

James, William, *The Varieties of Religious Experience,* New York: Collier Books, 1976. Originally published in 1902.

Lenski, Gerhard, *The Religious Factor,* Garden City, N.Y.: Doubleday, 1963.

Littleton, C. Scott, *The New Comparative Mythology,* Rev. ed. Berkeley, Calif.: University of California Press, 1973.

Malinowski, Bronislaw, *Magic, Science and Religion,* Garden City, N.Y.: Doubleday Anchor Books, 1954.

Marx, Karl, *Early Writings,* Trans. and ed. by T. B. Bottomore. London: C. A. Watts, 1963.

Waardenburg, Jacques, *Classical Studies to the Study of Religion,* Vol. 1. The Hague: Mouton, 1973.

Wach, Joachim, *Sociology of Religion,* Chicago: University of Chicago Press, 1944.

————, *The Comparative Study of Religions,* New York: Columbia University Press, 1958.

Weber, Max, *Ancient Judaism,* Trans. and ed. by H. H. Gerth and D. Martindale, Glencoe, Ill.: Free Press, 1952.

————, *The Protestant Ethic and the Spirit of Capitalism,* Trans. by Talcott Parsons. New York: Charles Scribner's Sons, 1958.

————, *The Religion of China,* Trans. and ed. by H. H. Gerth, Glencoe, Ill.: Free Press, 1962.

————, *The Religion of India,* Trans. and ed. by H. H. Gerth and D. Martindale, Glencoe, Ill.: Free Press, 1962.

Weber, Max, *Economy and Society,* Vol. 2. Ed. by Guenther Roth and Claus Wittich, New York: Bedminster Press, 1968, pp. 399–634.

Wilson, John, *Religion in American Society,* Englewood Cliffs, N.J.: Prentice-Hall, 1978.

CHAPTER ELEVEN

Abel, Theodore, *The Nazi Movement: Why Hitler Came to Power,* New York: Atherton Press, 1965. First published in 1938.

Ash, Roberta, *Social Movements in America,* Chicago: Markham, 1972.

Blackey, Robert, and Paynton, Clifford, *Revolution and the Revolutionary Ideal,* Cambridge, Mass.: Shenckman, 1976.

Brinton, Crane, *The Anatomy of Revolution,* New York: Vintage Books, 1965.

Ch'en, Jerome, *Mao and the Chinese Revolution,* London: Oxford University Press, 1965.

Ebenstein, William, *The Nazi State,* New York: Octagon Books, 1975.

Franklin, John Hope, *From Slavery to Freedom,* New York: Vintage Books, 1969.

Geschwender, James A., ed. *The Black Revolt,* Englewood Cliffs, N.J.: Prentice-Hall, 1971.

Grunberger, Richard, *A Social History of the Third Reich,* London: Weidenfeld and Nicholson, 1971.

Heberle, Rudolf, *Social Movements: An Introduction to Political Sociology,* New York: Appleton-Century-Crofts, 1951.

Houn, Franklin W., *A Short History of Chinese Communism,* Englewood Cliffs, N.J.: Prentice-Hall, 1973.

Killian, Lewis M., *The Impossible Revolution?* New York: Random House, 1968.

Muse, Benjamin, *The American Negro Revolution,* Bloomington, Ind.: Indiana University Press, 1968.

Neumann, Franz, *Behemoth,* New York: Harper Torchbooks, 1963. Originally published in 1942.

Pascal, Roy, *The Nazi Dictatorship,* London: Routledge & Sons, 1934.

Rauschning, Hermann, *The Beast From the Abyss,* London: Heinemann, 1941.

Rice, Edward E., *Mao's Way,* Berkeley, Calif.: University of California Press, 1972.

Schurmann, Franz, *Ideology and Organization in Communist China,* Berkeley, Calif.: University of California Press, 1973.

Shirer, William L., *The Rise and Fall of the Third Reich,* New York: Simon and Schuster, 1960.

Smelser, Neil, *Theory of Collective Behavior,* New York: Free Press of Glencoe, 1963.

Speer, Albert, *Inside the Third Reich,* New York: Macmillan, 1970.

Walton, Hanes, Jr. *The Political Philosophy of Martin Luther King, Jr.* Westport, Conn.: Greenwood, 1971.

CHAPTER TWELVE

Bayley, David H., and Mendelsohn, Harold, *Minorities and the Police: Confrontation in America,* New York: Free Press, 1969.

Becker, Howard S., *Outsiders: Studies in the Sociology of Deviance,* New York: Free Press, 1963.

Chambliss, William J., and Seidman, Robert B., *Law, Order, and Power,* Reading, Mass.: Addison-Wesley, 1971.

Cicourel, Aaron V., *The Social Organization of Juvenile Justice,* London: Heinemann, 1976. First published in 1968.

Cohen, Albert K., *Delinquent Boys: The Culture of the Gang,* New York: Free Press, 1955.

Cressey, Donald, and Ward, David A., eds. *Delinquency, Crime and Social Process,* New York: Harper & Row, 1969.

Curtis, Lynn A., *Violence, Race and Culture,* Lexington, Mass.: Lexington Books, 1975.

Durkheim, Emile, *The Rules of the Sociological Method,* Trans. by Sarah A. Solway and John H. Mueller, New York: Free Press, 1964.

———, *The Division of Labor in Society,* Trans. by George Simpson, New York: Free Press, 1965.

Erikson, Kai T., *Wayward Puritans: A Study in the Sociology of Deviance,* New York: John Wiley, 1966.

Ianni, Francis A. J., *Black Mafia: Ethnic Succession in Organized Crime,* New York: Simon and Schuster, 1974.

Lemert, Edwin M., *Human Deviance, Social Problems, and Social Control,* Englewood Cliffs, N.J.: Prentice-Hall, 1967.

Matza, David, *Delinquency and Drift,* New York: John Wiley, 1964.

———, *Becoming Deviant,* Englewood Cliffs, N.J.: Prentice-Hall, 1969.

McDonald, Lynn, *The Sociology of Law and Order,* London: Faber & Faber, 1976.

Owens, Charles E., and Bell, Jimmy, eds. *Blacks and Criminal Justice*, Lexington, Mass.: Lexington Books, 1977.

Quinney, Richard, *The Social Reality of Crime*, Boston: Little, Brown, 1970.

Schuessler, Karl, ed. *Edwin H. Sutherland on Analyzing Crime*, Chicago: University of Chicago Press, 1956.

Schur, Edwin M., *Labeling Deviant Behavior: Its Sociological Implications*, New York: Harper & Row, 1971.

Sutherland, Edwin H., *White Collar Crime*, New York: Holt, Rinehart and Winston, 1949.

Sutherland, Edwin H., and Cressey, Donald R., *Principles of Criminology*, 7th ed. Philadelphia: J. B. Lippincott, 1966.

Turk, Austin T., *Criminality and Legal Order*, Chicago: Rand McNally, 1969.

Winslow, Robert W., ed. *The Emergence of Deviant Minorities*, New Brunswick, N.J.: Transaction Books, 1972.

CHAPTER THIRTEEN

Andrews, Antony, *Greek Society*, London: Penguin Books, 1967.

Andrews, Richard B., *Urban Growth and Development*, New York: Simmons-Boardman, 1962.

Bartholomew, Harland, assisted by Jack Wood, *Land Uses in American Cities*, Cambridge, Mass.: Harvard University Press, 1955.

Berger, Bennett M., *Working-Class Suburbs: A Study of Auto Workers in Suburbia*, Berkeley, Calif.: University of California Press, 1968.

Blumstein, James F., and Martin, Eddie J., eds. *The Urban Scene in the Seventies*, Nashville, Tenn.: Vanderbilt University Press, 1974.

Burgess, Ernest W., and Bogue, Donald, eds. *Urban Sociology*, Chicago: University of Chicago Press, 1967.

Childe, V. Gordon, *Man Makes Himself*, New York: Mentor Book, 1957.

————, *What Happened in History*, Penguin Books, 1964.

Chinoy, Ely, ed. *The Urban Future*, New York: Lieber-Atherton, 1973.

Chudacoff, Howard P., *The Evolution of American Urban Society*, Englewood Cliffs, N.J.: Prentice-Hall, 1975.

Douglass, Harlan Paul, *The Suburban Trend*, New York: Century, 1925.

Editors of *Fortune* Magazine, *The Exploding Metropolis*, Garden City, N.Y.: Doubleday, 1958.

Ehrenberg, Victor, *The Greek State*, New York: W. W. Norton, 1960.

————, *From Solon to Socrates*, London: Methuen, 1967.

Finley, M. I., *The Ancient Economy*, Los Angeles: University of California Press, 1973.

————, *Economy and Society in Ancient Greece*, New York: Viking Press, 1982.

Fowler, W. Warde, *The City-State of the Greeks and Romans*, London: Macmillan, 1918.

Fustel de Coulanges, Numa Denis, *The Ancient City: A Study of the Re-*

ligion, Laws, and Institutions of Greece and Rome, Boston: Lee and Shepard, 1894.

Gans, Herbert J., *The Levittowners: Ways of Life and Politics in a New Suburban Community,* New York: Pantheon, 1967.

Glotz, G., *The Greek City and Its Institutions,* New York: Barnes & Noble, 1965. First published in 1929.

Gottman, Jean, *Megalopolis: The Urbanized Northeastern Seaboard of the United States,* Cambridge, Mass.: MIT Press, 1961

Green, Constance McLaughlin, *The Rise of Urban America,* New York: Harper & Row, 1965.

Halliday, William Reginald, *The Growth of the City State,* Liverpool: University Press of Liverpool, 1923.

Hammond, Mason, *The City in the Ancient World,* Cambridge, Mass.: Harvard University Press, 1972.

Kenyon, Kathleen, *Archeology in the Holy Land,* London: Ernest Benn, 1970.

Lapidus, Ira Marvin, *Musli: Cities in the Later Middle Ages,* Cambridge, Mass.: Harvard University Press, 1967.

McKenzie, Roderick D., *On Human Ecology,* Chicago and London: University of Chicago Press, 1968.

Mumford, Lewis, *The City in History,* New York: Harcourt, Brace & World, 1961.

Mundy, John H., and Riesenberg, Peter, *The Medieval Town,* Princeton, N.J.: Van Nostrand, 1958.

Park, Robert E., Burgess, Ernest W., and McKenzie, Roderick D., *The City,* Chicago: University of Chicago Press, 1925.

Pirenne, Henri, *Les Villes du Moyen Age,* Paris: Presses Universitaires de France, 1971.

Report of the Commission on the Cities in the 1970's, *The State of the Cities,* New York: Praeger, 1972.

Research and Policy Committee of the Committee for Economic Development, *An Approach to Federal Urban Policy,* New York: CED, 1977.

Schlesinger, Arthur Meier, *The Rise of the City,* Chicago: Quadrangle Books, 1961. Originally published in 1933.

Ste. Croix, G. E. M., *The Class Struggle in the Ancient Greek World,* Ithaca: Cornell University Press, 1982.

Toynbee, Arnold, *Cities on the Move,* New York: Oxford University Press, 1970.

Weber, Adna Ferrin, *The Growth of Cities in the Nineteenth Century,* Ithaca, N.Y.: Cornell University Press, 1965. Originally published in 1899.

Weber, Max, *The City,* Trans. and ed. by Dan Martindale and Gertrude Neuwirth, New York: Free Press, 1958.

CHAPTER FOURTEEN

Anstey, Vera, *The Economic Development of India,* London: Longman, Green, 1929.

Baran, Paul, *The Political Economy of Growth,* New York: Modern Reader Paperbacks, 1957.

Bernstein, Henry, ed. *Underdevelopment and Development: The Third World Today,* New York: Penguin Books, 1976.

Dobb, Maurice, *Studies in the Development of Capitalism,* New York: International Publishers, 1947.

Frank, Andre Gunder, *Capitalism and Underdevelopment in Latin America,* New York: Modern Reader Paperbacks, 1969.

Furtado, Celso, *Development and Underdevelopment,* Berkeley and Los Angeles: University of California Press, 1967.

Moore, Barrington, Jr., *Social Origins of Dictatorship and Democracy,* Boston: Beacon Press, 1966.

Moore, Wilbert E., *Social Change,* Englewood Cliffs, N.J.: Prentice-Hall, 1963.

Nehru, Jawaharlal, *The Discovery of India,* New York: J. Day, 1946.

Nisbet, Robert A., *Social Change and History,* New York: Oxford University Press, 1969.

Norman, E. H., *Japan's Emergence as a Modern State,* New York: Institute of Pacific Relations, 1940.

CHAPTER FIFTEEN

Brandt, Willy, ed., *North-South: A Program for Survival,* Cambridge, Mass.: MIT Press, 1980.

Clark, Rodney, *The Japanese Company,* New Haven: Yale University Press, 1979.

Hacker, Louis M., *The Triumph of American Capital,* New York: Simon and Schuster, 1940.

Hobsbawm, E. J., *The Age of Revolution: 1789–1848,* New York: Mentor, 1962.

Johnson, Chalmers, *MITI and the Japanese Miracle,* Stanford, Calif.: Stanford University Press, 1982.

Kennan, George F., "On Nuclear War," *The New York Review of Books,* January 21, 1982.

Moore, Jr., Barrington, *Social Origins of Dictatorship and Democracy,* Boston: Beacon Press, 1966.

Rohatyn, Felix, "Restructuring America," *New York Review of Books,* March 5, 1981.

Rostow, W. W., *The Stages of Economic Growth,* Cambridge: Cambridge University Press, 1964.

Rothschild, Emma, "Reagan and the Real America," *New York Review of Books,* February 5, 1981.

Schlesinger, Arthur M., *Political and Social History of the United States, 1821–1925,* New York: Macmillan, 1925.

Weber, Max, *General Economic History,* trans. by Frank H. Knight, New York: Collier Books, 1961.

Name Index

Subject Index

Subject Index